# GLOBAL ECOLOGY

Sinauer Associates Inc. • Publishers
Sunderland, Massachusetts

# GLOBAL ECOLOGY

Edited by Charles H. Southwick
UNIVERSITY OF COLORADO

**FRONTISPIECE**

Tropical deforestation in Brazil. (Photograph by James P. Blair. Copyright © 1983 National Geographic Society. Reprinted with permission.)

**THE COVER**

The Sand Mountains of Gansu Province in western China. According to the Global 2000 report, the world's deserts will expand 20 percent by the year 2000. The United Nations has identified 2 billion hectares of land where the risk of desertification is high or very high. (Photograph by James L. Stanfield. Copyright © 1982 National Geographic Society. Reprinted with permission.)

GLOBAL ECOLOGY



Printed in U.S.A.

**Library of Congress Cataloging in Publication Data**
Main entry under title:

Global ecology.

Includes bibliographies and index.
1. Ecology—Addresses, essays, lectures. 2. Pollution—Environmental aspects—Addresses, essays, lectures.
3. Man—Influence on nature—Addresses, essays, lectures.
I. Southwick, Charles H.
QH541.145.G584 1985 304.2′8 85-1800
ISBN 0-87893-810-9 (pbk.)

5 4 3 2

# Contents

# Preface



Although the concept of "One World" has been a popular idea among writers and politicians since the days of Wendell Wilkie in the 1930s, it did not become a visual reality until the space-age accomplishments of NASA in the 1960s. The astronauts on U.S. and Soviet space flights were impressed by the beauty and isolation of the earth. They saw its magnificent greens and blues and browns, surrounded by a swirling atmosphere of moisture-laden clouds. The impressions bordered on a religious experience for many of them. This was particularly true among the Apollo crews—as they receded from the life-giving earth to the lifeless moon, the contrasts were strong.

In more than twenty-five years of human activity in space, orbital flights have become routine, the use of space by science, industry, and human affairs has greatly expanded, and some impressions have changed as well. The astronauts of space shuttle *Challenger* reported in 1983 that they were shocked to see the amount of pollution surrounding the earth below them. Commander Paul Weitz stated, "Unfortunately, this world is becoming a grey planet . . . . Our environment apparently is going downhill . . . . We are fouling our own nest."

This realization is nothing new for most of us on earth. We have all experienced air and water pollution and the loss of a childhood field, forest, or pond to urban growth or suburban development. What is new is the fact that these changes are now visible from space, that they are affecting the entire planet.

These, and many scientific facts that come pouring at us from all directions, highlight the importance of global ecology. Global ecology is the study of ecological principles and problems on a worldwide basis. It has many components—some are simply the accumulation of local and regional events until they assume global importance, as may be true for acid rain, soil erosion, and coastal pollution, for example. Other events in global ecology are of such general planetary impact that they cannot be seen locally; new concepts and new approaches will be required to evaluate them. For example, chemical changes in the upper atmosphere or broad alterations in global heat balance can sometimes be measured locally, but their true significance involves more than the summation of local processes.

Thus, global ecology may be approached from many directions. One approach involves the physics and chemistry of the atmosphere and the interactions of the atmosphere with the oceans, land, and biota. Another is a more directly humanistic approach involving a look at our own populations and resources, our food supplies, our states of health and economics, and the conditions of our fellow travelers on planet Earth—a look at the condition of all the living organisms with which we share life support systems.

This collection of readings is based on the premise that these diverse approaches, from geophysics to world health, are all the valid domain of global ecology.

The book begins on the theme of the biosphere, its nature, extent, and some of its functional properties. This is followed by chapters with opposing views on the state of the world—the pessimistic projections of Global 2000 and the optimistic views of Julian Simon and Herman Kahn. These illustrate the range of opinions now available on global futures and human prospects.

The second section deals with ecological principles and trends—topics such as biogeochemical cycles, interactions of the atmosphere and hydrosphere, and measurable trends in global ecology. These chapters go beyond introductory textbooks, and they emphasize current areas of research. Chapter 8 provides one of the best real data assessments on worldwide environmental trends by which we can evaluate the pessimism and optimism expressed in preceding chapters.

The third section deals with human impacts on the biosphere—air and water pollution, land degradation, soil erosion, world food supplies, tropical deforestation, and desertification. These topics illustrate both ecologic and economic effects of human activities. Although most of the chapters in Sections II and III deal with the terrestrial environment, quite appropriately since this is where we live, Chapters 7 and 12 on marine ecology recognize that we do live on a watery planet, with more than 70% of the earth's surface covered by oceans, seas, and ice.

The fourth section focuses on human populations—demography, population trends, poverty, and world health, all representing a directly humanistic approach to global ecology. We sometimes forget that our own populations reflect global environmental conditions as well as alter them. This is especially true in the area of world health.

The book ends with a discussion of human prospects, biological diversity, environmental consequences of war, nuclear winter, and the roles of science and technology in guiding global futures.

I have chosen these readings because they represent the most important issues of our times—issues that deserve a broad audience among students in many fields; not only students in the biological sciences, but in the physical, behavioral, and social sciences as well, and certainly among students in the humanities, business, law, engineering, medicine, and public health. The issues of global ecology must receive the thoughtful and creative consideration of all of us in our collective search for solutions.

**Acknowledgments**

Many individuals and organizations have contributed to this collection of readings. I am indebted to my students and colleagues at the University of Colorado and Johns Hopkins University who have responded with keen interest and enthusiasm to discussions of global ecology, and I am grateful to all the authors and publishers of the articles in this book who have granted permission to use their materials.

I am also deeply indebted to many colleagues in India with whom I worked closely for more than 25 years and shared firsthand many aspects of global ecology.

My own research in ecology has been supported over a span of 35 years by various organizations, including the United States National Institutes of Health, the National Science Foundation, the U.S. Educational Foundation in India, and the National Geographic Society. During numerous trips through Asia, Africa, and Latin America I have enjoyed the encouragement, companionship, and insights of my wife, Heather, son, Steven, and daughter, Karen. They have shared with me the excitement and discouragements of viewing the world in ecological terms.

Practical assistance in typing these papers has been provided by Jeanie Cavanagh, Bryon Coe, Gretta Howell, Mary Marcotte, Barbara Miller, and Elizabeth Owen, all of the Department of Environmental, Population, and Organismic Biology of the University of Colorado. Finally, I am grateful to Andy Sinauer for stimulating discussions and editorial guidance on topics in global ecology.

CHARLES H. SOUTHWICK
Boulder, Colorado
January 1985

CHAPTER 1

# The Biosphere



CHARLES H. SOUTHWICK

The biosphere is that part of the earth in which living organisms exist. It is a thin and discontinuous film over the surface of the earth, varying in thickness, and quite incomplete in surface coverage. It extends above the surface of the earth to altitudes that may reach nearly 10,000 meters or even greater heights when insects and microorganisms are carried aloft by updrafts and wind currents. The biosphere extends below the ground to the deepest roots of plants, to the chambers of many subterranean caverns, and in the oceans to the depths of thermal vents. In a few rare instances such as the Mariana Trench, these may conceivably provide a habitat for living organisms over 10,000 meters beneath the ocean's surface, although no organisms have yet been found at this depth. These are the extreme limits of the biosphere on earth and they emphasize the need to distinguish between various zones within the biosphere.

The broad definition mentioned above includes any place on earth or in the atmosphere where living organisms can be found. The extent of the total biosphere is quite different from those areas of earth where organisms may actively reproduce. With a few exceptions, the zone of primary biological production is much narrower. Its greatest extent on land is in a tall forest, such as a redwood grove or a tropical forest, where the zone of biospheric primary production (BPP)* may be 100 meters thick. At the other extreme on land, the zone of BPP in a rice field or a potato field is only 1 or 2 meters thick, and in a mown lawn, only a few centimeters thick.

In aquatic environments, the zone of BPP may be several hundred meters thick, for example, in a very clear ocean or lake where sufficient light can penetrate to support photosynthesis well below the surface. This depth of light penetration would require unusually clear water, but such conditions do occur in some marine and freshwater environments. Conversely, in a typical lake or ocean coastal zone, photosynthesis occurs in a layer of water that extends only a few meters below the surface, and in turbid waters, only a few centimeters.

The major exceptions to the preceding statements are deep-sea thermal vents, discovered only in recent years. These vents are cracks in the earth's crust on the bottom of the sea where hot gases and heated water emerge, often several hundred degrees centigrade in temperature (Matthews, 1981). Such vents have an incredible assemblage of marine life—uniquely adapted clams, marine worms, and crustacea capable of living in complete darkness, under very high pressures, and at extreme temperatures. In such habitats, photosynthesis does not occur; rather, primary production is accomplished by bacteria capable of synthesizing organic compounds from hydrogen sulfide. This undersea world is totally different from the biosphere as we know it on the surface of the earth. It operates with energy sources and metabolic pathways different from those of life on the surface, and it radically alters our concepts of where and how living organisms can exist. Figure 1 portrays in a diagrammatic fashion some of the vertical dimensions of the biosphere. It may help us visualize the great variation that occurs in both the depth and the thickness of the biosphere and in its lateral extent as well.

Despite the extreme environments in the deep-sea thermal vents where living organisms occur, there are many places on earth where the biosphere does not exist; or when living organisms are present, they are so transient or so sparsely distributed that they do not constitute a permanent biotic community. For practical purposes, the biosphere does not extend into the extremes of the polar regions, into vast areas of the driest

*BPP is used here in the same sense as GPP (gross primary production) is used in traditional ecology. That is, it refers to total organic synthesis in plants by photosynthesis and in microorganisms by chemosynthesis.

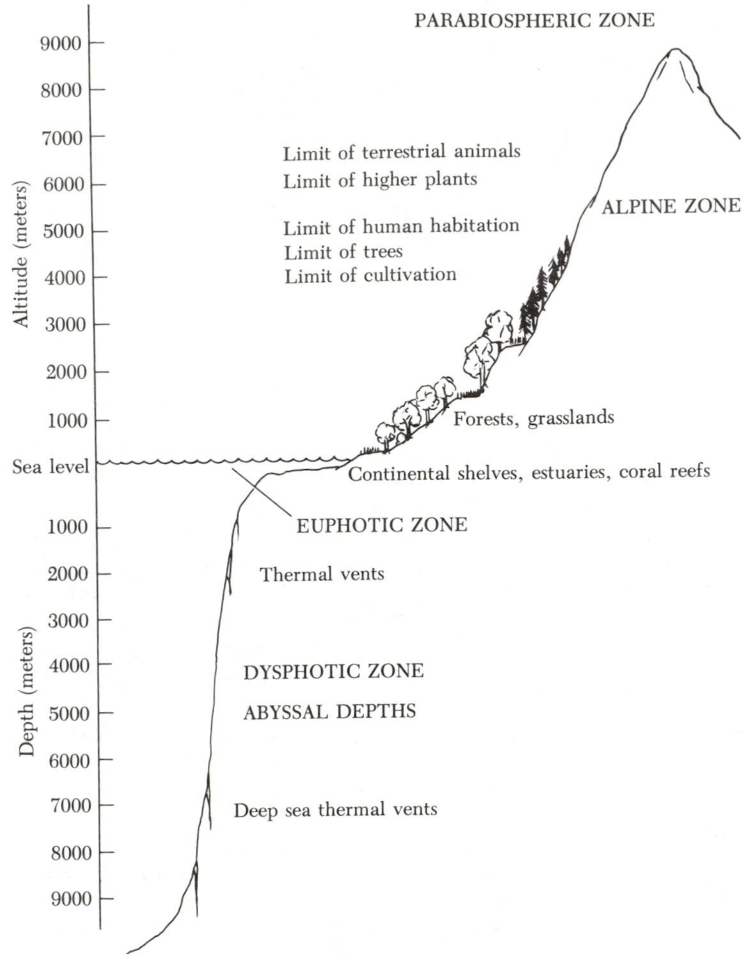


**Figure 1** The vertical extent of the biosphere. The parabiospheric zone represents altitudes where only dormant forms of life, such as bacterial and fungal spores, exist. The euphotic zone in aquatic environments is the zone of active photosynthesis. (From Southwick, 1976; redrawn from Hutchinson, 1970.)

deserts, onto most of the highest mountain peaks that have an environment of permanent ice and snow, into some land and water areas most highly polluted with toxic wastes, and throughout some of the deepest ocean volume in places other than thermal vents and upwellings. Such areas may have transient life forms, but they do not contribute significantly to the total picture of biospheric production.

The relative thickness of the biosphere may be visualized by considering an analogy with more familiar structures. If the diameter of the earth is represented by the height of an eight-story building (approximately 100 feet or 30 meters tall) the total thickness of the biosphere would be represented by the thickness of a two-by-four board (approximately 4 centimeters) on top of the building. On the same scale, the zone of active biological production, excluding deep-sea vents, would be represented by the thickness of a piece of paper (approximately 0.3 mm), and even this thickness would represent the most favorable habitats, such as a clear coral sea or a tropical rain forest.

The point of this descriptive exercise is to emphasize that the biosphere is surprisingly limited. Meaningful terrestrial biosphere occupies less than one-quarter of the earth's surface, and it is continually subject to alteration and insult at the hands of human populations. Yet this biosphere is our total life-support system. It generates our oxygen, produces our food, reprocesses our wastes, and makes all life possible. As Christensen (1984) has stated, it is our "grand oasis in space."

Although it is an elementary exercise in biology to enumerate what a natural ecosystem does, it is worth listing some of these properties in relation to a nonliving system. What, for example, can a natural grassland accomplish that Astroturf or a parking lot cannot? Some of these accomplishments are listed in Table 1, which is perhaps too simple a reminder for an educated reader, but nonetheless is a set of facts collectively forgotten when we express no concern that we have been paving the biosphere with concrete and asphalt, chopping down its trees, washing away its soil, and polluting its air and waters.

We can also think of the biosphere as a mosaic of biochemical processes, an infinitely complex biochemical system. It captures, converts, processes, and stores solar energy through an incredible diversity of organisms. Despite the diversity of hundreds of thousands of species of green plants (perhaps 500,000) and microorganisms and the even greater diversity of animal species (perhaps 5 to 10 million, though only 2 million are known to science at the present; Ehrlich and Ehrlich, 1981), the fundamental structures of living organ-

**Table 1.** Simplified comparison of some system properties between a natural ecosystem and a man-made structure.

| Natural ecosystem: pond, marsh, grassland, forest, etc. | Man-made system: house (non-solar) factory, parking lot, Astroturf, etc. |
|---|---|
| 1. Captures, converts, and stores energy from the sun | 1. Consumes energy from fossil or nuclear fuels |
| 2. Produces oxygen and consumes carbon dioxide | 2. Consumes oxygen and produces carbon dioxide |
| 3. Produces carbohydrates and proteins; accomplishes organic synthesis | 3. Cannot accomplish organic synthesis; produces only chemical degradation |
| 4. Filters and detoxifies pollutants and waste products | 4. Produces waste materials that must be treated elsewhere |
| 5. Is capable of self-maintenance and renewal | 5. Is not capable of self-maintenance and renewal |
| 6. Maintains beauty if not excessively disturbed | 6. Usually causes unsightly deterioration if not properly engineered and maintained |
| 7. Creates rich soil | 7. Destroys soil |
| 8. Stores and purifies water | 8. Often contributes to water pollution and loss |
| 9. Provides wildlife habitat | 9. Destroys wildlife habitat |

From Rodale, 1972, and Southwick, 1976.

isms show remarkable similarities in basic organization. The patterned structures of DNA, RNA, proteins, lipids, and carbohydrates form a blueprint for all life. Modern biology is stretching our understanding and amazement of this living world on both molecular and global scales.

As this remarkable biosphere of which we are a part recycles biogeochemical products between itself and the physical components of the earth, it counters the physical process of entropy (the increase of disorder and disorganization) by constantly organizing, structuring, and rebuilding the biochemical basis of living organisms. Life itself can be thought of as anti-entropic in its phases of normal growth, whereas it is entropic in its processes of catabolism and decomposition. Without life the world would indeed proceed to disorder; with life, there can be productive reorganization.

The biosphere restructures its components not only in a physical and biological sense; it does so in a behavioral and social sense as well. Groups of ants, bees, fish, mice, deer, monkeys, and people all tend to establish behavioral and social systems. Social breakdowns occur, but from each process of social entropy, reorganization begins. In this elemental sense, a corporation after Chapter 11 bankruptcy or a nation after war follows the same basic process as a herd of deer or a covey of quail after the hunting season: a corporation or a society must reorganize, achieve a new structure, find new leaders, develop new routes of communication, establish new systems of political process. This is simply a broad expression of our recognition that there is order in the living world, order that can be disassembled or shattered, but order that can also be reassembled providing we do not entirely destroy an ecosystem's ability to do so.

The biosphere may also be thought of as a great moderator or buffer of environmental conditions on earth. One need only compare the summer ground temperatures of a bare earth field at midday to those in the shade of a deep forest at the same time to realize how much a plant community moderates temperature. In a similar way, the biosphere moderates humidity, wind, precipitation, oxygen and carbon dioxide balances, and many aspects of atmospheric chemistry. Forests provide enough moisture to the air through transpiration to help maintain the rainfall necessary for their own survival (McCormick, 1959). Aspen forests in Colorado have a significant buffering effect on acid precipitation (Kling and Grant, 1984), and most biotic communities have various capacities to detoxify certain pollutants.

All of these qualities relate to ecosystem homeostasis—the ability of biotic communities to maintain environmental conditions favorable for the perpetuation of life. When the biosphere is destroyed, physical conditions are more likely to swing to extremes—reasonable balances can no longer be maintained. These principles have many ramifications, often of direct importance to human survival. We know that natural drought cycles can be tragically exacerbated when the vegetative cover is destroyed by overgrazing or excessive land misuse. Temperature differentials become much greater when forests are destroyed. Flood conditions become more dangerous when watersheds are denuded. Storms may become more violent. The maintenance of equitable climates on earth is intimately associated with intact ecological systems (Schneider and Londer, 1984).

Traditional courses in biology are organized around some aspect of the biosphere—its complexity, its taxonomy, and its central functions. Some textbooks of biology are entitled *Biosphere* (Jessop, 1970; Wallace et al., 1984). Within the study of the biosphere, the focus may be the cell, the organism, the population, or the ecosystem. At each of these levels, the interplay of diversity and unity is an impressive theme. At the most comprehensive biological level of all, the biosphere itself, this theme is also the logical one to emphasize. But perhaps it is even more important for us to realize what we are doing to the biosphere. How are we affecting the earth's ability to function as a life-support system? That question is a critical one, not only for ourselves, but for millions of other organisms. It is easy to see local effects at many points, but how important are these in a cumulative sense? Are numerous arenas of local pollution and countless scars of erosion, deforestation, and desertification altering the function of the global system? Or are they negligible? When we examine data on global increases in $CO_2$, accumulation of airborne lead in the Greenland icecap, DDT in the penguins of Antarctica, and expansions of the deserts of Africa, we have evidence of global influences.

Although the earth is a vast assemblage of infinitely complex and varied environments, we must also think of its total health. Global concepts are more essential than ever before—they can provide guidelines for local action and clues for what we might expect from local or regional developments when seen in broader perspective.

Hence this book. It cannot possibly cover the entire subject of global ecology, but it does provide a collection of outstanding articles, written by experts in their fields and covering an array of ecological topics from global atmospheric chemistry to world health. Collectively, these articles address various aspects of how we are impacting the biosphere, what some of the consequences might be, and what we might be able to do about it if we are sufficiently informed and concerned.

## References

Christensen, J. W. 1984. Global Science: Energy, Resources, Environment. Dubuque, Iowa: Kendall-Hunt Publishing Co. 355 pp.

Ehrlich, P. and A. Ehrlich. 1981. Extinction. New York: Random House. 305 pp.

Hutchinson, G. E. 1970. The Biosphere. Scientific American 223(3): 45–53.

Jessop, N. M. 1970. Biosphere: A Study of Life. Englewood Cliffs, N.J.: Prentice Hall. 954 pp.

Kling, G. W. and M. C. Grant. 1984. Acid Precipitation in the Colorado Front Range: An Overview with Time Predictions for Significant Effects. Arctic and Alpine Research 16(3): 321–329.

Matthews, S. W. 1981. New World of the Ocean. National Geographic 160: 792–832.

McCormick, J. 1959. Forest Ecology. New York: Harper and Brothers.

Rodale, R. 1972. Ecology and Luxury Living May Not Mix. Emmaus, Pa.: Rodale Press.

Schneider, S. and R. Londer. 1984. Co-Evolution of Climate and Life. San Francisco: Sierra Club. 563 pp.

Southwick, C. H. 1976. Ecology and the Quality of Our Environment, Second Edition. Boston: Willard Grant Press. 426 pp.

Wallace, R. A., J. L. King, and G. P. Sanders. 1984. Biosphere: The Realm of Life. Glenview, Ill.: Scott, Foresman. 699 pp.

# SECTION I

# DIVERGENT VIEWS ON THE STATE OF THE WORLD

C H A P T E R 2

# The Global 2000 Report



PREPARED BY THE COUNCIL ON ENVIRONMENTAL QUALITY AND THE DEPARTMENT OF STATE

## Letter of Transmittal

The President

Sir: In your Environmental Message to the Congress of May 23, 1977, you directed the Council on Environmental Quality and the Department of State, working with other federal agencies, to study the "probable changes in the world's population, natural resources, and environment through the end of the century." This endeavor was to serve as "the foundation of our longer-term planning."

The effort we then undertook to project present world trends and to establish a foundation for planning is now complete, and we are pleased to present our report to you. What emerges are not predictions but rather projections developed by U.S. Government agencies of what will happen to population, resources, and environment if present policies continue.

A report prepared by the Council on Environmental Quality and the Department of State for the President of the United States. Volume I, Summary: Entering the Twenty-First Century (1982). Gerald O. Barney, Study Director. U.S. Government Printing Office, Washington, DC 20402.

Our conclusions, summarized in the pages that follow, are disturbing. They indicate the potential for global problems of alarming proportions by the year 2000. Environmental, resource, and population stresses are intensifying and will increasingly determine the quality of human life on our planet. These stresses are already severe enough to deny many millions of people basic needs for food, shelter, health, and jobs, or any hope for betterment. At the same time, the earth's carrying capacity—the ability of biological systems to provide resources for human needs—is eroding. The trends reflected in the Global 2000 Study suggest strongly a progressive degradation and improverishment of the earth's natural resource base.

If these trends are to be altered and the problems diminished, vigorous, determined new initiatives will be required worldwide to meet human needs while protecting and restoring the earth's capacity to support life. Basic natural resources—farmlands, fisheries, forests, minerals, energy, air, and water—must be conserved and better managed. Changes in public policy are needed around the world before problems worsen and options for effective action are reduced.

A number of responses to global resource, environment, and population problems—responses only touched on in the Study—are underway. Heightened international concern is reflected in the "Megaconferences" convened by the United Nations during the last decade: Human Environment (1972), Population (1974), Food (1974), Human Settlements (1976), Water (1977), Desertification (1977), Science and Technology for Development (1979), and New and Renewable Sources of Energy, scheduled for August 1981 in Nairobi. The United States has contributed actively to these conferences, proposing and supporting remedial actions of which many are now being taken. We are also working with other nations bilaterally, building concern for population growth, natural resources, and environment into our foreign aid programs and cooperating with our immediate neighbors on common problems ranging from cleanup of air and water pollution to preservation of soils and development of new crops. Many nations around the world are adopting new approaches—replanting deforested areas, conserving energy, making family planning measures widely available, using natural predators and selective pesticides to protect crops instead of broadscale destructive application of chemicals.

Nonetheless, given the urgency, scope, and complexity of the challenges before us, the efforts now underway around the world fall far short of what is needed. An era of unprecedented global cooperation and commitment is essential.

The necessary changes go beyond the capability of any single nation. But our nation can itself take important and exemplary steps. Because of our preeminent position as a producer and consumer of food and energy, our efforts to conserve soil, farmlands, and energy resources are of global, as well as national, importance. We can avoid polluting our own environment, and we must take care that we do not degrade the global environment.

Beyond our borders we can expand our collaboration with both developed and developing nations in a spirit of generosity and justice. Hundreds of millions of the world's people are now trapped in a condition of abject poverty. People at the margin of existence must take cropland, grazing land, and fuel where they can find it, regardless of the effects upon the earth's resource base. Sustainable economic development, coupled with environmental protection, resource management, and family planning is essential. Equally important are better understanding and effective responses to such global problems as the buildup of carbon dioxide in the atmosphere and the threat of species loss on a massive scale.

Finally, to meet the challenges described in the Global 2000 Study our federal government requires a much stronger capability to project and analyze long-term trends. The Study clearly points to the need for improving the present foundation for long-term planning. On this foundation rest decisions that involve the future welfare of the Nation.

We wish to express our thanks to and our admiration for the Director of the Global 2000 Study, Dr. Gerald O. Barney and his staff. Their diligence, dedication, and ability to bring forth the best from a legion of contributors is much appreciated. Special thanks is also due to those of the Council on Environmental Quality and the Department of State who worked closely with the Study and to the 11 other agencies that contributed greatly to it.* Without the detailed knowledge provided by these agencies' experts, the Global 2000 Study would have been impossible.

Respectfully,

THOMAS R. PICKERING
*Assistant Secretary, Oceans and International Environmental and Scientific Affairs, Department of State*

GUS SPETH
*Chairman, Council on Environmental Quality*

*The Federal agencies that cooperated with us in this effort were the Departments of Agriculture, Energy, and the Interior, the Agency for International Development, the Central Intelligence Agency, the Environmental Protection Agency, the Federal Emergency Management Agency, the National Aeronautics and Space Administration, the National Science Foundation, the National Oceanic and Atmospheric Administration, and the Office of Science and Technology Policy.

# Preface



> Environmental problems do not stop at national boundaries. In the past decade, we and other nations have come to recognize the urgency of international efforts to protect our common environment.
>
> As part of this process, I am directing the Council on Environmental Quality and the Department of State, working in cooperation with the Environmental Protection Agency, the National Science Foundation, the National Oceanic and Atmospheric Administration, and other appropriate agencies, to make a one-year study of the probable changes in the world's population, natural resources, and environment through the end of the century. This study will serve as the foundation of our longer-term planning.

President Carter issued this directive in his Environmental Message to the Congress on May 23, 1977. It marked the beginning of what became a three-year effort to discover the long-term implications of present world trends in population, natural resources, and the environment and to assess the Government's foundation for long-range planning.

Government concern with trends in population, resources, and environment is not new. Indeed, study of these issues by Federal commissions and planning boards extends back at least 70 years.[1] The earlier studies, however, tended to view each issue without relation to the others, to limit their inquiries to the borders of this nation and the short-term future, and to have relatively little effect on policy.[2] What is new in more recent studies is a growing awareness of the interdependence of population, resources, and environment. The Global 2000 Study is the first U.S. Government effort to look at all three issues from a long-term global perspective that recognizes their interrelationships and attempts to make connections among them.

The Global 2000 Study is reported in three volumes. This Summary is the first volume. Volume II, the Technical Report, presents the Study in further detail and is referenced extensively in this Summary. The third volume provides technical documentation on the Government's global models. All three volumes are available from the U.S. Government Printing Office.

# Major Findings and Conclusions

If present trends continue, the world in 2000 will be more crowded, more polluted, less stable ecologically, and more vulnerable to disruption than the world we live in now. Serious stresses involving population, resources, and environment are clearly visible ahead. Despite greater material output, the world's people will be poorer in many ways than they are today.

For hundreds of millions of the desperately poor, the outlook for food and other necessities of life will be no better. For many it will be worse. Barring revolutionary advances in technology, life for most people on earth will be more precarious in 2000 than it is now—unless the nations of the world act decisively to alter current trends.

This, in essence, is the picture emerging from the U.S. Government's projections of probable changes in world population, resources, and environment by the end of the century, as presented in the Global 2000 Study. They do not predict what will occur. Rather, they depict conditions that are likely to develop if there are no changes in public policies, institutions, or rates of technological advance, and if there are no wars or other major disruptions. A keener awareness of the nature of the current trends, however, may induce changes that will alter these trends and the projected outcome.

## Principal Findings

Rapid growth in world population will hardly have altered by 2000. The world's population will grow from 4 billion in 1975 to 6.35 billion in 2000, an increase of more than 50 percent. The rate of growth will slow only marginally, from 1.8 percent a year to 1.7 percent. In terms of sheer numbers, population will be growing faster in 2000 than it is today, with 100 million people added each year compared with 75 million in 1975. Ninety percent of this growth will occur in the poorest countries.

While the economies of the less developed countries (LDCs) are expected to grow at faster rates than those of the industrialized nations, the gross national product per capita in most LDCs remains low. The average gross national product per capita is projected to rise substantially in some LDCs (especially in Latin America), but in the great populous nations of South Asia it remains below $200 a year (in 1975 dollars). The large existing gap between the rich and poor nations widens.

World food production is projected to increase 90 percent over the 30 years from 1970 to 2000. This translates into a global per capita increase of less than 15 percent over the same period. The bulk of that increase goes to countries that already have relatively high per capita food consumption. Meanwhile per capita consumption in South Asia, the Middle East, and the LDCs of Africa will scarcely improve or will actually decline below present inadequate levels. At the same time, real prices for food are expected to double.

Arable land will increase only 4 percent by 2000, so that most of the increased output of food will have to come from higher yields. Most of the elements that now contribute to higher yields—fertilizer, pesticides, power for irrigation, and fuel for machinery—depend heavily on oil and gas.

During the 1990s world oil production will approach geological estimates of maximum production capacity, even with rapidly increasing petroleum prices. The Study projects that the richer industrialized nations will be able to command enough oil and other commercial energy supplies to meet rising demands through 1990. With the expected price increases, many less developed countries will have increasing difficulties meeting energy needs. For the one-quarter of humankind that depends primarily on wood for fuel, the outlook is bleak. Needs for fuelwood will exceed available supplies by about 25 percent before the turn of the century.

While the world's finite fuel resources—coal, oil, gas, oil shale, tar sands, and uranium—are theoretically sufficient for centuries, they are not evenly distributed; they pose difficult economic and environmental problems; and they vary greatly in their amenability to exploitation and use.

Nonfuel mineral resources generally appear sufficient to meet projected demands through 2000, but further discoveries and investments will be needed to maintain reserves. In addition, production costs will increase with energy prices and may make some nonfuel mineral resources uneconomic. The quarter of the world's population that inhabits industrial countries will continue to absorb three-fourths of the world's mineral production.

Regional water shortages will become more severe. In the 1970–2000 period population growth alone will cause requirements for water to double in nearly half the world. Still greater increases would be needed to improve standards of living. In many LDCs, water supplies will become increasingly erratic by 2000 as a result of extensive deforestation. Development of new water supplies will become more costly virtually everywhere.

Significant losses of world forests will continue over the next 20 years as demand for forest products and fuelwood increases. Growing stocks of commercial-size timber are projected to decline 50 percent per capita. The world's forests are now disappearing at the rate of 18–20 million hectares a year (an area half the size of California), with most of the loss occurring in the humid tropical forests of Africa, Asia, and South America. The projections indicate that by 2000 some 40 percent of the remaining forest cover in LDCs will be gone.

Serious deterioration of agricultural soils will occur worldwide, due to erosion, loss of organic matter, desertification, salinization, alkalinization, and waterlogging. Already, an area of cropland and grassland approximately the size of Maine is becoming barren wasteland each year, and the spread of desert-like conditions is likely to accelerate.

Atmospheric concentrations of carbon dioxide and ozone-depleting chemicals are expected to increase at rates that could alter the world's climate and upper atmosphere significantly by 2050. Acid rain from increased combustion of fossil fuels (especially coal) threatens damage to lakes, soils, and crops. Radioactive and other hazardous materials present health and safety problems in increasing numbers of countries.

Extinctions of plant and animal species will increase dramatically. Hundreds of thousands of species—perhaps as many as 20 percent of all species on earth—will be irretrievably lost as their habitats vanish, especially in tropical forests.

The future depicted by the U.S. Government projections, briefly outlined above, may actually understate the impending problems. The methods available for carrying out the Study led to certain gaps and inconsistencies that tend to impart an optimistic bias. For example, most of the individual projections for the various sectors studied—food, minerals, energy, and so on—assume that sufficient capital, energy, water, and land will be available in each of these sectors to meet their needs, regardless of the competing needs of the other sectors. More consistent, better-integrated projections would produce a still more emphatic picture of intensifying stresses, as the world enters the twenty-first century.

## Conclusions

At present and projected growth rates, the world's population would reach 10 billion by 2030 and would approach 30 billion by the end of the twenty-first century. These levels correspond closely to estimates by the U.S. National Academy of Sciences of the maximum carrying capacity of the entire earth. Already the populations in sub-Saharan Africa and in the Himalayan hills of Asia have exceeded the carrying capacity of the immediate area, triggering an erosion of the land's capacity to support life. The resulting poverty and ill health have further complicated efforts to reduce fertility. Unless this circle of interlinked problems is broken soon, population growth in such areas will unfortunately be slowed for reasons other than declining birth rates. Hunger and disease will claim more babies and young children, and more of those surviving will be mentally and physically handicapped by childhood malnutrition.

Indeed, the problems of preserving the carrying capacity of the earth and sustaining the possibility of a decent life for the human beings that inhabit it are enormous and close upon us. Yet there is reason for hope. It must be emphasized that the Global 2000 Study's projections are based on the assumption that national policies regarding population stabilization, resource conservation, and environmental protection will remain essentially unchanged through the end of the century. But in fact, policies are beginning to change. In some areas, forests are being replanted after cutting. Some nations are taking steps to reduce soil losses and desertification. Interest in energy conservation is growing, and large sums are being invested in exploring alternatives to petroleum dependence. The need for family planning is slowly becoming better understood. Water supplies are being improved and waste treatment systems built. High-yield seeds are widely available and seed banks are being expanded. Some wildlands with their genetic resources are being protected. Natural predators and selective pesticides are being substituted for persistent and destructive pesticides.

Encouraging as these developments are, they are far from adequate to meet the global challenges projected in this Study. Vigorous, determined new initiatives are needed if worsening poverty and human suffering, environmental degradation, and international tension and conflicts are to be prevented. There are no quick fixes. The only solutions to the problems of population, resources, and environment are complex and long-term. These problems are inextricably linked to some of the most perplexing and persistent problems in the world—poverty, injustice, and social conflict. New and imaginative ideas—and a willingness to act on them—are essential.

The needed changes go far beyond the capability and responsibility of this or any other single nation. An era of unprecedented cooperation and commitment is essential. Yet there are opportunities—and a strong rationale—for the United States to provide leadership among nations. A high priority for this Nation must be a thorough assessment of its foreign and domestic policies relating to population, resources, and environment. The United States, possessing the world's largest economy, can expect its policies to have a significant influence on global trends. An equally important priority for the United States is to cooperate generously and justly with other nations—particularly in the areas of trade, investment, and assistance—in seeking solutions to the many problems that extend beyond our national boundaries. There are many unfulfilled opportunities to cooperate with other nations in efforts to relieve poverty and hunger, stabilize population, and enhance economic and environmental productivity. Further cooperation among nations is also needed to strengthen international mechanisms for protecting and utilizing the "global commons"—the oceans and atmosphere.

To meet the challenges described in this Study, the United States must improve its ability to identify emerging problems and assess alternative responses. In using and evaluting the Government's present capability for long-term global analysis, the Study found serious inconsistencies in the methods and assumptions employed by the various agencies in making their projections. The Study itself made a start toward resolving these inadequacies. It represents the Government's first attempt to produce an interrelated set of population, resource, and environmental projections, and it has brought forth the most consistent set of global projections yet achieved by U.S. agencies. Nevertheless, the projections still contain serious gaps and contradictions that must be corrected if the Government's analytic capability is to be improved. It must be acknowledged that at present the Federal agencies are not always capable of providing projections of the quality needed for long-term policy decisions.



While limited resources may be a contributing factor in some instances, the primary problem is lack of coordination. The U.S. Government needs a mechanism for continuous review of the assumptions and methods the Federal agencies use in their projection models and for assurance that the agencies' models are sound, consistent, and well documented. The improved analyses that could result would provide not only a clearer sense of emerging problems and opportunities, but also a better means for evaluating alternative responses, and a better basis for decisions of worldwide significance that the President, the Congress, and the Federal Government as a whole must make.

With its limitations and rough approximations, the Global 2000 Study may be seen as no more than a reconnaissance of the future; nonetheless its conclusions are reinforced by similar findings of other recent global studies that were examined in the course of the Global 2000 Study (see Appendix). All these studies are in general agreement on the nature of the problems and on the threats they pose to the future welfare of humankind. The available evidence leaves no doubt that the world—including this Nation—faces enormous, urgent, and complex problems in the decades immediately ahead. Prompt and vigorous changes in public policy around the world are needed to avoid or minimize these problems before they become unmanageable. Long lead times are required for effective action. If decisions are delayed until the problems become worse, options for effective action will be severely reduced.

# The Study in Brief



The President's directive establishing the Global 2000 Study called for a "study of the probable changes in the world's population, natural resources, and environment through the end of the century" and indicated that the Study as a whole was to "serve as the foundation of our longer-term planning."[3] The findings of the Study identify problems to which world attention must be directed. But because all study reports eventually become dated and less useful, the Study's findings alone cannot provide the foundation called for in the directive. The necessary foundation for longer-term planning lies not in study findings *per se*, but in the Government's continuing institutional capabilities—skilled personnel, data, and analytical models—for developing studies and analyses. Therefore, to meet the objectives stated in the President's directive, the Global 2000 Study was designed not only to assess probable changes in the world's population, natural resources, and environment, but also, through the study process itself, to identify and strengthen the Government's capability for longer-term planning and analysis.[4]

## Building the Study

The process chosen for the Global 2000 Study was to develop trend projections using, to the fullest extent possible, the long-term global data and models routinely employed by the Federal agencies. The process also included a detailed analysis of the Government's global modeling capabilities as well as a comparison of the Government's findings with those of other global analyses.

An executive group, established and co-chaired by the Council on Environmental Quality and the State Department, together with a team of designated agency coordinators, assisted in locating the agencies' experts, data, and analytical models. A number of Americans from outside Government and several people from other countries advised on the study structure. The agencies' expert met occasionally with some of these advisers to work out methods for coordinating data, models, and assumptions.

Overall, the Federal agencies have an impressive capability for long-term analyses of world trends in population, resources, and environment. Several agencies have extensive, richly detailed data bases and highly elaborate sectoral models. Collectively, the agencies' sectoral models and data constitute the Nation's present foundation for long-term planning and analysis.[5]

Currently, the principal limitation in the Government's long-term global analytical capability is that the models for various sectors were not designed to be used together in a consistent and interactive manner. The agencies' models were created at different times, using different methods, to meet different objectives. Little thought has been given to how the various sectoral models—and the institutions of which they are a part—can be related to each other to project a comprehensive, consistent image of the world. As a result, there has been little direct interaction among the agencies' sectoral models.[6]

With the Government's current models, the individual sectors addressed in the Global 2000 Study could be interrelated only by developing projections sequentially, that is, by using the results of some of the projections as inputs to others. Since population and gross national product (GNP) projections were required to estimate demand in the resource sector models, the population and GNP projections were developed first, in 1977. The resource projections followed in late 1977 and early 1978. All of the projections were linked to the environment projections, which were made during 1978 and 1979.[7]

The Global 2000 Study developed its projections in a way that furthered interactions, improved internal consistency, and generally strengthened the Government's global models. However, the effort to harmonize and integrate the Study's projections was only partially successful. Many internal contradictions and inconsistencies could not be resolved. Inconsistencies arose immediately from the fact that sequential projections are not as interactive as events in the real world, or as projections that could be achieved in an improved model. While the sequential process allowed some interaction among the model's sectors, it omitted the continuous influence that all the elements—population, resources, economic activity, environment—have upon each other. For example, the Global 2000 Study food projections assume that the catch from traditional fisheries will increase as fast as world population, while the fisheries projections indicate that this harvest will not increase over present levels on a sustainable basis. If it has been possible to link the fisheries and food projections, the expected fisheries contribution to the human food supply could have been realistically reflected in the food projections. This and other inconsistencies are discussed in detail in the Technical Report.[8]

Difficulties also arise from multiple allocation of resources. Most of the quantitative projections simply assume that resource needs in the sector they cover—needs for capital, energy, land, water, minerals—will be met. Since the needs for each sector are not clearly identified, they cannot be summed up and compared with estimates of what might be available. It is very likely that the same resources have been allocated to more than one sector.[9]

Equally significant, some of the Study's resource projections implicitly assume that the goods and services provided in the past by the earth's land, air, and water will continue to be available in larger and larger amounts, with no maintenance problems and no increase in costs. The Global 2000 Study projections for the environment cast serious doubt on these assumptions.[10]

Collectively, the inconsistencies and missing linkages that are unavoidable with the Government's current global models affect the Global 2000 projections in many ways. Analysis of the assumptions underlying the projections and comparisons with other global projections suggest that most of the Study's quantitative results understate the severity of potential problems the world will face as it prepares to enter the twenty-first century.[11]

The question naturally arises as to whether circumstances have changed significantly since the earliest projections were made in 1977. The answer is no. What changes have occurred generally support the projections and highlight the problems identified. The brief summaries of the projections (beginning on the next page) each conclude with comments on how the projections might be altered if redeveloped today.

The Global 2000 Study has three major underlying assumptions. First, the projections assume a general continuation around the world of present public policy relating to population stabilization, natural resource conservation, and environmental protection.* The projections thus point to the expected future if policies continue without significant changes.

The second major assumption relates to the effects of technological developments and of the market mechanism. The Study assumes that rapid rates of technological development and adoption will continue, and that the rate of development will be spurred on by efforts to deal with problems identified by this Study. Participating agencies were asked to use the technological assumptions they normally use in preparing long-term global projections. In general, the agencies assume a continuation of rapid rates of technological development and no serious social resistance to the adoption of new technologies. Agricultural technology, for example, is assumed to continue increasing crop yields as rapidly as during the past few decades, including the period of the Green Revolution (see Figure 1). The projections assume no revolutionary advances—such as immediate

---

*There are a few important exceptions to this rule. For example, the population projections anticipate shifts in public policy that will provide significantly increased access to family planning services. (See Chapter 14 of the Technical Report for further details.)

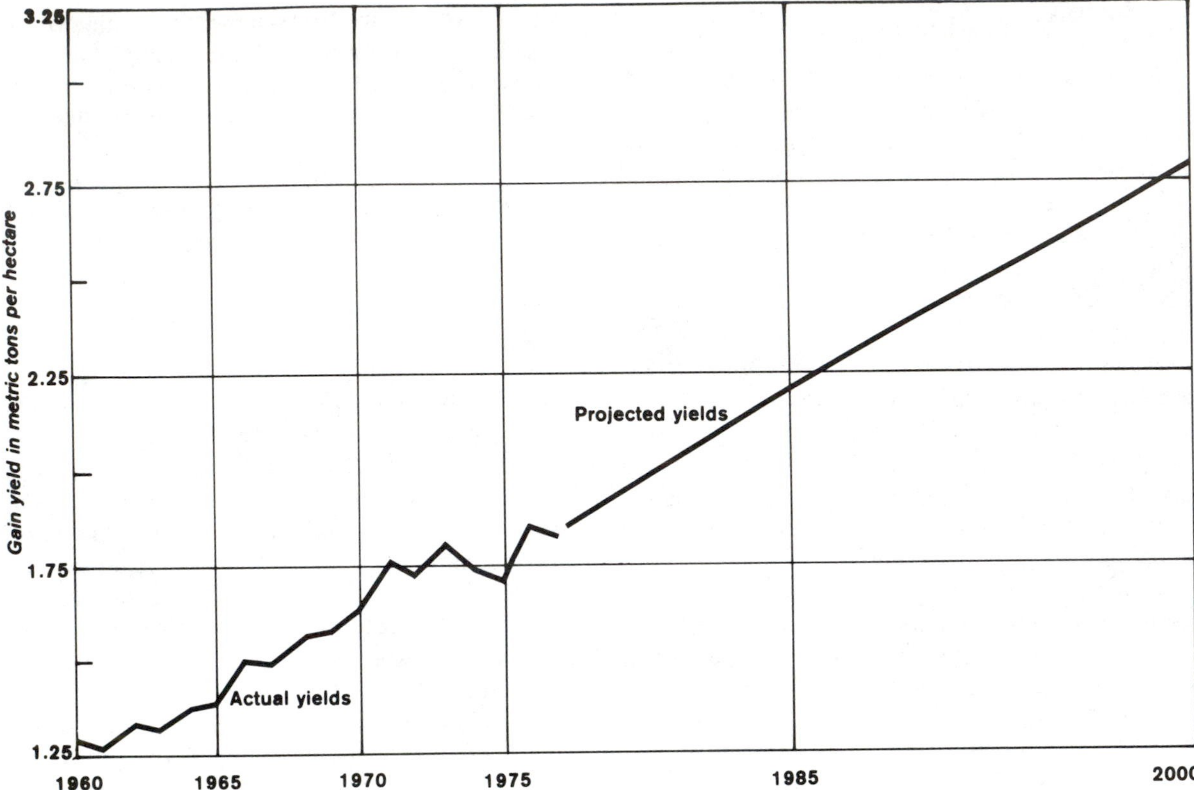


**Figure 1.** Historic and projected grain yields, 1960–2000. The food projections assume a continued rapid development and adoption of agricultural technology, much of it heavily dependent on fossil fuels.

wide-scale availability of nuclear fusion for energy production—and no disastrous setbacks—such as serious new health risks from widely used contraceptives or an outbreak of plant disease severely affecting an important strain of grain. The projections all assume that price, operating through the market mechanism, will reduce demand whenever supply constraints are encountered.[12]

Third, the Study assumes that there will be no major disruptions of international trade as a result of war, disturbance of the international monetary system, or political disruption. The findings of the Study do, however, point to increasing potential for international conflict and increasing stress on international financial arrangements. Should wars or a significant disturbance of the international monetary system occur, the projected trends would be altered in unpredictable ways.[13]

Because of the limitations outlined above, the Global 2000 Study is not the definitive study of future population, resource, and environment conditions. Nor is it intended to be a prediction. The Study does provide the most internally consistent and interrelated set of global projections available so far from the U.S. Government. Furthermore, its major findings are supported by a variety of nongovernmental global studies based on more highly interactive models that project similar trends through the year 2000 or beyond.[14]

## Population and Income

Population and income projections provided the starting point for the Study. These projections were used wherever possible in the resource projections to estimate demand.

### Population

One of the most important findings of the Global 2000 Study is that enormous growth in the world's population will occur by 2000 under any of the wide range of assumptions considered in the Study. The world's population increases 55 percent from 4.1 billion people in 1975 to 6.35 billion by 2000, under the Study's medium-growth projections.* While there is some uncertainty in these numbers, even the lowest-growth population projection shows a 46 percent increase—to 5.9 billion people by the end of the century.[15]

Another important finding is that the rapid growth of the world's population will not slow appreciably. The rate of growth per year in 1975 was 1.8 percent; the projected rate for 2000 is 1.7 percent. Even under the lowest growth projected, the number of persons being added annually to the world's population will be significantly greater in 2000 than today.[16]

Most of the population growth (92 percent) will occur in the less developed countries rather than in the industrialized countries. Of the 6.35 billion people in the world in 2000, 5 billion will live in LDCs. The LDCs' share of the world's population increased from 66 percent in 1950 to 72 percent in 1975, and is expected to reach 79 percent by 2000. LDC population growth rates will drop slightly, from 2.2 percent a year in 1975 to 2 percent in 2000, compared with 0.7 percent and 0.5 percent in developed countries. In some LDCs, growth rates will still be more than 3 percent a year in 2000. Table 1 summarizes the population projections. Figure 2 shows the distribution of the world's population in 1975 and 2000.[17]

Figure 3 shows the age structure of the population in less developed and industrialized nations for 1975 and 2000. While the structures shown for the industrialized nations become more column-shaped (characteristic of a mature and slowly growing population), the structures for the LDCs remain pyramid-shaped (characteristic of rapid growth). The LDC populations, predominantly young with their childbearing years ahead of them, have a built-in momentum for further growth. Because of this momentum, a world population of around 6 billion is a virtual certainty for 2000 even if fertility rates were somehow to drop quickly to replacement levels (assuming there are no disastrous wars, famine, or pestilence).[18]

The projected fertility rates and life expectancies, together with the age structure of the world's population, are extremely significant for later years since these factors influence how soon world population could cease to grow and what the ultimate stabilized global population could be. The Study's projections assume that world fertility rates will drop more than 20 percent over the 1975–2000 period, from an average of 4.3 children per fertile woman to 3.3. In LDCs, fertility rates are projected to drop 30 percent as a result of moderate progress in social and economic development and increased availability and use of contraceptive methods. The projections also assume that life expectancies at birth for the world will increase 11 percent, to 65.5 years, as a result of improved health. The projected increases in life expectancies and decreases in fertility rates produce roughly counterbalancing demographic effects.[19]

In addition to rapid population growth, the LDCs will experience dramatic movements of rural populations to cities and adjacent settlements. If present trends continue, many LDC cities will become almost inconceivably large and crowded. By 2000, Mexico City is projected to have more than 30 million people—roughly three times the present population of the New York metropolitan area. Calcutta will approach 20 million. Greater Bombay, Greater

*Most of the projections in the Technical Report—including the population projections—provide a high, medium, and low series. Generally, only the medium series are discussed in this Summary Report.

**TABLE 1**
**Population Projections for World, Major Regions, and Selected Countries**

| | 1975 | 2000 | Percent Increase by 2000 | Average Annual Percent Increase | Percent of World Population in 2000 |
|---|---|---|---|---|---|
| | *millions* | | | | |
| World | 4,090 | 6,351 | 55 | 1.8 | 100 |
| More developed regions | 1,131 | 1,323 | 17 | 0.6 | 21 |
| Less developed regions | 2,959 | 5,028 | 70 | 2.1 | 79 |
| Major regions | | | | | |
| Africa | 399 | 814 | 104 | 2.9 | 13 |
| Asia and Oceania | 2,274 | 3,630 | 60 | 1.9 | 57 |
| Latin America | 325 | 637 | 96 | 2.7 | 10 |
| U.S.S.R. and Eastern Europe | 384 | 460 | 20 | 0.7 | 7 |
| North America, Western Europe, Japan, Australia, and New Zealand | 708 | 809 | 14 | 0.5 | 13 |
| Selected countries and regions | | | | | |
| People's Republic of China | 935 | 1,329 | 42 | 1.4 | 21 |
| India | 618 | 1,021 | 65 | 2.0 | 16 |
| Indonesia | 135 | 226 | 68 | 2.1 | 4 |
| Bangladesh | 79 | 159 | 100 | 2.8 | 2 |
| Pakistan | 71 | 149 | 111 | 3.0 | 2 |
| Philippines | 43 | 73 | 71 | 2.1 | 1 |
| Thailand | 42 | 75 | 77 | 2.3 | 1 |
| South Korea | 37 | 57 | 55 | 1.7 | 1 |
| Egypt | 37 | 65 | 77 | 2.3 | 1 |
| Nigeria | 63 | 135 | 114 | 3.0 | 2 |
| Brazil | 109 | 226 | 108 | 2.9 | 4 |
| Mexico | 60 | 131 | 119 | 3.1 | 2 |
| United States | 214 | 248 | 16 | 0.6 | 4 |
| U.S.S.R. | 254 | 309 | 21 | 0.8 | 5 |
| Japan | 112 | 133 | 19 | 0.7 | 2 |
| Eastern Europe | 130 | 152 | 17 | 0.6 | 2 |
| Western Europe | 344 | 378 | 10 | 0.4 | 6 |

*Source:* Global 2000 Technical Report, Table 2-10.

Cairo, Jakarta, and Seoul are all expected to be in the 15–20 million range, and 400 cities will have passed the million mark.[20] Table 2 shows present and projected populations for 12 LDC cities.

Rapid urban growth will put extreme pressures on sanitation, water supplies, health care, food, shelter, and jobs. LDCs will have to increase urban services approximately two-thirds by 2000 just to stay even with 1975 levels of service per capita. The majority of people in large LDC cities are likely to live in "uncontrolled settlements"—slums and shantytowns where sanitation and other public services are minimal at best. In many large cities—for example, Bombay, Calcutta, Mexico City, Rio de Janeiro, Seoul, Taipei—a quarter or more of the population already lives in uncontrolled settlements, and the trend is sharply upward. It is not certain whether the trends projected

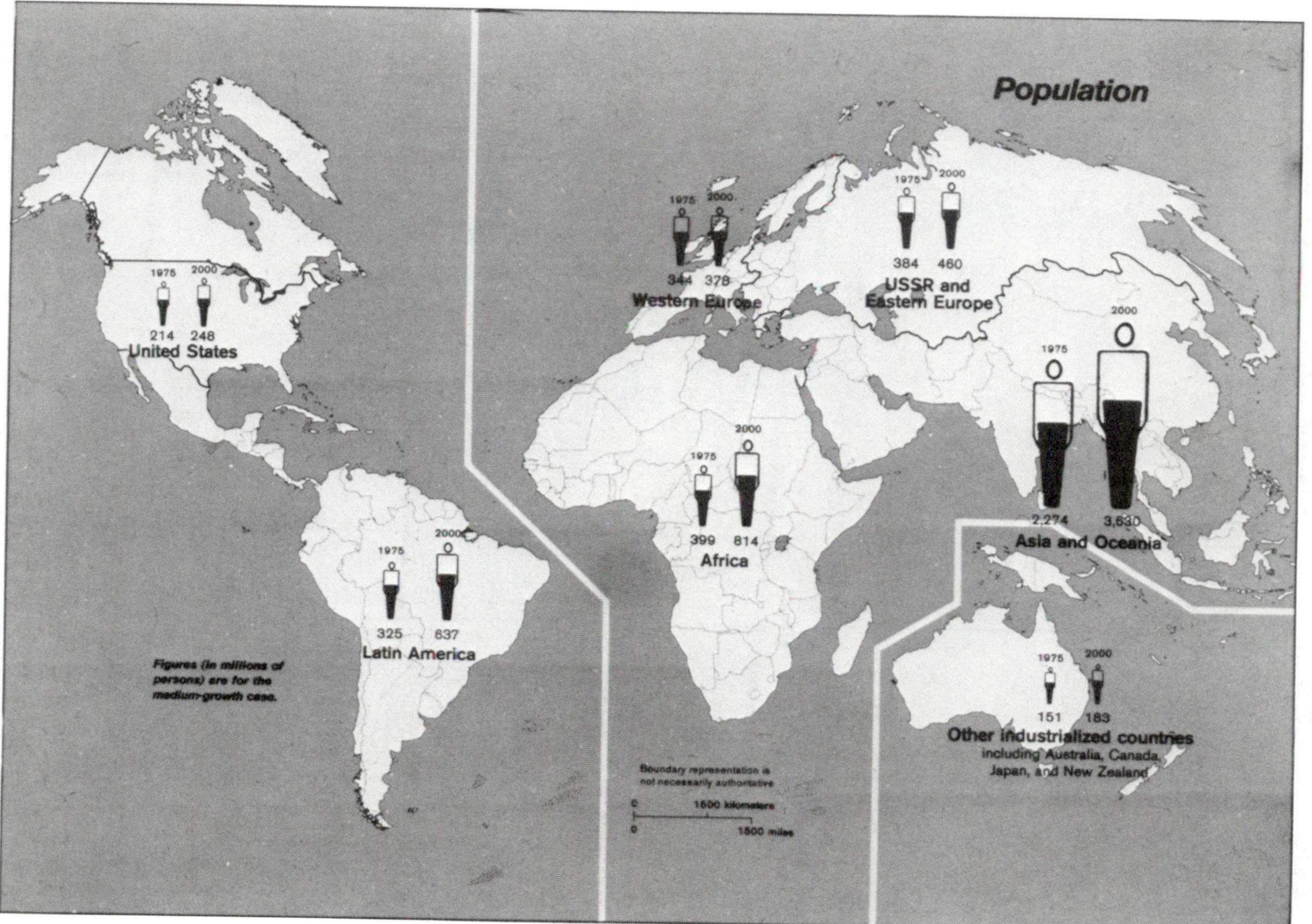


**Figure 2.** Distribution of the world's population, 1975 and 2000.

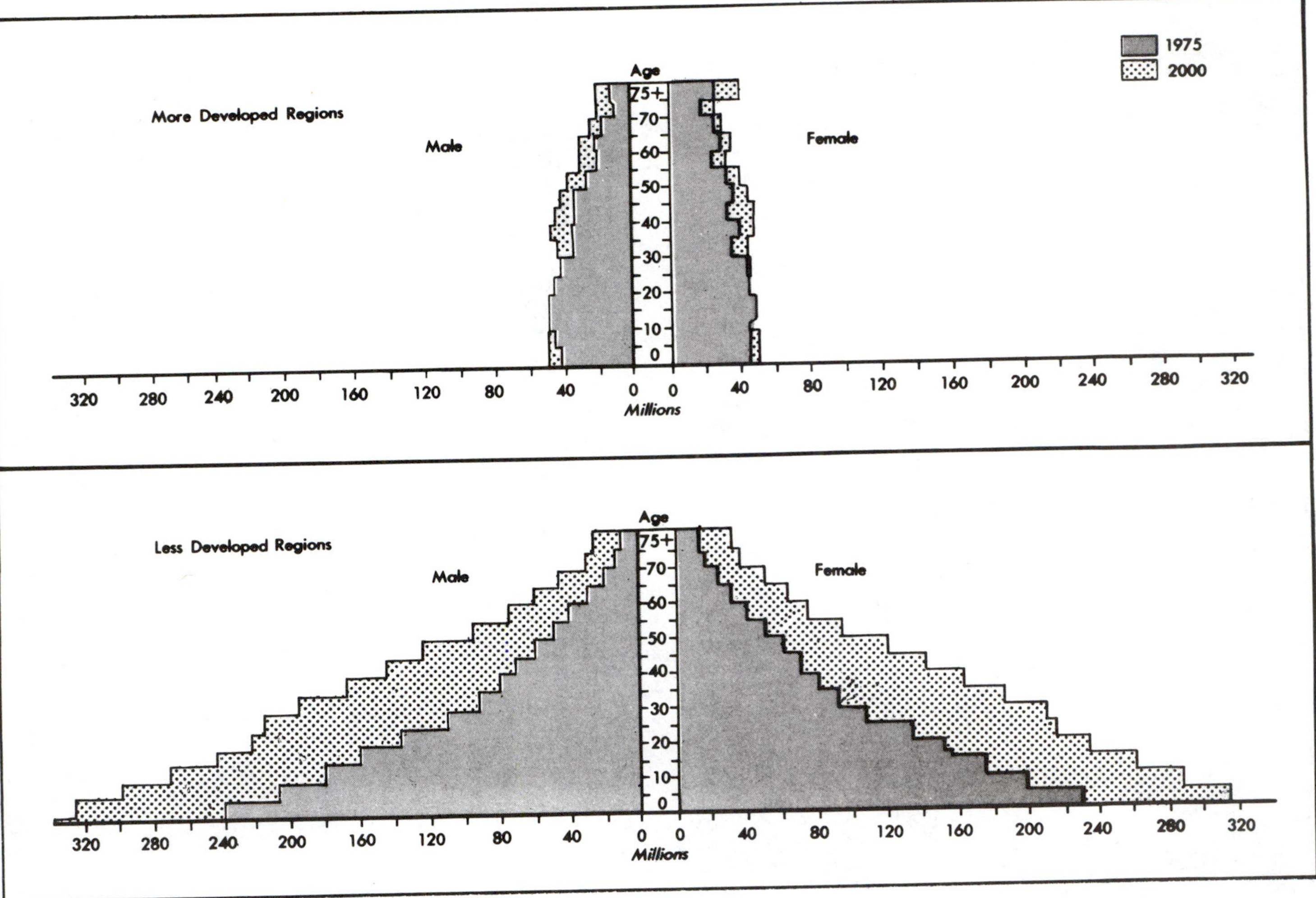


Figure 3. Age-sex composition of the world's population, medium series, 1975 and 2000.

**TABLE 2**
**Estimates and Rough Projections of Selected Urban Agglomerations in Developing Countries**

| | 1960 | 1970 | 1975 | 2000 |
|---|---|---|---|---|
| | *Millions of persons* | | | |
| Calcutta | 5.5 | 6.9 | 8.1 | 19.7 |
| Mexico City | 4.9 | 8.6 | 10.9 | 31.6 |
| Greater Bombay | 4.1 | 5.8 | 7.1 | 19.1 |
| Greater Cairo | 3.7 | 5.7 | 6.9 | 16.4 |
| Jakarta | 2.7 | 4.3 | 5.6 | 16.9 |
| Seoul | 2.4 | 5.4 | 7.3 | 18.7 |
| Delhi | 2.3 | 3.5 | 4.5 | 13.2 |
| Manila | 2.2 | 3.5 | 4.4 | 12.7 |
| Tehran | 1.9 | 3.4 | 4.4 | 13.8 |
| Karachi | 1.8 | 3.3 | 4.5 | 15.9 |
| Bogota | 1.7 | 2.6 | 3.4 | 9.5 |
| Lagos | 0.8 | 1.4 | 2.1 | 9.4 |

*Source*: Global 2000 Technical Report, Table 13-9.

for enormous increases in LDC urban populations will in fact continue for 20 years. In the years ahead, lack of food for the urban poor, lack of jobs, and increasing illness and misery may slow the growth of LDC cities and alter the trend.[21]

Difficult as urban conditions are, conditions in rural areas of many LDCs are generally worse. Food, water, health, and income problems are often most severe in outlying agricultural and grazing areas. In some areas rural-urban migration and rapid urban growth are being accelerated by deteriorating rural conditions.[22]

An updated medium-series population projection would show little change from the Global 2000 Study projections. World population in 2000 would be estimated at about 6.18 (as opposed to 6.35) billion, a reduction of less than 3 percent. The expectation would remain that, in absolute numbers, population will be growing more rapidly by the end of the century than today.[23]

The slight reduction in the population estimate is due primarily to new data suggesting that fertility rates in some areas have declined a little more rapidly than earlier estimates indicated. The new data indicate that fertility declines have occurred in some places even in the absence of overall socioeconomic progress.[24] Between 1970 and 1976, for example, in the presence of extreme poverty and malnutrition, fertility declines of 10–15 percent occurred in Indonesia and 15–20 percent in the poorest income classes in Brazil.[25]

## Income

Projected declines in fertility rates are based in part on anticipated social and economic progress, which is ultimately reflected in increased income. Income projections were not possible, and gross national product projections were used as surrogates. GNP, a rough and inadequate measure of social and economic welfare, is projected to increase worldwide by 145 percent over 25 years from 1975 to 2000. But because of population growth, per capita GNP increases much more slowly, from $1,500 in 1975 to $2,300 in 2000—an increase of 53 percent. For both the poorer and the richer countries, rates of growth in GNP are projected to decelerate after 1985.[26]

GNP growth is expected to be faster in LDCs (an average annual growth of 4.5 percent, or an approximate tripling over 25 years) than in developed regions (an average annual growth of 3.3 percent, or somewhat more than a doubling). However, the LDC growth in gross national product develops from a very low base, and population growth in the LDCs brings per capita increases in GNP down to very modest proportions. While parts of the LDC world, especially several countries in Latin America, are projected to improve significantly in per capita GNP by 2000, other countries will make little or no gains from their present low levels. India, Bangladesh, and Pakistan, for example, increase their per capita GNP by 31 percent, 8 percent, and 3 percent, respectively, but in all three countries GNP per capita remains below $200 (in 1975 dollars).[27] Figure 4 shows projected per capita gross national product by regions in 2000.

The present income disparities between the wealthiest and poorest nations are projected to widen. Assuming that present trends continue, the group of industrialized countries will have a per

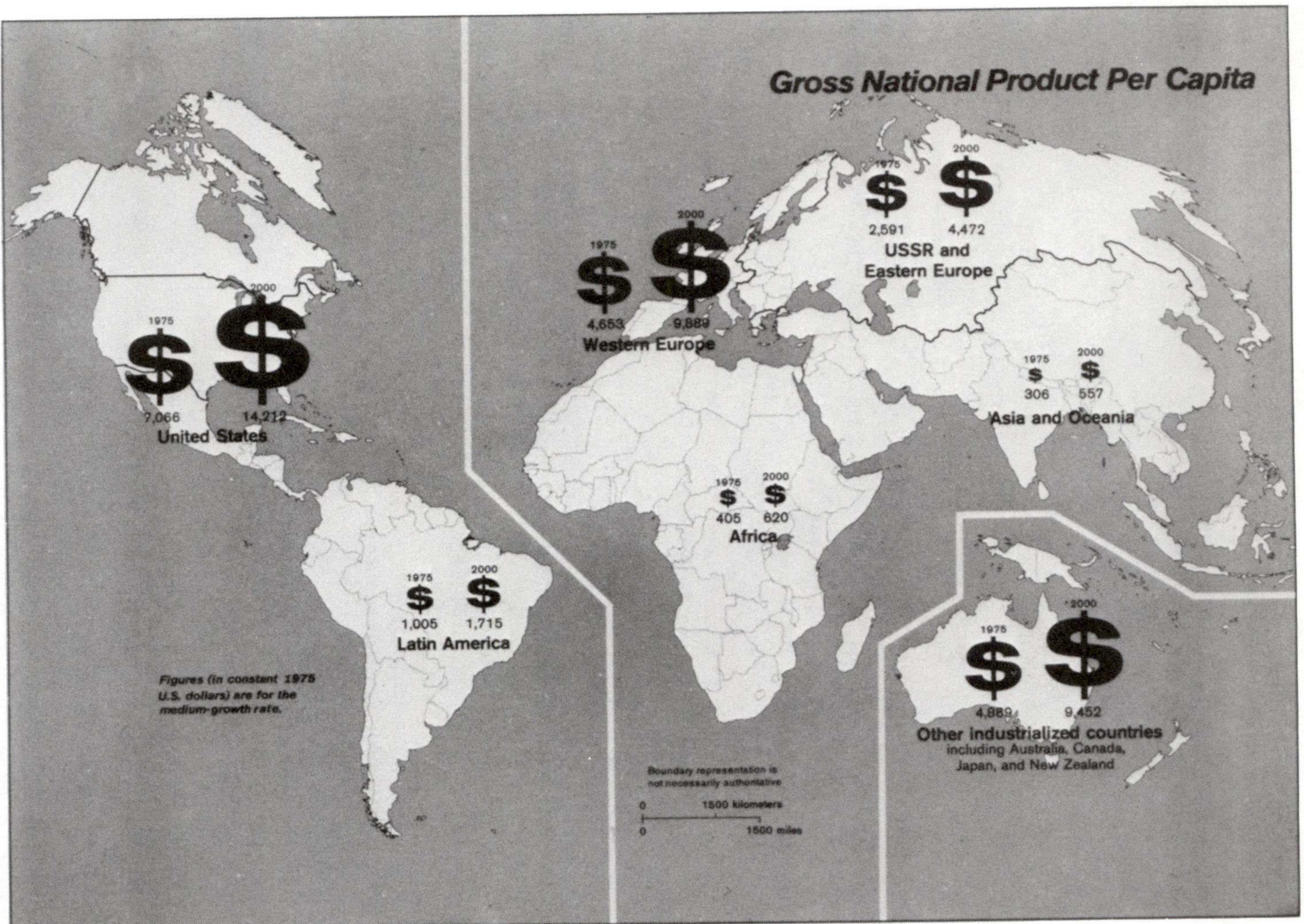


**Figure 4.** Per capita gross national product, by regions, 1975 and 2000.

capita GNP of nearly $8,500 (in 1975 dollars) in 2000, and North America, Western Europe, Australia, New Zealand, and Japan will average more than $11,000. By contrast, per capita GNP in the LDCs will average less than $600. For every $1 increase in GNP per capita in the LDCs, a $20 increase is projected for the industrialized countries.[28] Table 3 and 4 summarize the GNP

**TABLE 3**

**GNP Estimates (1975) and Projections and Growth Rates (1985, 2000) by Major Regions and Selected Countries and Regions**

*(Billions of constant 1975 dollars)*

| | 1975 GNP | 1975–85 Growth Rate | 1985 Projections[a] | 1985–2000 Growth Rate | 2000 Projections[a] |
|---|---|---|---|---|---|
| | | *percent* | | *percent* | |
| WORLD | 6,025 | 4.1 | 8,991 | 3.3 | 14,677 |
| More developed regions | 4,892 | 3.9 | 7,150 | 3.1 | 11,224 |
| Less developed regions | 1,133 | 5.0 | 1,841 | 4.3 | 3,452 |
| MAJOR REGIONS | | | | | |
| Africa | 162 | 5.2 | 268 | 4.3 | 505 |
| Asia and Oceania | 697 | 4.6 | 1,097 | 4.2 | 2,023 |
| Latin America[b] | 326 | 5.6 | 564 | 4.5 | 1,092 |
| U.S.S.R. and Eastern Europe | 996 | 3.3 | 1,371 | 2.8 | 2,060 |
| North America, Western Europe, Japan, Australia, and New Zealand | 3,844 | 4.0 | 5,691 | 3.1 | 8,996 |
| SELECTED COUNTRIES AND REGIONS[c] | | | | | |
| People's Republic of China | 286 | 3.8 | 413 | 3.8 | 718 |
| India | 92 | 3.6 | 131 | 2.8 | 198 |
| Indonesia | 24 | 6.4 | 45 | 5.4 | 99 |
| Bangladesh | 9 | 3.6 | 13 | 2.8 | 19 |
| Pakistan | 10 | 3.6 | 14 | 2.8 | 21 |
| Philippines | 16 | 5.6 | 27 | 4.4 | 52 |
| Thailand | 15 | 5.6 | 25 | 4.4 | 48 |
| South Korea | 19 | 5.6 | 32 | 4.4 | 61 |
| Egypt | 12 | 5.6 | 20 | 4.4 | 38 |
| Nigeria | 23 | 6.4 | 43 | 5.4 | 94 |
| Brazil | 108 | 5.6 | 185 | 4.4 | 353 |
| Mexico | 71 | 5.6 | 122 | 4.4 | 233 |
| United States[d] | 1,509 | 4.0 | 2,233 | 3.1 | 3,530 |
| U.S.S.R. | 666 | 3.3 | 917 | 2.8 | 1,377 |
| Japan | 495 | 4.0 | 733 | 3.1 | 1,158 |
| Eastern Europe (excluding U.S.S.R.) | 330 | 3.3 | 454 | 2.8 | 682 |
| Western Europe | 1,598 | 4.0 | 2,366 | 3.1 | 3,740 |

[a]Projected growth rates of gross national product were developed using complex computer simulation techniques described in Chapter 16 of the Global 2000 Technical Report. These projections represent the result of applying those projected growth rates to the 1975 GNP data presented in the 1976 World Bank Atlas. Projections shown here are for medium-growth rates.

[b]Includes Puerto Rico.

[c]In most cases, gross national income growth rates were projected for groups of countries rather than for individual countries. Thus the rates attributed to individual LDCs in this table are the growth rates applicable to the group with which that country was aggregated for making projections and do not take into account country specific characteristics.

[d]Does not include Puerto Rico.

*Source:* Global 2000 Technical Report, Table 3-3.

## TABLE 4

**Per Capita GNP Estimates (1975) and Projections and Growth Rates (1985, 2000) by Major Regions and Selected Countries and Regions**

*(Constant 1975 U.S. dollars)*

| | 1975 | Average Annual Growth Rate, 1975–85 | 1985 Projections[a] | Average Annual Growth Rate, 1985–2000 | 2000 Projections[a] |
|---|---|---|---|---|---|
| | | *percent* | | *percent* | |
| WORLD | 1,473 | 2.3 | 1,841 | 1.5 | 2,311 |
| More developed countries | 4,325 | 3.2 | 5,901 | 2.5 | 8,485 |
| Less developed countries | 382 | 2.8 | 501 | 2.1 | 587 |
| MAJOR REGIONS | | | | | |
| Africa | 405 | 2.2 | 505 | 1.4 | 620 |
| Asia and Oceania | 306 | 2.7 | 398 | 2.3 | 557 |
| Latin America[b] | 1,005 | 2.6 | 1,304 | 1.8 | 1,715 |
| U.S.S.R. and Eastern Europe | 2,591 | 2.4 | 3,279 | 2.1 | 4,472 |
| North America, Western Europe, Japan, Australia, and New Zealand | 5,431 | 3.4 | 7,597 | 2.6 | 11,117 |
| SELECTED COUNTRIES AND REGIONS[c] | | | | | |
| People's Republic of China | 306 | 2.3 | 384 | 2.3 | 540 |
| India | 148 | 1.5 | 171 | 0.8 | 194 |
| Indonesia | 179 | 4.1 | 268 | 3.1 | 422 |
| Bangladesh | 111 | 0.6 | 118 | 0.1 | 120 |
| Pakistan | 138 | 0.4 | 144 | −0.1 | 142 |
| Philippines | 368 | 3.2 | 503 | 2.3 | 704 |
| Thailand | 343 | 3.0 | 460 | 2.2 | 633 |
| South Korea | 507 | 3.5 | 718 | 2.7 | 1,071 |
| Egypt | 313 | 2.9 | 416 | 2.2 | 578 |
| Nigeria | 367 | 3.3 | 507 | 2.2 | 698 |
| Brazil | 991 | 2.2 | 1,236 | 1.6 | 1,563 |
| Mexico | 1,188 | 2.0 | 1,454 | 1.3 | 1,775 |
| United States[d] | 7,066 | 3.3 | 9.756 | 2.5 | 14,212 |
| U.S.S.R. | 2,618 | 2.3 | 3,286 | 2.1 | 4,459 |
| Japan | 4,437 | 3.1 | 6,023 | 2.5 | 8,712 |
| Eastern Europe | 2,539 | 2.6 | 3,265 | 2.2 | 4,500 |
| Western Europe | 4,653 | 3.7 | 6,666 | 2.7 | 9,889 |

[a]The medium-series projections of gross national product and population presented in Tables 3-3 and 3-4 of the Global 2000 Technical Report were used to calculate the 1975, 1985, and 2000 per capita gross national product figures presented in this table.

[b]Includes Puerto Rico.

[c]In most cases, gross national product growth rates were projected for groups of countries rather than for individual countries. Thus, the rates attributed to individual LDCs in this table are the growth rates applicable to the group with which that country was aggregated for making projections and do not take into account country-specific characteristics.

[d]Does not include Puerto Rico.

*Source*: Global 2000 Technical Report, Table 3-5.

projections. The disparity between the developed countries and the less developed group is so marked that dramatically different rates of change would be needed to reduce the gap significantly by the end of the century.* Disparities between the rich and poor of many LDCs are equally striking.

Updated GNP projections would indicate somewhat lower economic growth than shown in the Global 2000 projections. Projections for the member nations of the Organization for Economic Cooperation and Development (OECD) have been revised downward over the past 2–3 years because of the effects of increasing petroleum prices and because of anticipated measures to reduce inflation. In turn, depressed growth in the OECD economies is expected to lead to slowed growth in LDC economies. For example, in 1976 the World Bank projected that the industrialized nations' economies would expand at 4.9 percent annually over the 1980–85 period; by 1979 the Bank had revised these projections downward to 4.2 percent annually over the 1980–90 period. Similarly, between 1976 and 1979 Bank projections for LDC economies dropped from 6.3 percent (1980-85 period) to 5.6 percent (1980-90 period).[29]

## Resources

The Global 2000 Study resource projections are based to the fullest extent possible on the population and GNP projections presented previously. The resource projections cover food, fisheries, forests, nonfuel minerals, water, and energy.

### Food

The Global 2000 Study projects world food production to increase at an average annual rate of about 2.2 percent over the 1970–2000 period. This rate of increase is roughly equal to the record growth rates experienced during the 1950s, 1960s, and early 1970s, including the period of the so-called Green Revolution. Assuming no deterioration in climate or weather, food production is projected to be 90 percent higher in 2000 than in 1970.[30]

The projections indicate that most of the increase in food production will come from more intensive use of yield-enhancing, energy-intensive inputs and technologies such as fertilizer, pesticides, herbicides, and irrigation—in many cases with diminishing returns. Land under cultivation is projected to increase only 4 percent by 2000 because most good land is already being cultivated. In the early 1970s one hectare of arable land supported an average of 2.6 persons; by 2000 one hectare will have to support 4 persons. Because of this tightening land constraint, food production is not likely to increase fast enough to meet rising demands unless world agriculture becomes significantly more dependent on petroleum and petroleum-related inputs. Increased petroleum dependence also has implications for the cost of food production. After decades of generally falling prices, the real price of food is projected to increase 95 percent over the 1970–2000 period, in significant part as a result of increased petroleum dependence.[31] If energy prices in fact rise more rapidly than the projections anticipate, then the effect on food prices could be still more marked.

On the average, world food production is projected to increase more rapidly than world population, with average per capita consumption increasing about 15 percent between 1970 and 2000. Per capita consumption in the industrialized nations is projected to rise 21 percent from 1970 levels, with increases of from 40 to more than 50 percent in Japan, Eastern Europe, and the U.S.S.R., and 28 percent in the United States.* In

*The gap would be significantly smaller—in some cases it would be reduced by about one half—if the comparison were based on purchasing power considerations rather than exchange rates, but a large gap would remain. (See I. B. Kravis et al., *International Comparisons of Real Product and Purchasing Power*, Baltimore: Johns Hopkins University Press. 1978.)

*"Consumption" statistics are based on the amount of food that leaves the farms and does not leave the country and therefore include transportation and processing losses. Projected increases in per capita consumption in countries like the United States, where average consumption is already at least nutritionally adequate, reflect increasing losses of food during transportation and processing and might also be accounted for by increased industrial demand for grain, especially for fermentation into fuels.

the LDCs, however, rising food output will barely keep ahead of population growth.[32]

An increase of 9 percent in per capita food consumption is projected for the LDCs as a whole, but with enormous variations among regions and nations. The great populous countries of South Asia—expected to contain 1.3 billion people by 2000—improve hardly at all, nor do large areas of low-income North Africa and the Middle East. Per capita consumption in the sub-Saharan African LDCs will actually decline, according to the projections. The LDCs showing the greatest per capita growth (increases of about 25 percent) are concentrated in Latin America and East Asia.[33] Table 5 summarizes the projections for food production and consumption, and Table

**TABLE 5**
**Grain Production, Consumption, and Trade, Actual and Projected, and Percent Increase in Total Food Production and Consumption**

| | Grain *(million metric tons)* | | | Food *(Percent increase over the 1970-2000 period)* |
|---|---|---|---|---|
| | 1969–71 | 1973–75 | 2000 | |
| Industrialized countries | | | | |
| Production | 401.7 | 434.7 | 679.1 | 43.7 |
| Consumption | 374.3 | 374.6 | 610.8 | 47.4 |
| Trade | +32.1 | +61.6 | +68.3 | |
| United States | | | | |
| Production | 208.7 | 228.7 | 402.0 | 78.5 |
| Consumption | 169.0 | 158.5 | 272.4 | 51.3 |
| Trade | +39.9 | +72.9 | +129.6 | |
| Other developed exporters | | | | |
| Production | 58.6 | 61.2 | 106.1 | 55.6 |
| Consumption | 33.2 | 34.3 | 65.2 | 66.8 |
| Trade | +28.4 | +27.7 | +40.9 | |
| Western Europe | | | | |
| Production | 121.7 | 132.9 | 153.0 | 14.6 |
| Consumption | 144.2 | 151.7 | 213.1 | 31.6 |
| Trade | −21.8 | −19.7 | −60.1 | |
| Japan | | | | |
| Production | 12.7 | 11.9 | 18.0 | 31.5 |
| Consumption | 27.9 | 30.1 | 60.1 | 92.8 |
| Trade | −14.4 | −19.3 | −42.1 | |
| Centrally planned countries | | | | |
| Production | 401.0 | 439.4 | 722.0 | 74.0 |
| Consumption | 406.6 | 472.4 | 758.5 | 79.9 |
| Trade | −5.2 | −24.0 | −36.5 | |
| Eastern Europe | | | | |
| Production | 72.1 | 89.4 | 140.0 | 83.2 |
| Consumption | 78.7 | 97.7 | 151.5 | 81.7 |
| Trade | −6.1 | −7.8 | −11.5 | |
| U.S.S.R. | | | | |
| Production | 165.0 | 179.3 | 290.0 | 72.7 |
| Consumption | 161.0 | 200.7 | 305.0 | 85.9 |
| Trade | +3.9 | −10.6 | −15.0 | |
| People's Republic of China | | | | |
| Production | 163.9 | 176.9 | 292.0 | 69.0 |
| Consumption | 166.9 | 180.8 | 302.0 | 71.4 |
| Trade | −3.0 | −3.9 | −10.0 | |

**TABLE 5** (Cont.)

| | Grain (*million metric tons*) | | | Food (*Percent increase over the 1970-2000 period)* |
|---|---|---|---|---|
| | 1969–71 | 1973–75 | 2000 | |
| Less developed countries | | | | |
| Production | 306.5 | 328.7 | 740.6 | 147.7 |
| Consumption | 326.6 | 355.0 | 772.4 | 142.8 |
| Trade | −18.5 | −29.5 | −31.8 | |
| Exporters[a] | | | | |
| Production | 30.1 | 34.5 | 84.0 | 125.0 |
| Consumption | 18.4 | 21.5 | 36.0 | 58.0 |
| Trade | +11.3 | +13.1 | +48.0 | |
| Importers[b] | | | | |
| Production | 276.4 | 294.2 | 656.6 | 149.3 |
| Consumption | 308.2 | 333.5 | 736.4 | 148.9 |
| Trade | −29.8 | −42.6 | −79.8 | |
| Latin America | | | | |
| Production | 63.8 | 72.0 | 185.9 | 184.4 |
| Consumption | 61.2 | 71.2 | 166.0 | 165.3 |
| Trade | +3.2 | +0.2 | +19.9 | |
| North Africa/Middle East | | | | |
| Production | 38.9 | 42.4 | 89.0 | 157.8 |
| Consumption | 49.5 | 54.1 | 123.7 | 167.3 |
| Trade | −9.1 | −13.8 | −29.7 | |
| Other African LDCs | | | | |
| Production | 32.0 | 31.3 | 63.7 | 104.9 |
| Consumption | 33.0 | 33.8 | 63.0 | 96.4 |
| Trade | −1.0 | −2.4 | +0.7 | |
| South Asia | | | | |
| Production | 119.1 | 127.7 | 259.0 | 116.8 |
| Consumption | 125.3 | 135.1 | 275.7 | 119.4 |
| Trade | −6.2 | −9.3 | −16.7 | |
| Southeast Asia | | | | |
| Production | 22.8 | 21.4 | 65.0 | 210.0 |
| Consumption | 19.3 | 17.9 | 47.0 | 163.6 |
| Trade | +3.4 | +3.7 | +18.0 | |
| East Asia | | | | |
| Production | 29.9 | 34.0 | 73.0 | 155.3 |
| Consumption | 38.3 | 42.9 | 97.0 | 164.9 |
| Trade | −8.8 | −9.7 | −24.0 | |
| World | | | | |
| Production/Consumption | 1,108.0 | 1,202.0 | 2,141.7 | 91.0 |

*Note:* In grade figures, plus sign indicates export, minus sign indicates import.

[a]Argentina and Thailand.

[b]All others, including several countries that export in some scenarios (e.g., Brazil, Indonesia, and Colombia).

*Source:* Global 2000 Technical Report, Table 6-5.

6 and Figure 5 show per capita food consumption by regions.

The outlook for improved diets for the poorest people in the poorest LDCs is sobering. In the 1970s, consumption of calories in the LDCs averaged only 94 percent of the minimum requirements set by the U.N. Food and Agriculture Organization (FAO).† Moreover, income and

†The FAO standard indicates the *minimum* consumption that will allow normal activity and good health in adults and will permit children to reach normal body weight and intelligence in the absence of disease.

**TABLE 6**
**Per Capita Grain Production, Consumption, and Trade, Actual and Projected, and Percent Increase in Per Capita Total Food Production and Consumption**

| | Grain (*kilograms per capita*) | | | Food (*Percent increase over the 1970-2000 period)* |
|---|---|---|---|---|
| | 1969–71 | 1973–75 | 2000 | |
| Industrialized countries | | | | |
| Production | 573.6 | 592.6 | 769.8 | 18.4 |
| Consumption | 534.4 | 510.7 | 692.4 | 21.2 |
| Trade | +45.8 | +84.0 | +77.4 | |
| United States | | | | |
| Production | 1,018.6 | 1,079.3 | 1,640.3 | 51.1 |
| Consumption | 824.9 | 748.0 | 1,111.5 | 28.3 |
| Trade | +194.7 | +344.0 | +528.8 | |
| Other developed exporters | | | | |
| Production | 1,015.6 | 917.0 | 915.6 | −11.3 |
| Consumption | 575.4 | 514.0 | 562.6 | −5.7 |
| Trade | +492.2 | +415.0 | +353.0 | |
| Western Europe | | | | |
| Production | 364.9 | 388.4 | 394.0 | 1.0 |
| Consumption | 432.4 | 443.3 | 548.8 | 15.5 |
| Trade | −65.4 | −57.6 | −154.8 | |
| Japan | | | | |
| Production | 121.7 | 108.5 | 135.4 | 6.1 |
| Consumption | 267.5 | 274.4 | 452.3 | 54.2 |
| Trade | −138.1 | −175.9 | −316.7 | |
| Centrally planned countries | | | | |
| Production | 356.1 | 368.0 | 451.1 | 29.6 |
| Consumption | 361.0 | 395.6 | 473.9 | 35.8 |
| Trade | −4.6 | −20.1 | −22.8 | |
| Eastern Europe | | | | |
| Production | 574.0 | 693.0 | 921.9 | 53.3 |
| Consumption | 626.6 | 757.4 | 997.6 | 52.1 |
| Trade | −48.6 | −60.5 | −75.8 | |
| U.S.S.R. | | | | |
| Production | 697.6 | 711.2 | 903.2 | 28.1 |
| Consumption | 663.1 | 796.1 | 949.9 | 41.4 |
| Trade | +16.1 | −42.0 | −46.7 | |
| People's Republic of China | | | | |
| Production | 216.3 | 217.6 | 259.0 | 17.4 |
| Consumption | 220.2 | 222.4 | 267.8 | 19.1 |
| Trade | −4.0 | −4.8 | −8.8 | |

food distribution within individual LDCs is so skewed that national average caloric consumption generally must be 10–20 percent above minimum levels before the poorest are likely to be able to afford a diet that meets the FAO minimum standard. Latin America is the only major LDC region where average caloric consumption is projected to be 20 percent or more above the FAO minimum standard in the year 2000. In the other LDC regions—South, East, and Southeast Asia, poor areas of North Africa and the Middle East, and especially Central Africa, where a calamitous drop

**TABLE 6** (Cont.)

| | Grain (kilograms per capita) | | | Food Percent increase over the 1970–2000 period |
|---|---|---|---|---|
| | 1969–71 | 1973–75 | 2000 | |
| Less developed countries | | | | |
| Production | 176.7 | 168.7 | 197.1 | 10.8 |
| Consumption | 188.3 | 182.2 | 205.5 | 8.6 |
| Trade | −10.7 | −15.1 | −8.4 | |
| Exporters[a] | | | | |
| Production | 491.0 | 521.9 | 671.7 | 10.4 |
| Consumption | 300.1 | 325.3 | 287.8 | −22.6 |
| Trade | +184.3 | +198.2 | +383.9 | |
| Importers[b] | | | | |
| Production | 159.4 | 173.8 | 180.7 | 10.8 |
| Consumption | 177.7 | 193.6 | 202.7 | 10.8 |
| Trade | −17.2 | −24.1 | −21.9 | |
| Latin America | | | | |
| Production | 236.1 | 241.0 | 311.4 | 33.7 |
| Consumption | 226.5 | 238.3 | 278.1 | 25.1 |
| Trade | +11.8 | +2.7 | +33.3 | |
| North Africa/Middle East | | | | |
| Production | 217.1 | 214.6 | 222.5 | −1.8 |
| Consumption | 276.2 | 273.8 | 292.8 | 2.2 |
| Trade | −50.8 | −69.8 | −70.3 | |
| Other African LDCs | | | | |
| Production | 134.9 | 118.3 | 113.2 | −15.5 |
| Consumption | 139.1 | 127.7 | 112.0 | −19.1 |
| Trade | −4.2 | −9.1 | +1.2 | |
| South Asia | | | | |
| Production | 161.6 | 162.4 | 170.0 | 4.6 |
| Consumption | 170.0 | 171.8 | 181.0 | 5.8 |
| Trade | −8.4 | −11.8 | −11.0 | |
| Southeast Asia | | | | |
| Production | 244.7 | 214.5 | 316.5 | 35.9 |
| Consumption | 207.2 | 182.6 | 228.5 | 14.6 |
| Trade | +37.5 | +31.9 | +87.5 | |
| East Asia | | | | |
| Production | 137.3 | 136.0 | 163.5 | 22.8 |
| Consumption | 176.2 | 171.5 | 217.3 | 27.3 |
| Trade | −40.4 | −38.8 | −53.8 | |
| World | | | | |
| Production/Consumption | 311.5 | 313.6 | 343.2 | 14.5 |

*Note:* In trade figures, plus sign indicates export, minus sign indicates import.

[a]Argentina and Thailand.

[b]All others, including several countries that export in some scenarios (e.g., Brazil, Indonesia, and Colombia).

*Source:* Global 2000 Technical Report, Table 6-6.

in food per capita is projected—the quantity of food available to the poorest groups of people will simply be insufficient to permit children to reach normal body weight and intelligence and to permit normal activity and good health in adults. Consumption in the LDCs of central Africa is projected to be more than 20 percent below the FAO minimum standard, assuming no recurrence of severe drought. In South Asia (primarily India, Pakistan, and Bangladesh), average caloric intake is projected to remain below the FAO minimum standard, although increasing slightly—from 12

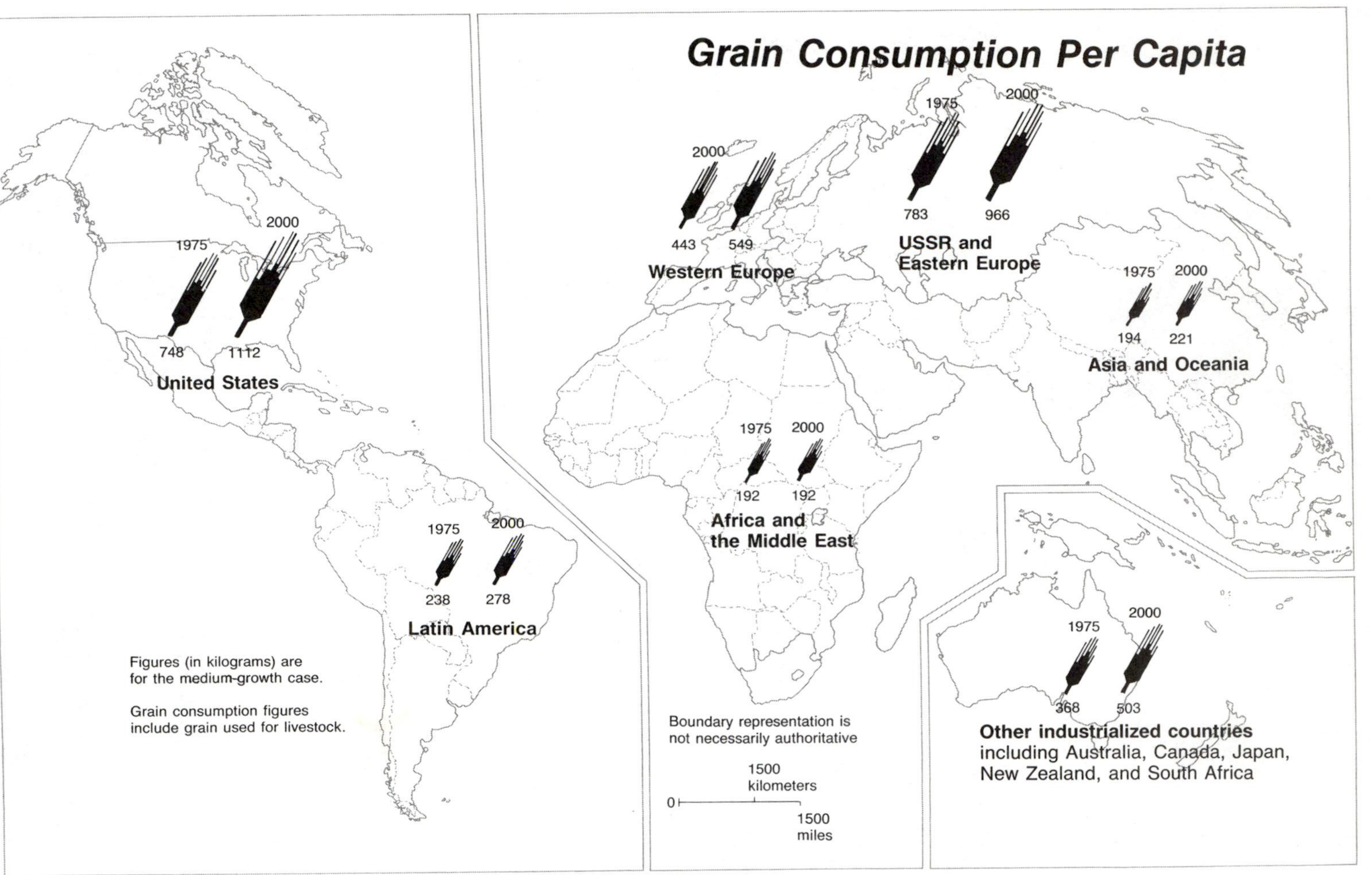


**Figure 5.** Per capita grain consumption, by regions, 1975 and 2000.

percent below the FAO standard in the mid-1970s to about 3 percent below the standard in 2000. In East Asia, Southeast Asia, and affluent areas of North Africa and the Middle East, average per capita caloric intakes are projected to be 6-17 percent above FAO minimum requirement levels, but because the great majority of people in these regions are extremely poor, they will almost certainly continue to eat less than the minimum. The World Bank has estimated that the number of malnourished people in LDCs could rise from 400-600 million in the mid-1970s to 1.3 billion in 2000.[34]

The projected food situation has many implications for food assistance and trade. In the developing world, the need for imported food is expected to grow. The most prosperous LDCs will turn increasingly to the world commercial markets. In the poorest countries, which lack the wherewithal to buy food, requirements for international food assistance will expand. LDC exporters (especially Argentina and Thailand) are projected to enlarge food production for export because of their cost advantage over countries dependent on energy-intensive inputs. LDC grain-exporting countries, which accounted for only a little more than 10 percent of the world grain market in 1975, are projected to capture more than 20 percent of the market by 2000. The United States is expected to continue its role as the world's principal food exporter. Moreover, as the year 2000 approaches and more marginal, weather-sensitive lands are brought into production around the world, the United States is likely to become even more of a residual world supplier than today; that is, U.S. producers will be responding to widening, weather-related swings in world production and foreign demand.[35]

Revised and updated food projections would reflect reduced estimates of future yields, increased pressure on the agricultural resource base, and several changes in national food policies.

Farmers' costs of raising—and even maintaining—yields have increased rapidly in recent years. The costs of energy-intensive, yield-enhancing inputs—fertilizer, pesticides, and fuels—have risen very rapidly throughout the world, and where these inputs are heavily used, increased applications are bringing diminishing returns. In the United States, the real cost of producing food increased roughly 10 percent in both 1978 and 1979.[36] Other industrialized countries have experienced comparable production cost increases. Cost increases in the LDCs appear to be lower, but are still 2-3 times the annual increases of the 1960s and early 1970s. While there have been significant improvements recently in the yields of selected crops, the diminishing returns and rapidly rising costs of yield-enhancing inputs suggest that yields overall will increase more slowly than projected.

Since the food projections were made, there have been several important shifts in national food and agricultural policy concerns. In most industrialized countries, concern with protecting agricultural resources, especially soils, has increased as the resource implications of sustained production of record quantities of food here become more apparent. Debate on the 1981 U.S. farm bill, for example, will certainly include more consideration of "exporting top soil" than was foreseeable at the time the Global 2000 Study's food projections were made.[37] The heightened concern for protection of agricultural resources is leading to a search for policies that encourage improved resource management practices. Still further pressure on the resource base can be expected, however, due to rising industrial demand for grain, especially for fermentation into alcohol-based fuels. Accelerated erosion, loss of natural soil fertility and other deterioration of the agricultural resource base may have more effect in the coming years than is indicated in the Global 2000 food projections.

In the LDCs, many governments are attempting to accelerate investment in food production capacity. This policy emphasis offers important long-term benefits. Some LDC governments are intervening more frequently in domestic food markets to keep food prices low, but often at the cost of low rural incomes and slowed development of agricultural production capacity.[38]

Worldwide, the use of yield-enhancing inputs is likely to be less, and soil deterioration greater, than expected. As a result, revised food projections would show a tighter food future—somewhat less production and somewhat higher prices—than indicated in the Global 2000 projections.



## Fisheries

Fish is an important component of the world's diet and has sometimes been put forth as a possible partial solution to world food shortages. Unfortunately, the world harvest of fish is expected to rise little, if at all, by the year 2000.* The world catch of naturally produced fish leveled off in the 1970s at about 70 million metric tons a year (60 million metric tons for marine fisheries, 10 million metric tons for freshwater species). Harvests of traditional fisheries are not likely to increase on a sustained basis, and indeed to maintain them will take good management and improved protection of the marine environment. Some potential for greater harvests comes from aquaculture and from nontraditional marine species, such as Antarctic krill, that are little used at present for direct human consumption.[39]

Traditional freshwater and marine species might be augmented in some areas by means of aquaculture. The 1976 FAO World Conference on Aquaculture concluded that a five- to tenfold increase in production from aquaculture would be possible by 2000, given adequate financial and technical support. (Aquaculture contributed an estimated 6 million metric tons to the world's total catch in 1975.) However, limited investment and technical support, as well as increasing pollution of freshwater ponds and coastal water, are likely to be a serious impediment to such growth.[40]

While fish is not a solution to the world needs for calories, fish does provide an important source of protein. The 70 million metric tons caught and raised in 1975 is roughly equivalent to 14 million metric tons of protein, enough to supply 27 percent of the minimum protein requirements of 4 billion people. (Actually since more than one-third of the fish harvest is used for animal feed, not food for humans, the contribution of fish to human needs for protein is lower than these figures suggest.[41]) A harvest of about 115 million metric tons would be required to supply 27 percent of the protein needs of 6.35 billion people in 2000. Even assuming that the catch of marine and freshwater fish rises to the unlikely level of 100 million metric tons annually, and that yields from aquaculture double, rising to 12 million tons, the hypothetical total of 112 metric tons would not provide as much protein per capita as the catch of the mid-1970s. Thus, on a per capita basis, fish may well contribute less to the world's nutrition in 2000 than today.

Updated fisheries projections would show little change from the Global 2000 Study projections. FAO fisheries statistics are now available for 1978 and show a world catch of 72.4 million metric tons. (The FAO statistics for the 1970-78 period have been revised downward somewhat to reflect improved data on the catch of the People's Republic of China.) While there has been some slight recovery of the anchovy and menhaden fisheries, traditional species continue to show signs of heavy pressure. As indicated in the Global 2000 projections, the catch of nontraditional species is filling in to some extent. Perhaps the biggest change in updated fisheries projections would stem from a careful analysis of the effects of the large increase in oil prices that occurred in 1979. Scattered observations suggest that fishing fleets throughout the world are being adversely affected except where governments are keeping oil prices to fishing boats artificially low.[42]

*The food projections assumed that the world fish catch would increase at essentially the same rate as population and are therefore likely to prove too optimistic on this point. (See Chapters 6, 14, and 18 of the Global 2000 Technical Report for further discussion of this point.)

## Forests

If present trends continue, both forest cover and growing stocks of commercial-size wood in the less developed regions (Latin America, Africa,

Asia, and Oceania) will decline 40 percent by 2000. In the industrialized regions (Europe, the U.S.S.R., North America, Japan, Australia, New Zealand) forests will decline only 0.5 percent and growing stock about 5 percent. Growing stock per capita is expected to decline 47 percent worldwide and 63 percent in LDCs.[43] Table 7 shows projected forest cover and growing stocks by region for 1978 and 2000.

Deforestation is projected to continue until about 2020, when the total world forest area will stabilize at about 1.8 billion hectares. Most of the loss will occur in the tropical forests of the developing world. About 1.45 billion hectares of forest in the industrialized nations has already stabilized and about 0.37 billion hectares of forest in the LDCs is physically or economically inaccessible. By 2020, virtually all of the physically accessible forest in the LDCs is expected to have been cut.[44]

The real prices of wood products—fuelwood, sawn lumber, wood panels, paper, wood-based chemicals, and so on—are expected to rise considerably as GNP (and thus also demand) rises and world supplies tighten. In the industrialized nations, the effects may be disruptive, but not catastrophic. In the less developed countries, however, 90 percent of wood consumption goes for cooking and heating, and wood is a necessity of life. Loss of woodlands will force people in many LDCs to pay steeply rising prices for fuelwood and charcoal or to spend much more effort collecting wood—or else to do without.[45]

Updated forest projections would present much the same picture as the Global 2000 Study projections. The rapid increase in the price of crude oil will probably limit the penetration of kerosene sales into areas now depending on fuelwood and dung and, as a result, demand for fuelwood may be somewhat higher than expected. Some replanting of cut tropical areas is occurring, but only at low rates similar to those assumed in the Global 2000 Study projections. Perhaps the most encouraging developments are those associated with heightened international awareness of the seriousness of current trends in world forests.[46]

**TABLE 7**
**Estimates of World Forest Resources, 1978 and 2000**

| | Closed Forest[a] *(millions of hectares)* | | Growing Stock *(billions cu m overbark)* | |
|---|---|---|---|---|
| | 1978 | 2000 | 1978 | 2000 |
| U.S.S.R. | 785 | 775 | 79 | 77 |
| Europe | 140 | 150 | 15 | 13 |
| North America | 470 | 464 | 58 | 55 |
| Japan, Australia, New Zealand | 69 | 68 | 4 | 4 |
| Subtotal | 1,464 | 1,457 | 156 | 149 |
| Latin America | 550 | 329 | 94 | 54 |
| Africa | 188 | 150 | 39 | 31 |
| Asia and Pacific LDCs | 361 | 181 | 38 | 19 |
| Subtotal (LDCs) | 1,099 | 660 | 171 | 104 |
| Total (world) | 2,563 | 2,117 | 327 | 253 |
| | | | Growing Stock per Capita *(cu m biomass)* | |
| Industrial countries | | | 142 | 114 |
| LDCs | | | 57 | 21 |
| Global | | | 76 | 40 |

[a]Closed forests are relatively dense and productive forests. They are defined variously in different parts of the world. For further details, see Global 2000 Technical Report, footnote, p. 117.

*Source:* Global 2000 Technical Report, Table 13-29.

## Water

The Global 2000 Study population, GNP, and resource projections all imply rapidly increasing demands for fresh water.[47] Increases of at least 200-300 percent in world water withdrawals are expected over the 1975-2000 period. By far the largest part of the increase is for irrigation. The United Nations has estimated that water needed for irrigation, which accounted for 70 percent of human uses of water in 1967, would double by 2000. Moreover, irrigation is a highly consumptive use, that is, much of the water withdrawn for this purpose is not available for immediate reuse because it evaporates, is transpired by plants, or becomes salinated.[48]

Regional water shortages and deterioration of water quality, already serious in many parts of the world, are likely to become worse by 2000. Estimates of per capita water availability for 1971 and 2000, based on population growth alone, *without allowance for other causes of increased demand,* are shown in Figures 6 and 7. As indicated in these maps, population growth alone will cause demands for water at least to double relative to 1971 in nearly half the countries of the world. Still greater increases would be needed to improve standards of living.[49]

Much of the increased demand for water will be in the LDCs of Africa, South Asia, the Middle East, and Latin America, where in many areas fresh water for human consumption and irrigation is already in short supply. Although the data are sketchy, it is known that several nations in these areas will be approaching their maximum developable water supply by 2000, and that it will be quite expensive to develop the water remaining. Moreover, many LDCs will also suffer destabilization of water supplies following extensive loss of forests. In the industrialized countries competition among different uses of water—for increasing food production, new energy systems (such as production of synthetic fuels from coal and shale), increasing power generation, expanding food production, and increasing needs of other industry—will aggravate water shortages in many areas.[50]

Updated water projections would present essentially the same picture. The only significant change that has occurred since the projections were developed is that the price of energy (especially oil) has increased markedly. Increased energy costs will adversely affect the economics of many water development projects, and may reduce the amount of water available for a variety of uses. Irrigation, which usually requires large amounts of energy for pumping, may be particularly affected.

### Nonfuel Minerals

The trends for nonfuel minerals, like those for the other resources considered in the Global 2000 Study, show steady increases in demand and consumption. The global demand for and consumption of most major nonfuel mineral commodities is projected to increase 3-5 percent annually, slightly more than doubling by 2000. Consumption of all major steelmaking mineral commodities is projected to increase at least 3 percent annually. Consumption of all mineral commodities for fertilizer production is projected to grow at more than 3 percent annually, with consumption of phosphate rock growing at 5.2 percent per year—the highest growth rate projected for any of the major nonfuel mineral commodities. The nonferrous metals show widely varying projected growth rates; the growth rate for aluminum, 4.3 percent per year, is the largest.[51]

The projections suggest that the LDC's share of nonfuel mineral use will increase only modestly. Over the 1971-75 period, Latin America, Africa, and Asia used 7 percent of the world's aluminum production, 9 percent of the copper, and 12 percent of the iron ore. The three-quarters of the world's population living in these regions in 2000 are projected to use only 8 percent of aluminum production, 13 percent of copper production, and 17 percent of iron ore production. The one-quarter of the world's population that inhabits industrial countries is projected to continue absorbing more than three-fourths of the world's nonfuel minerals production.[52] Figure 8 shows the geographic distribution of per capita consumption of nonfuel minerals for 1975 and 2000.

The projections point to no mineral exhaustion problems but, as indicated in Table 8, further discoveries and investments will be needed to maintain reserves and production of several mineral commodities at desirable levels. In most cases, however, the resource potential is still large (see Table 9), especially for low grade ores.[53]

Updated nonfuel minerals projections would need to give further attention to two factors affecting investment in mining. One is the shift over the past decade in investment in extraction and processing away from the developing countries toward industrialized countries (although this trend may now be reversing). The other factor is

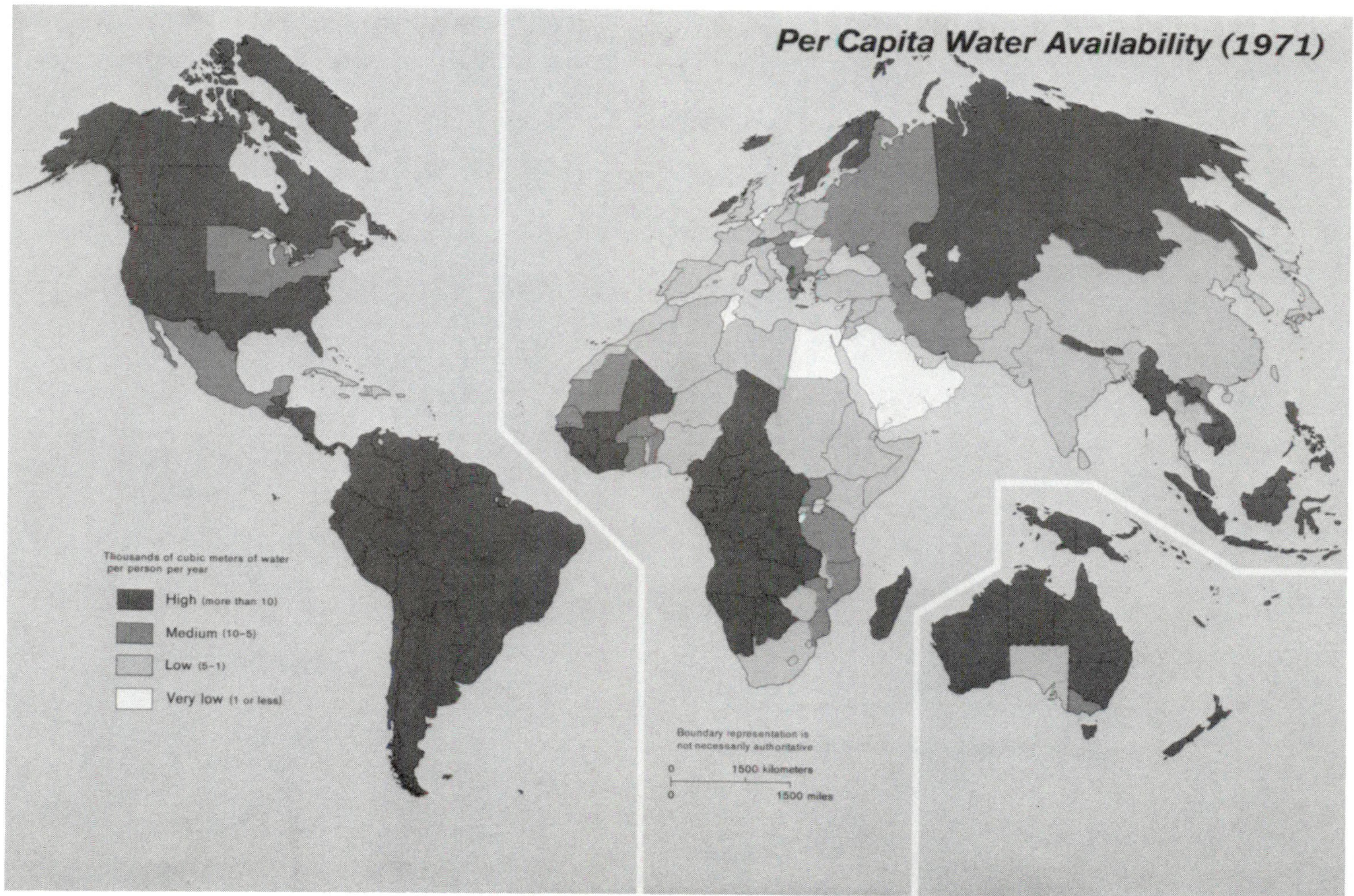


**Figure 6.** Per capita water availability, 1971.

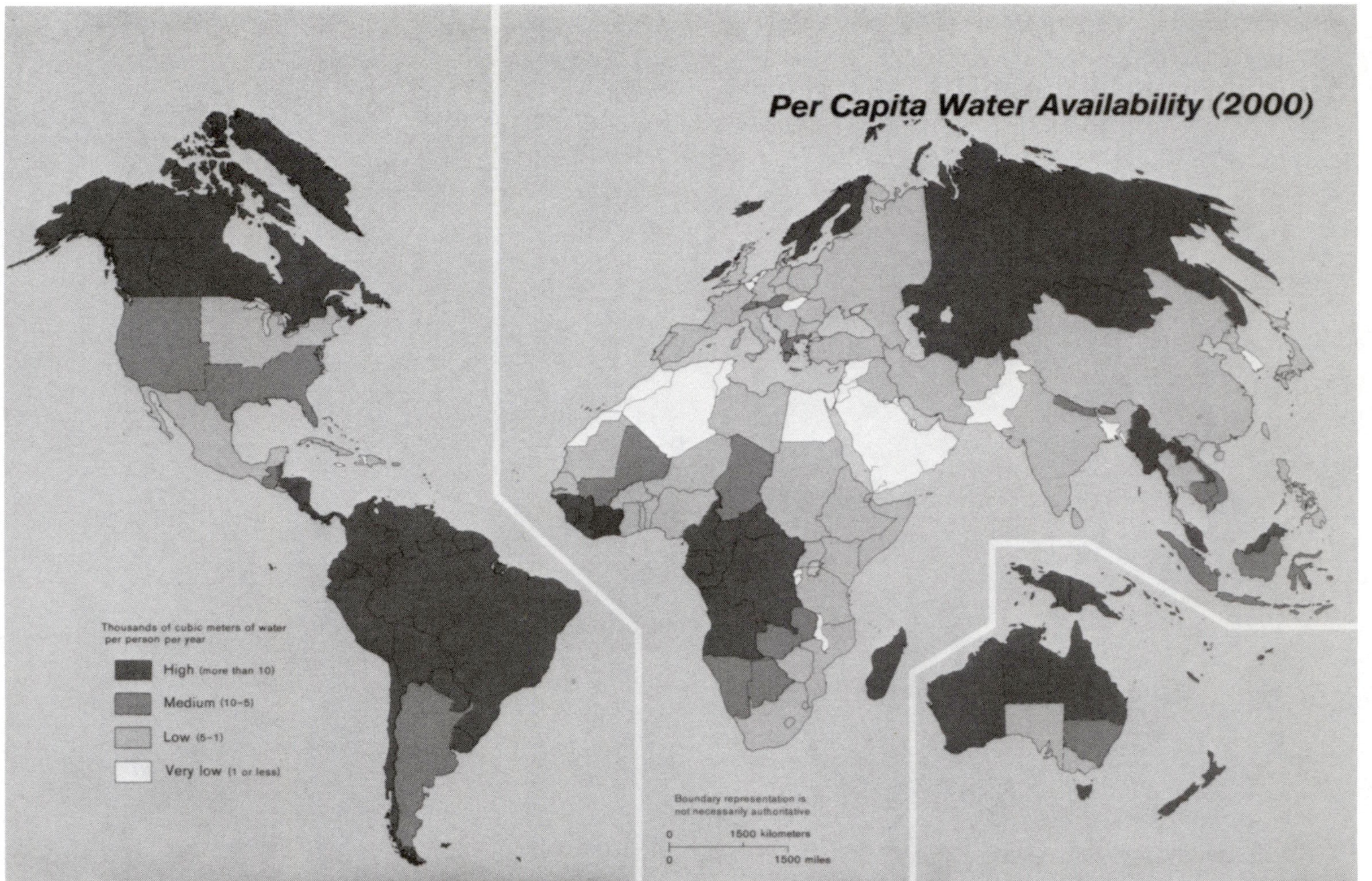


**Figure 7.** Projected per capita water availability, 2000.

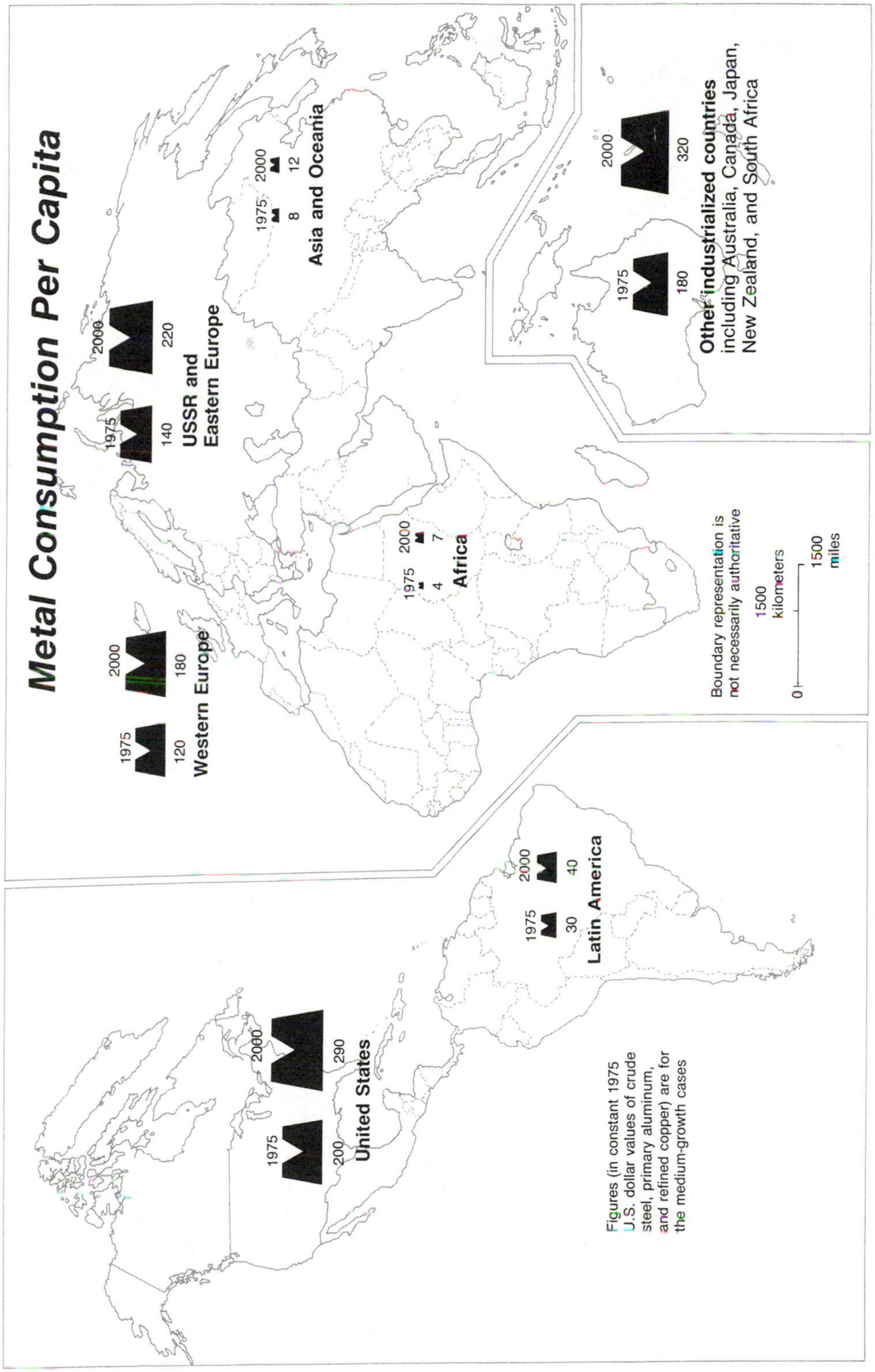

**Metal Consumption Per Capita**

Western Europe
1975 120
2000 180

USSR and Eastern Europe
1975 140
2000 220

Asia and Oceania
1975 8
2000 12

Africa
1975 4
2000 7

Other industrialized countries including Australia, Canada, Japan, New Zealand, and South Africa
1975 180
2000 320

United States
1975 200
2000 290

Latin America
1975 30
2000 40

Boundary representation is not necessarily authoritative

0 1500 kilometers

1500 miles

Figures (in constant 1975 U.S. dollar values of crude steel, primary aluminum, and refined copper) are for the medium-growth cases

**Figure 8.** Distribution of per capita consumption of nonfuel minerals, 1975 and 2000.

**TABLE 8**
**Life Expectancies of 1976 World Reserves of Selected Mineral Commodities at Two Different Rates of Demand**

| | 1976 Reserves | 1976 Primary Demand | Projected Demand Growth Rate | Life Expectancy in Years[a] Static at 1976 Level | Life Expectancy in Years[a] Growing at Projected Rates |
|---|---|---|---|---|---|
| | | | *percent* | | |
| Fluorine (*million short tons*) | 37 | 2.1 | 4.58 | 18 | 13 |
| Silver (*million troy ounces*) | 6,100 | 305 | 2.33 | 20 | 17 |
| Zinc (*million short tons*) | 166 | 6.4 | 3.05 | 26 | 19 |
| Mercury (*thousand flasks*) | 5,210 | 239 | 0.50 | 22 | 21 |
| Sulfur (*million long tons*) | 1,700 | 50 | 3.16 | 34 | 23 |
| Lead (*million short tons*) | 136 | 3.7 | 3.14 | 37 | 25 |
| Tungsten (*million pounds*) | 4,200 | 81 | 3.26 | 52 | 31 |
| Tin (*thousand metric tons*) | 10,000 | 241 | 2.05 | 41 | 31 |
| Copper (*million short tons*) | 503 | 8.0 | 2.94 | 63 | 36 |
| Nickel (*million short tons*) | 60 | 0.7 | 2.94 | 86 | 43 |
| Platinum (*million troy ounces*) | 297 | 2.7 | 3.75 | 110 | 44 |
| Phosphate rock (*million metric tons*) | 25,732 | 107 | 5.17 | 240 | 51 |
| Manganese (*million short tons*) | 1,800 | 11.0 | 3.36 | 164 | 56 |
| Iron in ore (*billion short tons*) | 103 | 0.6 | 2.95 | 172 | 62 |
| Aluminum in bauxite (*million short tons*) | 5,610 | 18 | 4.29 | 312 | 63 |
| Chromium (*million short tons*) | 829 | 2.2 | 3.27 | 377 | 80 |
| Potash (*million short tons*) | 12,230 | 26 | 3.27 | 470 | 86 |

*Note:* Corresponding data for helium and industrial diamonds not available.
[a]Assumes no increase to 1976 reserves.
*Source:* After Global 2000 Technical Report, Table 12-4, but with updated and corrected entries. Updated reserves and demand data from U.S. Bureau of Mines, *Mineral Trends and Forecasts*, 1979. Projected demand growth rates are from Global 2000 Technical Report, Table 12-2.

the rapid increase in energy prices. Production of many nonfuel minerals is highly energy-intensive, and the recent and projected increases in oil prices can be expected to slow the expansion of these mineral supplies.[54]

### Energy

The Global 2000 Study's energy projections show no early relief from the world's energy problems. The projections point out that petroleum production capacity is not increasing as rapidly as demand. Furthermore, the rate at which petroleum reserves are being added per unit of exploratory effort appears to be falling. Engineering and geological considerations suggest that world petroleum production will peak before the end of the century. Political and economic decisions in the OPEC countries could cause oil production to level off even before technological constraints come into play. A world transition away from petroleum dependence must take place, but there is still much uncertainty as to how this transition will occur. In the face of this uncertainty, it was not possible at the time the Global 2000 energy projections were made—late 1977—for the Department of Energy (DOE) to develop meaningful energy projections beyond 1990.[55] Updated DOE analyses, discussed at the end of this section, extend the global energy projections available from the U.S. Government to 1995.

**TABLE 9**

**World Production and Reserves in 1977 (Estimated), Other Resources in 1973–77 (as Data Available), Resource Potential, and Resource Base of 17 Elements**

*(Millions of metric tons)*

| | Production | Reserves | Other Resources | Resource Potential (Recoverable) | Resource Base (Crustal Mass) |
|---|---|---|---|---|---|
| Aluminum | 17[a] | 5,200[a] | 2,800[a] | 3,519,000 | 1,990,000,000,000 |
| Iron | 495[b] | 93,100 | 143,000[c] | 2,035,000 | 1,392,000,000,000 |
| Potassium | 22 | 9,960 | 103,000 | n.a. | 408,000,000,000 |
| Manganese | 10[d] | 2,200 | 1,100[e] | 42,000 | 31,200,000,000 |
| Phosphorus | 14[f] | 3,400[f] | 12,000[f] | 51,000 | 28,800,000,000 |
| Fluorine | 2[g] | 72 | 270 | 20,000 | 10,800,000,000 |
| Sulfur | 52 | 1,700 | 3,800[h] | | 9,600,000,000 |
| Chromium | 3[i] | 780[i] | 6,000[i] | 3,260 | 2,600,000,000 |
| Zinc | 6 | 159 | 4,000 | 3,400 | 2,250,000,000 |
| Nickel | 0.7 | 54 | 103[e] | 2,590 | 2,130,000,000 |
| Copper | 8 | 456 | 1,770[j] | 2,120 | 1,510,000,000 |
| Lead | 4 | 123 | 1,250 | 550 | 290,000,000 |
| Tin | 0.2 | 10 | 27 | 68 | 40,800,000 |
| Tungsten | 0.04 | 1.8 | 3.4 | 51 | 26,400,000 |
| Mercury | 0.008 | 0.2 | 0.4 | 3.4 | 2,100,000 |
| Silver | 0.010 | 0.2 | 0.5 | 2.8 | 1,800,000 |
| Platinum group[k] | 0.0002 | 0.02 | 0.05[l] | 1.2[m] | 1,100,000 |

[a]In bauxite, dry basis, assumed to average 21 percent recoverable aluminum.
[b]In ore and concentrates assumed to average 58 percent recoverable iron.
[c]In ore and concentrates assumed to average 26 percent recoverable iron.
[d]In ore and concentrates assumed to average 40 percent manganese.
[e]Excludes metal in deep-sea nodules and, in the case of nickel, unidentified resources.
[f]In phosphate rock ore and concentrates assumed to average 13 percent phosphorus.
[g]In fluorspar and phosphate rock ore and concentrates assumed to average 44 percent fluorine.
[h]Excludes unidentified sulfur resources, enormous quantities of sulfur in gypsum and anhydrite, and some 600 billion tons of sulfur in coal, oil shale, and in shale that is rich in organic matter.
[i]In ore and concentrates assumed to average 32 percent chromium.
[j]Includes 690 million tons in deep-sea nodules.
[k]Platinum, palladium, iridium, cesium, rhodium, and ruthenium.
[l]Approximate midpoint of estimated range of 0.03–0.06 million metric tons.
[m]Platinum only.

*Source:* Global 2000 Technical Report, Table 12-7.

DOE projections prepared for the Study show large increases in demand for all commercial sources over the 1975-90 period (see Table 10). World energy demand is projected to increase 58 percent, reaching 384 quads (quadrillion British thermal units) by 1990. Nuclear and hydro sources (primarily nuclear) increase most rapidly (226 percent by 1990), followed by oil (58 percent), natural gas (43 percent), and coal (13 percent). Oil is projected to remain the world's leading energy source, providing 46-47 percent of the world's total energy through 1990, assuming that the real price of oil on the international market increases 65 percent over the 1975-90 period. The energy projections indicate that there is considerable potential for reductions in energy consumption.[56]

Per capita energy consumption is projected to increase everywhere. The largest increase—72 percent over the 1975-90 period—is in industrialized countries other than the United States. The smallest increase, 12 percent, is in the centrally planned economies of Eastern Europe. The percentage increases for the United States and for the LDCs are the same—27 percent—but actual per capita energy consumption is very different. By 2000, U.S. per capita energy consumption is projected to be about 422 million Btu (British thermal units) annually. In the LDCs, it will be only 14 million Btu, up from 11 million in 1975[57] (see Table 11 and Figure 9).

While prices for oil and other commercial energy sources are rising, fuelwood—the poor person's oil—is expected to become far less available than it is today. The FAO has estimated that the demand for fuelwood in LDCs will increase at 2.2 percent per year, leading to local fuelwood shortages in 1994 totaling 650 million cubic meters—approximately 25 percent of the projected need.

## TABLE 10
## Global Primary[a] Energy Use, 1975 and 1990, by Energy Type

| | 1975 | | 1990 | | Percent Increase (1975-90) | Average Annual Percent Increase |
|---|---|---|---|---|---|---|
| | $10^{15}$ Btu | Percent of Total | $10^{15}$ Btu[b] | Percent of Total | | |
| Oil | 113 | 46 | 179 | 47 | 58 | 3.1 |
| Coal | 68 | 28 | 77 | 20 | 13 | 0.8 |
| Natural gas | 46 | 19 | 66 | 17 | 43 | 2.4 |
| Nuclear and hydro | 19 | 8[c] | 62 | 16[c] | 226 | 7.9 |
| Solar (other than conservation/ and hydro)[d] | — | — | — | — | — | — |
| Total | 246 | 100 | 384 | 100 | 56 | 3.0 |

[a]All of the nuclear and much of the coal primary (i.e., input) energy is used thermally to generate electricity. In the process, approximately two-thirds of the primary energy is lost as waste heat. The figures given here are primary energy.

[b]The conversions from the DOE projections in Table 10-8 were made as follows: *Oil* $84.8 \times 10^6$ bbl/day $\times$ 365 days $\times 5.8 \times 10^6$ Btu/bbl $= 179 \times 10^6$ Btu. *Coal:* $5{,}424 \times 10^6$ short tons/yr $\times 14.1 \times 10^6$ Btu/short ton [DOE figure for world average grade coal] $= 77 \times 10^{15}$ Btu. *Natural gas:* $64.4 \times 10^{12}$ft$^3$/yr $\times$ 1,032 Btu/ft$^3$ $= 66 \times 10^{15}$ Btu. *Nuclear and Hydro:* $6{,}009 \times 10^{12}$ Wh [output]/yr $\times$ 3,412 Btu/Wh $\times$ 3 input Btu/output Btu $= 62 \times 10^{12}$ Btu.

[c]After deductions for lost (waste) heat (see note a), the corresponding figures for output energy are 2.7 percent in 1975 and 6.0 in 1990.

[d]The IIES projection model is able to include solar only as conservation or hydro.

*Source:* Global 2000 Technical Report, Table 13-32.

## TABLE 11
## Per Capita Global Primary Energy Use, Annually, 1975 and 1990

| | 1975 | | 1990 | | Percent Increase (1975-90) | Average Annual Percent Increase |
|---|---|---|---|---|---|---|
| | $10^6$ Btu | Percent of World Average | $10^6$ Btu | Percent of World Average | | |
| United States | 332 | 553 | 422 | 586 | 27 | 1.6 |
| Other industrialized countries | 136 | 227 | 234 | 325 | 72 | 3.6 |
| Less developed countries[a] | 11 | 18 | 14 | 19 | 27 | 1.6 |
| Centrally planned economies | 58 | 97 | 65 | 90 | 12 | 0.8 |
| World | 60 | 100 | 72 | 100 | 20 | 1.2 |

[a]Since population projections were not made separately for the OPEC countries, those countries have been included here in LDC category.

*Source:* Global 2000 Technical Report, Table 13-34.

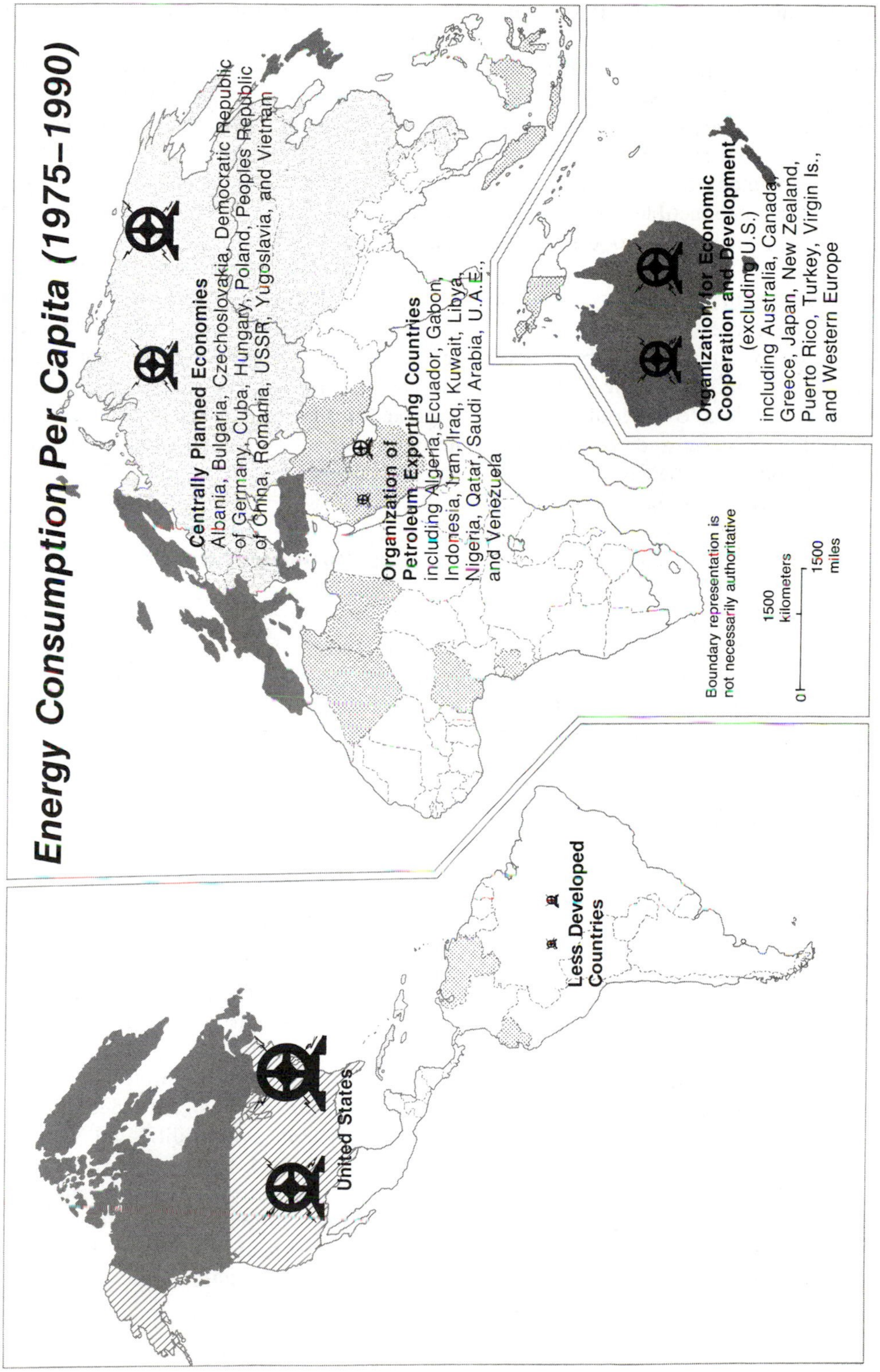

**Figure 9.** Energy consumption per capita, 1975–90.

Scarcities are now local but expanding. In the arid Sahel of Africa, fuelwood gathering has become a full-time job requiring in some places 360 person-days of work per household each year. When demand is concentrated in cities, surrounding areas have already become barren for considerable distances—50 to 100 kilometers in some places. Urban families, too far from collectible wood, spend 20 to 30 percent of their income on wood in some West African cities.[58]

The projected shortfall of fuelwood implies that fuel consumption for essential uses will be reduced, deforestation expanded, wood prices increased, and growing amounts of dung and crop residues shifted from the field to the cooking fire. No explicit projections of dung and crop residue combustion could be made for the Study, but it is known that a shift toward burning these organic materials is already well advanced in the Himalayan hills, in the treeless Ganges plain of India, in other parts of Asia, and in the Andean region of South America. The FAO reports that in 1970 India burned 68 million tons of cow dung and 39 million tons of vegetable waste, accounting for roughly a third of the nation's total noncommercial energy consumption that year. Worldwide, an estimated 150-400 million tons of dung are burned annually for fuel.[59]

Updated energy projections have been developed by the Department of Energy based on new price scenarios that include the rapid 1979 increase in the price of crude oil. The new price scenarios are not markedly different from the earlier estimates for the 1990s. The new medium-scenario price for 1995 is $40 per barrel (in 1979 dollars), which is about 10 percent higher than the $36 price (1979 dollars) implied by the earlier scenario. However, the prices for the early 1980s are almost 100 percent higher than those in the projections made by DOE in late 1977 for the Study.[60] The sudden large increase in oil prices of 1979 is likely to have a more disruptive effect on other sectors than would the gradual increase assumed in the Global 2000 Study projections.

DOE's new projections differ in several ways from those reported in this Study. Using the higher prices, additional data, and a modified model, DOE is now able to project supply and demand for an additional five years, to 1995. Demand is projected to be lower because of the higher prices and also because of reduced estimates of economic growth. Coal is projected to provide a somewhat larger share of the total energy supply. The nuclear projections for the OECD countries are lower, reflecting revised estimates of the speed at which new nuclear plants will be built. Updated estimates of OPEC maximum production are lower than earlier estimates, reflecting trends toward resource conservation by the OPEC nations. The higher oil prices will encourage the adoption of alternative fuels and technologies, including solar technology and conservation measures.[61]

## Environmental Consequences

The population, income, and resource projections all imply significant consequences for the quality of the world environment. Virtually every aspect of the earth's ecosystems and resource base will be affected.[62]

### Impacts on Agriculture

Perhaps the most serious environmental development will be an accelerating deterioration and loss of the resources essential for agriculture. This overall development includes soil erosion; loss of nutrients and compaction of soils; increasing salinization of both irrigated land and water used for irrigation; loss of high-quality cropland to urban development; crop damage due to increasing air and water pollution; extinction of local and wild crop strains needed by plant breeders for improving cultivated varieties; and more frequent and more severe regional water shortages—especially where energy and industrial developments compete for water supplies, or where forest losses are heavy and the earth can no longer absorb, store, and regulate the discharge of water.

Deterioration of soils is occurring rapidly in LDCs, with the spread of desert-like conditions in drier regions, and heavy erosion in more humid areas. Present global losses to desertification are estimated at around 6 million hectares a year (an area about the size of Maine), including 3.2 million hectares of rangeland, 2.5 million hectares of rainfed cropland, and 125 thousand hectares of irrigated farmland. Desertification does not necessarily mean the creation of Sahara-like sand deserts, but rather it includes a variety of ecological changes that destroy the cover of vegetation and fertile soil in the earth's drier regions, rendering the land useless for range or crops. Principal direct causes are overgrazing, destructive cropping practices, and use of woody plants for fuel.

At presently estimated rates of desertification, the world's desert areas (now some 800 million hectares) would expand almost 20 percent by 2000. But there is reason to expect that losses to desertification will accelerate, as increasing numbers of people in the world's drier regions put more pressures on the land to meet their needs for livestock range, cropland, and fuelwood. The United Nations has identified about 2 billion hectares of lands (Figure 10) where the risk of desertification is "high" or "very high." These lands at risk total about two and one-half times the area now classified as desert.

Although soil loss and deterioration are especially serious in many LDCs, they are also affecting agricultural prospects in industrialized nations. Present rates of soil loss in many industrialized nations cannot be sustained without serious implications for crop production. In the United States, for example, the Soil Conservation Service, looking at wind and water erosion of U.S. soils, has concluded that to sustain crop production indefinitely at even present levels, soil losses must be cut in half.

The outlook for making such gains in the United States and elsewhere is not good. The food and forestry projections imply increasing pressures on soils throughout the world. Losses due to improper irrigation, reduced fallow periods, cultivation of steep and marginal lands, and reduced vegetative cover can be expected to accelerate, especially in North and Central Africa, the humid and high-altitude portions of Latin America, and much of South Asia. In addition, the increased burning of dung and crop wastes for domestic fuel will deprive the soil of nutrients and degrade the soil's ability to hold moisture by reducing its organic content. For the world's poor, these organic materials are often the only source of the nutrients needed to maintain the productivity of farmlands. It is the poorest people—those least able to afford chemical fertilizers—who are being forced to burn their organic fertilizers. These nutrients will be urgently needed for food production in the years ahead, since by 2000 the world's croplands will have to feed half again as many people as in 1975.[63] In the industrialized regions, increasing use of chemical fertilizers, high-yield plant varieties, irrigation water, and herbicides and pesticides have so far compensated for basic declines in soil conditions. However, heavy dependence on chemical fertilizers also leads to losses of soil organic matter, reducing the capacity of the soil to retain moisture.

Damage and loss of irrigated lands are especially significant because these lands have yields far above average. Furthermore, as the amount of arable land per capita declines over the next two decades, irrigated lands will be counted upon increasingly to raise per capita food availability. As of 1975, 230 million hectares—15 percent of the world's arable area—were being irrigated; an additional 50 million hectares are expected to be irrigated by 1990. Unfortunately there is great difficulty in maintaining the productivity of irrigated lands. About half of the world's irrigated land has already been damaged to some degree by salinity, alkalinity, and waterlogging, and much of the additional land expected to be irrigated by 1990 is highly vulnerable to irrigation-related damage.

Environmental problems of irrigation exist in industrialized countries (for example, in the San Joaquin Valley in California) as well as in LDCs (as in Pakistan, where three-quarters of the irrigated lands are damaged). It is possible, but slow

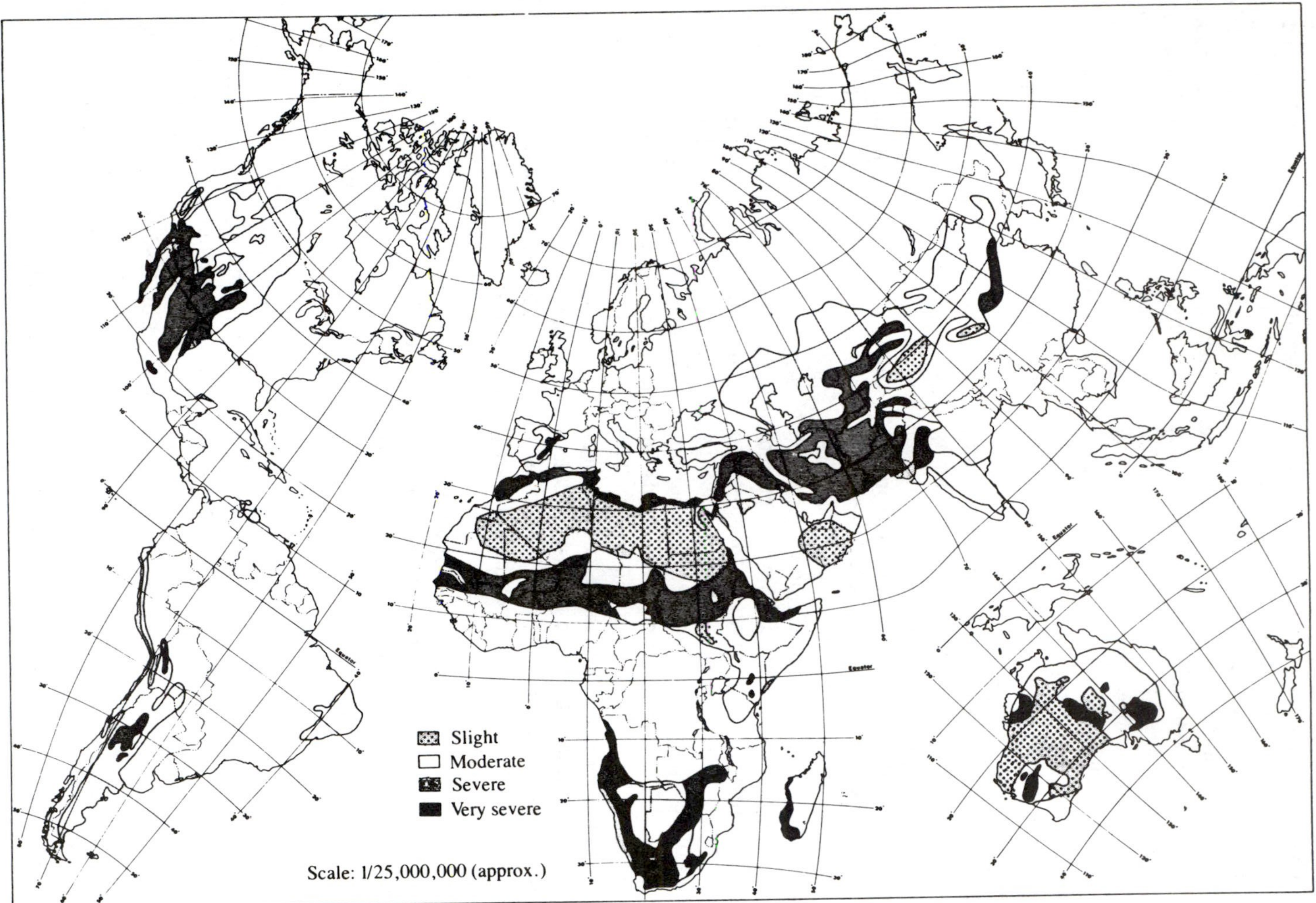


**Figure 10**. Desertification map (*U.N. Desertification Conference, 1977*).

and costly, to restore damaged lands. Prevention requires careful consideration of soils and attention to drainage, maintenance, and appropriate water-saving designs.

Loss of good cropland to urban encroachment is another problem affecting all countries. Cities and industries are often located on a nation's best agricultural land—rich, well-watered alluvial soils in gently sloping river valleys. In the industrialized countries that are members of the OECD, the amount of land devoted to urban uses has been increasing twice as fast as population. The limited data available for LDCs point to similar trends. In Egypt, for example, despite efforts to open new lands to agriculture, the total area of irrigated farmland has remained almost unchanged in the past two decades. As fast as additional acres are irrigated with water from the Aswan Dam, old producing lands on the Nile are converted to urban uses.

The rising yields assumed by the Global 2000 food projections depend on wider adoption of existing high-yield agricultural technology and on accelerating use of fertilizers, irrigation, pesticides, and herbicides. These yield-enhancing inputs, projected to more than double in use worldwide and to quadruple in LDCs, are heavily dependent on fossil fuels. Even now, a rapid escalation of fossil fuel prices or a sudden interruption of supply could severely disturb world agricultural production, raise food prices, and deprive larger numbers of people of adequate food. As agriculture becomes still more dependent on energy-intensive inputs, the potential for disruption will be even greater.

Accelerating use of pesticides is expected to raise crop yields quickly and substantially, especially in LDCs. Yet, many of these chemicals produce a wide range of serious environmental consequences, some of which adversely affect agricultural production. Destruction of pest predator populations and the increasing resistance of pests to heavily used pesticides have already proved to be significant agricultural problems. On California farms, for example, 17 of 25 major agricultural pests are now resistant to one or more types of pesticides, and the populations of pest predators have been severely reduced. Many millions of dollars in crop damage are now caused annually in California by resistant pests whose natural predators have been destroyed.

Crop yields are expected to be increased significantly by much wider use of high-yield strains of grains. Unfortunately, large monocultures of genetically identical crops pose increased risks of catastrophic loss from insect attacks or crop epidemics. The corn blight that struck the U.S. corn belt in 1970 provided a clear illustration of the vulnerability of genetically identical monocultures.

### Impacts on Water Resources

The quality of the world's water resources is virtually certain to suffer from the changes taking place between now and the year 2000. Water pollution from heavy application of pesticides will cause increasing difficulties. In the industrialized countries, shifts from widespread use of long-lived chemicals such as DDT are now underway, but in the LDCs—where the largest increases in agricultural chemical use is projected—it is likely that the persistent pesticides will continue to be used. Pesticide use in LDCs is expected to at least quadruple over the 1975–2000 period (a sixfold increase is possible if recent rates of increase continue). Pollution from the persistent pesticides in irrigation canals, ponds, and rice paddies is already a worrisome problem. Farmers in some parts of Asia are reluctant to stock paddies and ponds because fish are being killed by pesticides. This means a serious loss of high-quality protein for the diets of rural families.

In addition to the potential impacts on soils discussed above, irrigation adversely affects water quality by adding salt to the water returning to streams and rivers. Downstream from extensive irrigation projects the water may become too saline for further use, unless expensive desalinization measures are undertaken. As the use of water for irrigation increases, water salinity problems are certain to increase as well.

Water pollution in LDCs is likely to worsen as the urban population soars and industry expands. Already the waters below many LDC cities are heavily polluted with sewage and wastes from pulp and paper mills, tanneries, slaughterhouses, oil refineries, and chemical plants.

River basin development that combines flood control, generation of electricity, and irrigation is likely to increase in many less developed regions, where most of the world's untapped hydropower potential lies. While providing many benefits, large-scale dams and irrigation projects can also cause highly adverse changes in both freshwater and coastal ecosystems, creating health problems (including schistosomiasis, river blindness, malaria), inundating valuable lands, and displacing populations. In addition, if erosion in the watersheds of these projects is not controlled, siltation and buildup of sediments may greatly reduce the useful life of the projects.

Virtually all of the Global 2000 Study's projections point to increasing destruction or pollution of coastal ecosystems, a resource on which the commercially important fisheries of the world depend heavily. It is estimated that 60-80 percent of commercially valuable marine fishery species use estuaries, salt marshes, or mangrove swamps for habitat at some point in their life cycle. Reef habitats also provide food and shelter for large numbers of fish and invertebrate species. Rapidly expanding cities and industry are likely to claim coastal wetland areas for development; and increasing coastal pollution from agriculture, industry, logging, water resources development, energy systems, and coastal communities is anticipated in many areas.

### Impacts of Forest Losses

The projected rapid, widespread loss of tropical forests will have severe adverse effects on water and other resources. Deforestation—especially in South Asia, the Amazon basin, and central Africa—will destabilize water flows, leading to siltation of streams, reservoirs behind hydroelectric dams, and irrigation works, to depletion of ground water, to intensified flooding, and to aggravated water shortages during dry periods. In South and Southeast Asia approximately one billion people live in heavily farmed alluvial basins and valleys that depend on forested mountain watersheds for their water. If present trends continue, forests in these regions will be reduced by about half in 2000, and erosion, siltation, and erratic streamflows will seriously affect food production.

In many tropical forests, the soils, land forms, temperatures, patterns of rainfall, and distribution of nutrients are in precarious balance. When these forests are disturbed by extensive cutting, neither trees nor productive grasses will grow again. Even in less fragile tropical forests, the great diversity of species is lost after extensive cutting.

### Impacts on the World's Atmosphere and Climate

Among the emerging environmental stresses are some that affect the chemical and physical nature of the atmosphere. Several are recognized as problems; others are more conjectural but nevertheless of concern.

Quantitative projections of urban air quality around the world are not possible with the data and models now available, but further pollution in LDCs and some industrial nations is virtually certain to occur under present policies and practices. In LDC cities, industrial growth projected for the next 20 years is likely to worsen air quality. Even now, observations in scattered LDC cities show levels of sulfur dioxide, particulates, nitrogen dioxide, and carbon monoxide far above levels considered safe by the World Health Organization. In some cities, such as Bombay and Caracas, recent rapid increases in the numbers of cars and trucks have aggravated air pollution.

Despite recent progress in reducing various types of air pollution in many industrialized countries, air quality there is likely to worsen as increased amounts of fossil fuels, especially coal, are burned. Emissions of sulfur and nitrogen oxides are particularly troubling because they combine with water vapor in the atmosphere to form acid rain or produce other acid deposition. In large areas of Norway, Sweden, southern Canada, and the eastern United States, the pH value of rainfall has dropped from 5.7 to below 4.5, well into the acidic range. Also, rainfall has almost certainly become more acid in parts of Germany, Eastern Europe, and the U.S.S.R., although available data are incomplete.

The effects of acid rain are not yet fully understood, but damage has already been observed in lakes, forests, soils, crops, nitrogen-fixing plants, and building materials. Damage to lakes has been studied most extensively. For example, of 1,500 lakes in southern Norway with a pH below 4.3, 70 percent had no fish. Similar damage has been observed in the Adirondack Mountains of New York and in parts of Canada. River fish are also severely affected. In the last 20 years, first salmon and then trout disappeared in many Norwegian rivers as acidity increased.

Another environmental problem related to the combustion of fossil fuels (and perhaps also to the global loss of forests and soil humus) is the increasing concentration of carbon dioxide in the earth's atmosphere. Rising $CO_2$ concentrations are of concern because of their potential for causing a warming of the earth. Scientific opinion differs on the possible consequences, but a widely held view is that highly disruptive effects on world agriculture could occur before the middle of the twenty-first century. The $CO_2$ content of the world's atmosphere has increased about 15 percent in the last century and by 2000 is expected to be nearly a third higher than preindustrial levels. If the projected rates of increase in fossil fuel combustion (about 2 percent per year) were to continue, a doubling of the $CO_2$ content of the atmosphere could be expected after the middle of the next century; and if deforestation substantially reduces tropical forests (as projected), a doubling of atmosphereic $CO_2$ could occur sooner. The result could be significant alterations of precipitation patterns around the world, and a 2°–3 °C rise in temperatures in the middle latitudes of the earth. Agriculture and other human endeavors would have great difficulty in adapting to such large, rapid changes in climate. Even a 1 °C increase in average global temperatures would make the earth's climate warmer than it has been any time in the last 1,000 years.

A carbon dioxide-induced temperature rise is expected to be 3 or 4 times greater at the poles than in the middle latitudes. An increase of 5°–10°C in polar temperatures could eventually lead to the melting of the Greenland and Antarctic ice caps and a gradual rise in sea level, forcing abandonment of many coastal cities.

Ozone is another major concern. The stratospheric ozone layer protects the earth from damaging ultraviolet light. However, the ozone layer is being threatened by chlorofluorocarbon emissions from aerosol cans and refrigeration equipment, by nitrous oxide ($N_2O$) emissions from the denitrification of both organic and inorganic nitrogen fertilizers, and possibly by the effects of high-altitude aircraft flights. Only the United States and a few other countries have made serious efforts to date to control the use of aerosol cans. Refrigerants and nitrogen fertilizers present even more difficult challenges. The most widely discussed effect of ozone depletion and the resulting increase in ultraviolet light is an increased incidence of skin cancer, but damage to food crops would also be significant and might actually prove to be the most serious ozone related problem.

### Impacts of Nuclear Energy

The problems presented by the projected production of increasing amounts of nuclear power are different from but no less serious than those related to fossil fuel combustion. The risk of radioactive contamination of the environment due to nuclear power reactor accidents will be increased, as will the potential for proliferation of nuclear weapons. No nation has yet conducted a demonstration program for the satisfactory disposal of radioactive wastes, and the amount of wastes is increasing rapidly. Several hundred thousand tons of highly radioactive spent nuclear fuel will be generated over the lifetimes of the nuclear plants likely to be constructed through the year 2000. In addition, nuclear power production will create millions of cubic meters of low-level radioactive wastes, and uranium mining and processing will lead to the production of hundreds of millions of tons of low-level radioactive tailings. It has not yet been demonstrated that all of these high- and low-level wastes from nuclear power production can be safely stored and disposed of without incident. Some of the by-products of reactors, it should be noted, have half-lives approximately five times as long as the period of recorded history.

**Species Extinctions**

Finally, the world faces an urgent problem of loss of plant and animal genetic resources. An estimate prepared for the Global 2000 Study suggests that between half a million and 2 million species—15 to 20 percent of all species on earth—could be extinguished by 2000, mainly because of loss of wild habitat but also in part because of pollution. Extinction of species on this scale is without precedent in human history.[63]

One-half to two-thirds of the extinctions projected to occur by 2000 will result from the clearing or degradation of tropical forests. Insect, other invertebrate, and plant species—many of them unclassified and unexamined by scientists—will account for most of the losses. The potential value of this genetic reservoir is immense. If preserved and carefully managed, tropical forest species could be a sustainable source of new foods (especially nuts and fruits), pharmaceutical chemicals, natural predators of pests, building materials, speciality woods, fuel, and so on. Even careful husbandry of the remaining biotic resources of the tropics cannot compensate for the swift, massive losses that are to be expected if present trends continue.

Current trends also threaten freshwater and marine species. Physical alterations—damming, channelization, siltation—and pollution by salts, acid rain, pesticides, and other toxic chemicals are profoundly affecting freshwater ecosystems throughout the world. At present 274 freshwater vertebrate taxa are threatened with extinction, and by the year 2000 many may have been lost.

Some of the most important genetic losses will involve the extinction not of species but of subspecies and varieties of cereal grains. Four-fifths of the world's food supplies are derived from less than two dozen plant and animal species. Wild and local domestic strains are needed for breeding resistance to pests and pathogens into the high-yield varieties now widely used. These varietal stocks are rapidly diminishing as marginal wild lands are brought into cultivation. Local domesticated varieties, often uniquely suited to local conditions, are also being lost as higher-yield varieties displace them. And the increasing practice of monoculture of a few strains—which makes crops more vulnerable to disease epidemics or plagues of pests—is occurring at the same time that the genetic resources to resist such disasters are being lost.

# Entering the Twenty-First Century

The preceding sections have presented individually the many projections made by U.S. Government agencies for the Global 2000 Study. How are these projections to be interpreted collectively? What do they imply about the world's entry into the twenty-first century?[64]

The world in 2000 will be different from the world today in important ways. There will be more people. For every two persons on the earth in 1975 there will be three in 2000. The number of poor will have increased. Four-fifths of the world's population will live in less developed countries. Furthermore, in terms of persons per year added to the world, population growth will be 40 percent *higher* in 2000 than in 1975.[65]

The gap between the richest and the poorest will have increased. By every measure of material welfare the study provides—per capita GNP and consumption of food, energy, and minerals—the gap will widen. For example, the gap between the GNP per capita in the LDCs and the industrialized countries is projected to grow from about $4,000 in 1975 to about $7,900 in 2000.[66] Great disparities within countries are also expected to continue.

There will be fewer resources to go around. While on a worldwide average there was about four-tenths of a hectare of arable land per person in 1975, there will be only about one-quarter hectare per person in 2000[67] (see Figure 11 below). By 2000 nearly 1,000 billion barrels of the world's total original petroleum resource of approximately 2,000 billion barrels will have been consumed. Over just the 1975-2000 period, the world's remaining petroleum resources per capita can be expected to decline by at least 50 percent.[68] Over the same period world per capita water supplies will decline by 35 percent because of greater population alone; increasing competing demands will put further pressure on available water supplies.[69] The world's per capita growing stock of wood is projected to be 47 percent lower in 2000 than in 1978[70].

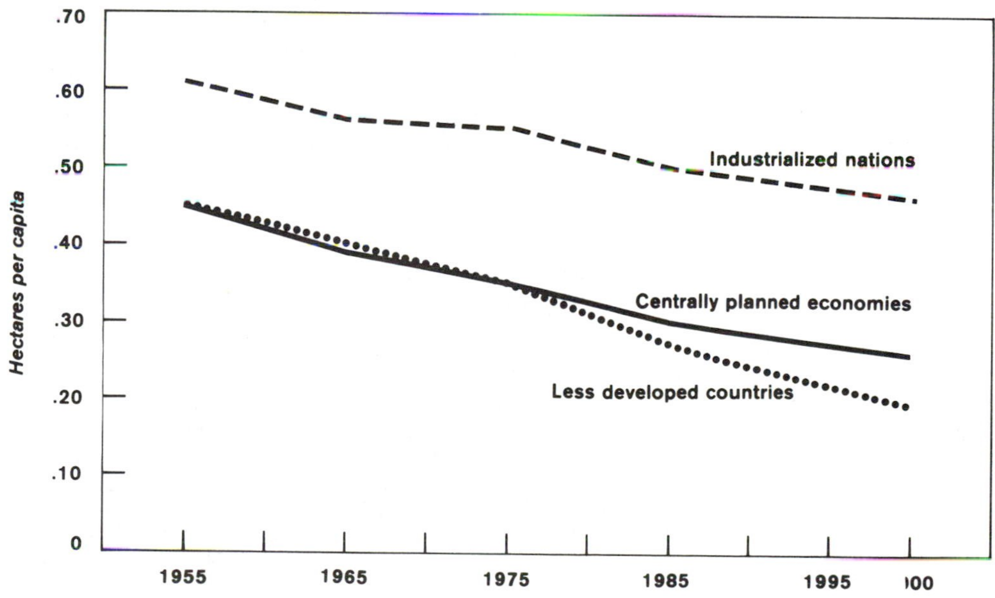


**Figure 11.** Arable land per capita, 1955, 1975, 2000.

The environment will have lost important life-supporting capabilities. By 2000, 40 percent of the forests still remaining in the LDCs in 1978 will have been razed.[71] The atmospheric concentration of carbon dioxide will be nearly one-third higher than preindustrial levels.[72] Soil erosion will have removed, on the average, several inches of soil from croplands all over the world. Desertification (including salinization) may have claimed a significant fraction of the world's rangeland and cropland. Over little more than two decades, 15-20 percent of the earth's total species of plants and animals will have become extinct—a loss of at least 500,000 species.[73]

Prices will be higher. The price of many of the most vital resources is projected to rise in real terms—that is, over and above inflation. In order to meet projected demand, a 100 percent increase in the real price of food will be required.[74] To keep energy demand in line with anticipated supplies, the real price of energy is assumed to rise more than 150 percent over the 1975-2000 period.[75] Supplies of water, agricultural land, forest products, and many traditional marine fish species are projected to decline relative to growing demand at current prices,[76] which suggests that real price rises will occur in these sectors too. Collectively, the projections suggest that resource-based inflationary pressures will continue and intensify, especially in nations that are poor in resources or are rapidly depleting their resources.

The world will be more vulnerable both to natural disaster and to disruptions from human causes. Most nations are likely to be still more dependent on foreign sources of energy in 2000 than they are today.[77] Food production will be more vulnerable to disruptions of fossil fuel energy supplies and to weather fluctuations as cultivation expands to more marginal areas. The loss of diverse germ plasm in local strains and wild progenitors of food crops, together with the increase of monoculture, could lead to greater risks of massive crop failures.[78] Larger numbers of people will be vulnerable to higher food prices or even famine when adverse weather occurs.[79] The world will be more vulnerable to the disruptive effects of war. The tensions that could lead to war will have multiplied. The potential for conflict over fresh water alone is underscored by the fact that out of 200 of the world's major river basins, 148 are shared by two countries and 52 are shared by three to ten countries. Long standing conflicts over shared rivers such as the Plata (Brazil, Argentina), Euphrates (Syria, Iraq), or Ganges (Bangladesh, India) could easily intensify.[80]

Finally, it must be emphasized that if public policy continues generally unchanged the world will be different as a result of lost opportunities. The adverse effects of many of the trends discussed in this Study will not be fully evident until 2000 or later; yet the actions that are necessary to change the trends cannot be postponed without foreclosing important options. The opportunity to stabilize the world's population below 10 billion, for example, is slipping away; Robert McNamara, President of the World Bank, has noted that for every decade of delay in reaching replacement fertility, the world's ultimately stabilized population will be about 11 percent greater.[81] Similar losses of opportunity accompany delayed perceptions or action in other areas. If energy policies and decisions are based on yesterday's (or even today's) oil prices, the opportunity to wisely invest scarce capital resources will be lost as a consequence of undervaluing conservation and efficiency. If agricultural research continues to focus on increasing yields through practices that are highly energy-intensive, both energy resources and the time needed to develop alternative practices will be lost.

The full effects of rising concentrations of carbon dioxide, depletion of stratospheric ozone, deterioration of soils, increasing introduction of complex persistent toxic chemicals into the environment, and massive extinction of species may not occur until well after 2000. Yet once such global environmental problems are in motion they are very difficult to reverse. In fact, few if any of the problems addressed in the Global 2000 Study are amenable to quick technological or policy fixes; rather, they are inextricably mixed with the world's most perplexing social and economic problems.

Perhaps the most troubling problems are

those in which population growth and poverty lead to serious long-term declines in the productivity of renewable natural resource systems. In some areas the capacity of renewable resource systems to support human populations is already being seriously damaged by efforts of present populations to meet desperate immediate needs, and the damage threatens to become worse.[82]

Examples of serious deterioration of the earth's most basic resources can already be found today in scattered places in all nations, including the industrialized countries and the better-endowed LDCs. For instance, erosion of agricultural soil and salinization of highly productive irrigated farmland is increasingly evident in the United States,[83] and extensive deforestation, with more or less permanent soil degradation, has occurred in Brazil, Venezuela, and Colombia.[84] But problems related to the decline of the earth's carrying capacity are most immediate, severe, and tragic in those regions of the earth containing the poorest LDCs.

Sub-Saharan Africa faces the problem of exhaustion of its resource base in an acute form. Many causes and effects have come together there to produce excessive demands on the environment, leading to expansion of the desert. Overgrazing, fuelwood gathering, and destructive cropping practices are the principal immediate causes of a series of transitions from open woodland, to scrub, to fragile semiarid range, to worthless weeds and bare earth. Matters are made worse when people are forced by scarcity of fuelwood to burn animal dung and crop wastes. The soil, deprived of organic matter, loses fertility and the ability to hold water—and the desert expands. In Bangladesh, Pakistan, and large parts of India, efforts by growing numbers of people to meet their basic needs are damaging the very cropland, pasture, forests, and water supplies on which they must depend for a livelihood.[85] To restore the lands and soils would require decades—if not centuries—*after* the existing pressures on the land have diminished. But the pressures are growing, not diminishing.

There are no quick or easy solutions, particularly in those regions where population pressure is already leading to a reduction of the carrying capacity of the land. In such regions a complex of social and economic factors (including very low incomes, inequitable land tenure, limited or no educational opportunities, a lack of non-agricultural jobs, and economic pressures toward higher fertility) underlies the decline in the land's carrying capacity. Furthermore, it is generally believed that social and economic conditions must improve before fertility levels will decline to replacement levels. Thus a vicious circle of causality may be at work. Environmental deterioration caused by large populations creates living conditions that make reductions in fertility difficult to achieve; all the while, continuing population growth increases further the pressures on the environment and land.[86]

The declines in carrying capacity already being observed in scattered areas around the world point to a phenomenon that could easily be much more widespread by 2000. In fact, the best evidence now available—even allowing for the many beneficial effects of technological developments and adoptions—suggests that by 2000 the world's human population may be within only a few generations of reaching the entire planet's carrying capacity.

The Global 2000 Study does not estimate the earth's carrying capacity, but it does provide a basis for evaluating an earlier estimate published in the U.S. National Academy of Sciences' report, *Resources and Man*. In this 1969 report, the Academy concluded that a world population of 10 billion "is close to (if not above) the maximum that an *intensively managed* world might hope to support with some degree of comfort and individual choice." The Academy also concluded that even with the sacrifice of individual freedom and choice, and even with chronic near starvation for the great majority, the human population of the world is unlikely to ever exceed 30 billion.[87]

Nothing in the Global 2000 Study counters the Academy's conclusions. If anything, data gathered over the past decade suggest the Academy may have underestimated the extent of some problems, especially deforestation and the loss and deterioration of soils.[88]

At present and projected growth rates, the world's population would rapidly approach the Academy's figures. If the fertility and mortality rates projected for 2000 were to continue unchanged into the twenty-first century, the world's population would reach 10 billion by 2030. Thus anyone with a present life expectancy of an additional 50 years could expect to see the world population reach 10 billion. This same rate of growth would produce a population of nearly 30 billion before the end of the twenty-first century.[89]

Here it must be emphasized that, unlike most of the Global 2000 Study projections, the population projections assume extensive policy changes and developments to reduce fertility rates. Without the assumed policy changes, the projected rate of population growth would be still more rapid.

Unfortunately population growth may be slowed for reasons other than declining birth rates. As the world's populations exceed and reduce the land's carrying capacity in widening areas, the trends of the last century or two toward improved health and longer life may come to a halt. Hunger and disease may claim more lives—especially lives of babies and young children. More of those surviving infancy may be mentally and physically handicapped by childhood malnutrition.

The time for action to prevent this outcome is running out. Unless nations collectively and individually take bold and imaginative steps toward improved social and economic conditions, reduced fertility, better management of resources, and protection of the environment, the world must expect a troubled entry into the twenty-first century.

# APPENDIX

## The Global 2000 Study Compared with Other Global Studies

In the course of the Global 2000 Study, the Government's several models (here referred to collectively as the "Government's global model") and their projections were compared with those of five other global studies.[90] The purpose was not only to compare the results of different projections, but also to see whether and how different assumptions and model structures may have led to different projections and findings.

The Global 2000 Study's principal findings are generally consistent with those of the five other global studies despite considerable differences in models and assumptions. On the whole, the other studies and their models lack the richness of detail that the Government's global model provides for the various individual sectors—food and agriculture, forests, water, energy, and so on. However, the linkages among the sectors in the other models are much more complete. Many apparent inconsistencies and contradictions in the Global 2000 projections are due to the weakness of the linkages among sectors of the Government's global model.

Another important difference is that the Government's projections stop at the year 2000 or before, while the other global studies project well into the twenty-first century. The most dramatic developments projected in the other studies—serious resource scarcities, population declines due to rising death rates, severe environmental deterioration—generally occur in the first half of the twenty-first century and thus cannot be compared with the Government's projections. Up to the turn of the century, all of the analyses, including the Government's, indicate more or less similar trends: continued economic growth in most areas, continued population growth everywhere, reduced energy growth, an increasingly tight and expensive food situation, increasing water problems, and growing environmental stress.

The most optimistic of the five models is the Latin American World Model. Instead of projecting future conditions on the basis of present policies and trends, this model asks: "How can global resources best be used to meet basic human needs for all people?" The model allocates labor and capital to maximize life expectancy. It assumes that personal consumption is sacrificed to maintain very high investment rates (25 percent of GNP per year), and it posits an egalitarian, nonexploitative, wisely managed world society that avoids pollution, soil depletion, and other forms of environmental degradation. Under these assumptions it finds that in little more than one generation basic human needs could be adequately satisfied in Latin America and in Africa. Thereafter, GNP would grow steadily and population growth would begin to stabilize.

But in Asia, even assuming these near-utopian social conditions and high rates of investment, the system collapses. The model projects an Asian food crisis beginning by 2010, as land runs out; food production cannot rise fast enough to keep up with population growth, and a vicious circle begins that leads to starvation and economic collapse by midcentury. The modelers suggest that an Asian food crisis could be avoided by such means as food imports from other areas with more cropland, better crop yields, and effective family planning policies. Nonetheless, it is striking that this model, which was designed to show that the fundamental constraints on human welfare were social, not physical, does project catastrophic food shortages in Asia due to land scarcity.

The World 2 and World 3 models, which were the basis of the 1972 Club of Rome report *The Limits to Growth,* give much attention to environmental factors—the only models in the group of five to do so. The World models, like the Global 2000 Study, considered trends in population, resources, and environment. However, these models are highly aggregated, looking at the world as a whole and omitting regional differences. In the cases that assume a continuation of present policies, the World 2 and 3 models project large global increases in food and income per capita until 2020, at which time either food scarcity or resource depletion would cause a downturn. The two models do suggest that major changes of policy can significantly alter these trends.

The World Integrated Model, a later effort sponsored by the Club of Rome, is much more detailed than the World 2 and 3 models in its treatment of regional differences, trade, economics, and shifts from one energy source to another, but it is less inclusive in its treatment of the environment. This complex model has been run under many different assumptions of conditions and policies. Almost invariably the runs project a long-term trend of steeply rising food prices. Under a wide range of policies and conditions the runs indicate massive famine in Asia and, to a lesser degree, in non-OPEC Africa, before the turn of the century.

The United Nations World Model found that to meet U.N. target rates for economic growth, developing countries would have to make great sacrifices in personal consumption, saving and investing at unprecedented rates. Personal consumption would not exceed 63 percent of income in any developing region, and none would have a level of private investment of less than 20 percent. To meet food requirements, global agricultural production would have to rise fourfold by 2000, with greater increases required in many places (500 percent, for example, in low-income Asia and Latin America).

The Model of International Relations in Agriculture (MOIRA) confines itself to agriculture; it takes into account the effects of agriculture policies but not those of environmental degradation. Its results are more optimistic than the Global 2000 projections: world food production more than doubles from 1975 to 2000, and per capita consumption rises 36 percent. Even so, because of unequal distribution, the number of people subsisting on two-thirds or less of the biological protein requirement rises from 350 million in 1975 to 740 million in 2000.

The Global 2000 Study conducted an experiment with two of the more integrated nongovernment models to answer the question: "How would projections from the Government's global model be different if the model were more integrated and included more linkages and feedback?" The linkages in the two nongovernment models were severed so that they bore some resemblance to the unconnected and inconsistent structure of the Government's global model. Chosen for the experiment were the World 3 model and the World Integrated Model.

In both models, severing the linkages led to distinctly more favorable outcomes. On the basis of results with World 3, the Global 2000 Study concluded that a more integrated Government model would project that:

- Increasing competition among agriculture, industry, and energy development for capital would lead to even higher resource cost inflation and significant decreases in real GNP growth (this assumes no major technological advances).
- The rising food prices and regional declines in food consumption per capita that are presently projected would be intensified by competition for capital and by degradation of the land.
- Slower GNP and agricultural growth would lead to higher death rates from widespread hunger—or from outright starvation—and to higher birth rates, with greater numbers of people trapped in absolute poverty.
- A decisive global downturn in incomes and food per capita would probably not take place until a decade or two after 2000 (this assumes no political disruptions).

When links in the World Integrated Model (WIM) were cut, outcomes again were more favorable. The results of the unlinked version were comparable to the Global 2000 quantitative projections for global GNP, population, grain production, fertilizer use, and energy use. But in the original integrated version of WIM, gross world product was 21 percent lower than in the unlinked version—$11.7 trillion instead of $14.8 trillion in 2000. In the linked version, world agricultural production rose 85 percent instead of 107 percent; grain available for human consumption rose less than 85 percent because some of the grain was fed to animals for increased meat production. Population rose only to 5.9 billion rather than 6.2 billion, in part because of widespread starvation (158 million deaths cumulatively by 2000) and in part because of lower birth rates in the industrialized countries. The effects of severing the linkages are much less in lightly populated regions with a wealth of natural resources, such as North America, than in regions under stress, where great numbers of people are living at the margin of existence. In North America, the difference in GNP per capita was about 5 percent; in South Asia, about 30 percent.

The inescapable conclusion is that the omission of linkages imparts an optimistic bias to the Global 2000 Study's (and the U.S. Government's) quantitative projections.[91] This appears to be particularly true of the GNP projections. The experiments with the World Integrated Model suggest that the Study's figure for gross world product in 2000 may be 15—20 percent too high.



## REFERENCES


1. *The Global 2000 Report to the President: Entering the Twenty-First Century,* vol. 2, *Technical Report,* Gerald O. Barney, Study Director, Washington: Government Printing Office, 1980, App. A.
2. Ibid.
3. Jimmy Carter, *The President's Environmental Program, 1977,* Washington: Government Printing Office, May 1977, p. M-11.
4. A more detailed discussion of the Global 2000 Study process is provided in *The Global 2000 Report to the President,* vol. 2, *Technical Report,* "Preface and Acknowledgments" and Ch. 1, "Introduction to the Projections."
5. *The Global 2000 Report to the President: Entering the Twenty-First Century,* vol. 2, Ch. 14–23; and *The Global 2000 Report to the President: Entering the Twenty-First Century,* vol. 3, *The Government's Global Model,* Gerald O. Barney, Study Director, Washington: Government Printing Office, 1980.

### NOTE

Unless otherwise indicated, the following chapter citations refer to the various chapters of *The Global 2000 Report to the President: Entering the Twenty-First Century,* vol. 2, *Technical Report,* Washington: Government Printing Office, 1980.

6. Ch. 13, 14, and 31.
7. Ch. 1.
8. "Closing the Loops," Ch. 13; Ch. 14.
9. Ibid.
10. Ch. 13, especially "Closing the Loops."
11. "Closing the Loops," Ch. 13; Ch. 14 and 31.
12. Ch. 1, 5, 14, and 23.
13. Ch. 1 and 14.
14. Ch. 30 and 31.
15. Ch. 2.
16. Ibid.
17. Ibid.
18. Ibid.
19. Ibid.; "Population Projections and the Environment," Ch. 13.
20. "Population Projections and the Environment." Ch. 13.
21. Ibid.; "Closing the Loops," Ch. 13.
22. Ibid.
23. *The Global 2000 Report to the President,* vol. 3, *The Government's Global Model,* Chapter on population models and the population projections update.
24. Ronald Freedman, "Theories of Fertility Decline: A Reappraisal," in Philip M. Hauser, ed., *World Population and Development,* Syracuse, N.Y.: Syracuse University Press, 1979; John C. Caldwell, "Toward a Restatement of Demographic Transition Theory," *Population and Development Review,* Sept./Dec., 1976.
25. For a discussion of Indonesia, see Freeman, op. cit.; for a discussion of Brazil, see "Demographic Projections Show Lower Birth Rate for the Poor" (in Portuguese), *VEJA,* Oct. 24, 1979, p. 139, citing the research of Elza Berquo of the Brazilian Analysis and Planning Center.
26. Ch. 3.
27. Ibid.
28. Ibid.
29. World Bank, *Prospects for Developing Countries, 1977–85,* Washington, Sept. 1976, Statistical Appendix, Table 1; World Bank, *World Development Report, 1979,* Washington, 1979, p. 13.
30. Ch. 6.
31. Ibid., "Food and Agriculture Projections and the Environment," Ch. 13.
32. Ch. 6.
33. Ibid.
34. "Food and Agriculture Projections and the Environment," Ch. 13.
35. Ch. 6.
36. U.S. Department of Agriculture, *Farm Income Statistics,* Washington: Economics, Statistics, and Cooperative Services, U.S.D.A., 1978 and 1979.
37. J. B. Penn, "The Food and Agriculture Policy Challenge

of the 1980's," Washington: Economics, Statistics and Cooperative Service, U.S.D.A., Jan. 1980.

38. P. Osam, *Accelerating Foodgrain Production in Low-Income Food-Deficit Countries—Progress, Potentials and Paradoxes,* Hawaii: East-West Center, May 1978; J. Gravan, *The Calorie Energy Gap in Bangladesh, and Strategies for Reducing It,* Washington: International Food Policy Research Institute, Aug. 1977.
39. Ch. 7.
40. Ibid.; "The Projections and the Marine Environment" and "Water Projections and the Environment," Ch. 13.
41. Ch. 7.
42. Food and Agriculture Organization, *Fisheries Statistics Yearbook, 1978.* Rome, 1979; Richard Hennemuth, Deputy Director, Northeast Fisheries Center, National Oceanic and Atmospheric Administration, personal communication, 1980.
43. Ch. 8 and App. C.
44. Ibid.
45. Ibid.; "Population Projections and the Environment," "Forestry Projections and the Environment," and "Energy Projections and the Environment," Ch. 13.
46. Norman Myers, *The Sinking Ark,* New York: Pergamon Press, 1979; U.S. Department of State, *Proceedings of the U.S. Strategy Conference on Tropical Deforestation,* Washington, Oct. 1978.
47. See especially Ch. 2, 6, 9, 10, and 12.
48. Ch. 9.
49. Ibid.; "Water Projections and the Environment," Ch. 13.
50. Ch. 9 and 13.
51. Ch. 12.
52. Ibid.
53. Ibid.
54. Ibid.; "Nonfuel Minerals Projections and the Environment," Ch. 13.
55. Ch. 10; "Energy Projections and the Environment," Ch. 13.
56. Ibid.
57. Ibid.
58. Ch. 8; "Population Projections and the Environment," "Forestry Projections and the Environment," and "Energy Projections and the Environment," Ch. 13.
59. Ch. 8; "Forestry Projections and the Environment," "Energy Projections and the Environment," and "Closing the Loops," Ch. 13.
60. See the IEES Model Projections reported in "International Energy Assessment," in Energy Information Administration, *Annual Report to Congress, 1978,* vol. 3, Washington: Department of Energy, 1979, pp. 11-34; Energy Information Administration, *Annual Report to the Congress, 1979,* forthcoming.
61. See Energy Information Administration, *Annual Report, 1979,* op. cit.; John Pearson and Derriel Cato, personal communication, Mar. 13, 1980.
62. The discussion of "Environmental Consequences" that follows is based on Ch. 13.
63. Thomas E. Lovejoy, "A Projection of Species Extinctions," Ch. 13, pp. 328-31.
64. This section is based largely on material contained in "Closing the Loops," Ch. 13.
65. Ch. 2.
66. Ch. 3.
67. Ch. 6.
68. Ch. 10; "Energy Projections and the Environment," Ch. 13.
69. Ch. 9.
70. Ch. 8.
71. Ibid.; "Forestry Projections and the Environment," Ch. 13.
72. Ch. 4; "Climate Projections and the Environment," Ch. 13.
73. "Food and Agriculture Projections and the Environment," "Forestry Projections and the Environment," and "Closing the Loops," Ch. 13.
74. Ch. 6.
75. Extrapolating from Ch. 10, which assumes a 5 percent per year increase over the 1980-90 period.
76. Ch. 6-9.
77. Ch. 10 and 11.
78. Ch. 6; "Food and Agriculture Projections and the Environment," Ch. 13.
79. Ibid.; Ch. 4; "Climate Projections and the Environment," Ch. 13.
80. Ch. 9.
81. Robert S. McNamara, President, World Bank, "Address to the Board of Governors," Belgrade, Oct. 2, 1979, pp. 9, 10.
82. Ch. 13.
83. "Food and Agriculture Projections and the Environment," Ch. 13.
84. "Forestry Projections and the Environment," Ch. 13, and Peter Freeman, personal communication, 1980, based on field observations in 1973.
85. "Population Projections and the Environment," "Food and Agriculture Projections and the Environment," "Forestry Projections and the Environment," and "Water Projections and the Environment," Ch. 13.
86. "Population Projections and the Environment," and "Closing the Loops," Ch. 13; Erik Eckholm, *The Dispossessed of the Earth: Land Reform and Sustainable Development,* Washington: Worldwatch Institute, June 1979.
87. National Academy of Sciences, Committee on Resources and Man, *Resources and Man,* San Francisco: Freeman, 1969, p. 5; "Closing the Loops," Ch. 13.
88. "Closing the Loops," Ch. 13.
89. Projection by the U.S. Bureau of the Census communicated in a personal letter, Feb. 26, 1980, from Dr. Samuel Baum, Chief, International Demographic Data Center. This letter and projection are presented in vol. 3 of the *Global 2000 Report to the President,* Population section.
90. The discussion in this Appendix is based on the detailed analyses in Ch. 24-31 and on two papers by Jennifer Robinson (author of those chapters) presented at the International Conference on Modeling Related to Environment, sponsored by the Polish Academy of Sciences: "The Global 2000 Study: An Attempt to Increase Consistency in Government Forecasting" and "Treatment of the Environment in Global Models."
91. Further discussion of these and other potential biases in the Government's projections are provided in Ch. 14-23, and App. B.

CHAPTER 3

# Simon and Kahn versus *Global 2000*

CONSTANCE HOLDEN

Futurist Herman Kahn, who died of a heart attack on 7 July, left behind a partially completed book that has attracted a good deal of attention even though it will not appear until some time next year. Coedited by Kahn and economist Julian Simon, who shares Kahn's optimistic vision of the future of the globe, the book is an attempt to refute the conclusions of *Global 2000*, an influential report issued by the Carter Administration in 1980.

*Global 2000* said that if present policies continue, the future in terms of population, resources, and the environment does not look good. Simon, who teaches at the University of Illinois, and Kahn, who headed the Hudson Institute, have argued, in contrast, that the trends by and large look fine and that the world will sort itself out if left to its own devices.

The two schools of thought have met in some preliminary skirmishes, notably at the AAAS annual meeting in Detroit. Presumably the debate will heat up when the book, christened *Global 2000 Revised* and financed by the Heritage Foundation, comes out. But whether it will instruct or further confuse the interested public is open to question.

*Global 2000* is, in its way, confused enough. The three-volume study, which is the government's first attempt at a coordinated analysis of the global environment/resources picture, used the resources of 14 agencies and several outside sources to analyze 11 selected "elements," such as food, water, and energy. Although teeming with qualifications and alternate scenarios, it came out with a general picture which, while it steers clear of apocalyptic visions, is not too happy. It predicts that the world is likely to be confronted with ever higher prices for food, oil, minerals, and fertilizer. In less-developed countries (LDC's) it sees increasing soil erosion, little room for expansion of cropland, water shortages, deforestation, loss of species, more overcrowding, and more pollution.

Recommendations based on the report, *Global Future: A Time to Act*, came out in the last days of the Carter Administration and sank out of sight with Carter. As for the original report, it inspired the formation of a coalition called Global Tomorrow (chaired by Russell Peterson, chairman of the National Audubon Society), which recently held a conference in Washington on the report; and a Year 2000 Committee of prominent men (chaired by Russell Train, head of the World Wildlife Fund–U.S.), which is pushing global foresight legislation and doing studies of private sector global data use.

*Global 2000* may have a "juggernaut" behind it (Simon's term), but Kahn and Simon have tried to balance it by gathering what Simon calls a group of "world class" authors for their book. The executive summary, which has been widely circulated, is far more provocative than the contributions. Written by Simon and Kahn, it explicitly contradicts the wording in *Global 2000*, saying, "If present trends continue, the world in 2000 will be less crowded, less polluted, more stable ecologically, and less vulnerable to resource-supply disruption than the world we live in now." Based on historical trends, it predicts declining scarcity, lowering prices and increased wealth. Trends in forests are "not worrisome," and "there is no evidence for the rapid loss of species." Simon frequently extrapolates from United States trends to predict developments in LDC's: For example, he says that as people get richer they will have more floor space in their homes. They will also have better roads and more vehicles. So they will have more room just as Americans have more than they did at the turn of the century.

As for the papers by the 23 authors,

---

Reprinted from *Science 221:* 341–343 (1983). 

Constance Holden is a member of the editorial staff of *Science*.

Simon and Kahn did not insist they toe the line and most of them eschew extreme positions. Nonetheless, despite the fact that *Global 2000* and *Global 2000 Revised* draw on many of the same original data sources, many findings are distinctly opposed.

Take, for comparison, the food and agriculture paper by D. Gale Johnson of the University of Chicago and the *Global 2000* food section. The two analyses differ markedly in their assessment of the role of fuel prices and environmental disruption in agricultural production. Johnson says that, according to the Food and Agriculture Organization (FAO), there will be an annual rise of 2.8 percent in food production in LDC's—more than enough for nutritional improvement. There is little need to bring new land into production because high yield practices (fertilizer, pesticides, and irrigation) are more efficient. Fertilizer prices, which have remained low, are not necessarily tied to petroleum prices, he says. Increases in per capita income make it implausible that an increasing share of resources will be into food production—as predicted by *Global 2000*—particularly since the percentage of the world's labor force engaged in agriculture has been declining. Prices of basic commodities such as grain and vegetable oils will stay low and may even decline. Increases in life expectancy indicate that malnutrition is declining. "Unavailability of food is no longer an important source of famine" (rather, it is war and strife). Johnson says that his projections are likely to prove valid if hindrances like trade restrictions and artifically low farm prices are removed.

*Global 2000* used the government's grain-oilseed-livestock model to conclude that food production will increase at a 2.2 percent annual rate. Because of rising petroleum costs, however, it predicts a 95 percent increase in food prices. It sees rapid rises in costs of fertilizer, pesticides and fuel, and diminishing returns because of accelerated erosion, loss of soil fertility and irrigation damage. Cropland may increase by only 4 percent because the good land is already cultivated and quality land is being lost to urbanization. It says that the World Bank estimates the number of malnourished people in LDC's could rise to 1.3 billion in 2000 and a substantial increase in the share of the world's resources devoted to food production will be needed to meet demand.

There are similar conflicts between the two studies in their analysis of world fisheries. John Wise says that the world haul, now at 70 million tonnes a year, will probably continue to increase for the next two decades. It will reach the FAO-predicted total of 100 to 120 million tonnes by 2000. Primary gains could come from improved management and harvesting or lightly exploited stocks. Possible further gains could come from finding new ways to fish krill and other unconventional species; developing ways to use fish meal directly as human food, and reducing discards at sea, and spoilage. Although overfishing has been a problem, pollution has had little effect on large-scale marine fisheries.

*Global 2000* asserts that traditional marine fish populations are now fully exploited, and the generally accepted annual potential of 100 million tonnes is unlikely on a sustained basis. Even if that figure is reached it would supply slightly less protein per capita—for a population of 6.35 billion—than it does now. (Wise, using the same figures, finds an increase in per capita protein.) Increasing ocean pollution is likely to effect significant reduction in yields, and improved technology has already masked real declines in fish populations. Increased harvest from lightly exploited areas and nontraditional species are inhibited by severe economic (that is, oil prices), technological, and management constraints.

An indication of the problems in forecasting fish catches is that the two analyses even differ on trends over the past decade, although they use the same set of figures. Wise, for example, maintains that the global haul rose by 10 percent in the 1970's, while *Global 2000* says the harvest leveled off in 1970.

Why the radical discrepancies, not only in future projections but in assessment of the current situation? There appear to be at least two explanations, one relating to the methodology, and the other to underlying assumptions.

In the Kahn-Simon book, the subjects are pretty much treated in isolation, with no reference to what may be pertinent trends outside the author's field. The creators of *Global 2000*, on the other hand, went through an agonizing process trying to integrate the data on each topic with data on everything else. This was extremely difficult because the computerized models used by each agency are generally devoted to narrow sectoral concerns or designed to justify particular policies. The energy model, for example, was intended to prove Project Independence would work by 1985.

Modelers often make assumptions about resources availability without referring to related efforts in other departments. Someone engaged in crop forecasting, for example, will assume the availability of a certain amount of water, which is also needed by an energy planner. So the two analysts may end up assuming 150 percent of the available water. Individual sectoral models thus often have to be modified in the light of other sets of projections.*

Nonetheless, after coordinating all the diverse sources, the creators of *Global*

---

*Here is an example from the report of efforts to correlate various assumptions on food, population, environment, and gross national product (GNP): " . . . the food projections show that . . . there will be some declines in food per capita. . . . This reinforces the finding of the per capita GNP projections that social and economic conditions will not improve throughout the world.

"The food projections also assume there will be no constraints on water development for agriculture. But this is contradicted by the water, forestry, and environmental projections, implying that a downward adjustment should be made to the food projections.

"The food projections also assume that land deterioration . . . will not occur. But this is contradicted by the environmental projections, implying the need for further downward adjustments to the food projections. . . .

"Downward adjustments to the per capita food projections would necessitate increases in the population projections. . . .

"Higher population projections would in turn probably increase the severity of water problems . . . and the rate of land deterioration . . . further lowering the food projections . . .

". . . if one wanted to adjust the food, GNP or population projections for consistency . . . where is one to begin—or end? Where is the lever and where is the fulcrum?"

*2000* felt they had come up with the best data available.

Kahn and Simon are inclined to think *Global 2000*'s struggles were nothing but a huge waste of time. "Our philosophy is totally different," Kahn said last month. "We are hostile to big models . . . any attempt to have a global model to integrate everything becomes uncontrollable" and is "of dubious value." What about resolving inconsistencies? "If you find inconsistencies the model is better off without them." Simon agrees that the number of factors calibrated into an analysis has to be reasonably small or "you'll never get on with your work." For example, the Kahn-Simon summary says: "The future price of energy is not a key input for estimating the future price and quantity of food." They believe in trend analysis: extrapolating "simplistically with ruler and pencil" produces better results, said Kahn. All the global modelers get is *gigo*—"garbage in, gospel out."

Methodology, then, is one of the main areas where the two works differ. Another is in their concepts of the overall direction of human history. *Global 2000* depicts a time of historical discontinuity in which traditional ways of doing things and the old supply-demand equations will lead eventually to pillage and desecration of natural resources and increasing human misery. *Global 2000 Revised* reflects a belief in humanity's continuing ability to sort everything out for the best.

*Global 2000 Revised* also seems to put a lot more faith in man's ingenuity and the rate of technological advance than does *Global 2000*. For example, if the oil runs out, the former believes new substitutes will be found, whereas the latter is more likely to see higher prices and more pressures on the environment.

The free play of market forces—including natural ones—is fundamental to the Simon-Khan vision, whereas the problems as *Global 2000* sees them would seem to call for government policy changes on every level.

The faith in the ability of market forces always to promote equilibrium is apparently why it was not deemed crucial to have a discussion of population growth in *Global 2000 Revised*. There is a paper by Mark Perlman of the University of Pittsburgh on the difficulties of making population projections, but according to Simon, the editors decided not to go into the implications of such growth because they did not want to divide the authors. Anyway, he said, populations level off by themselves when they reach a certain stage of economic well-being. Needless to say, *Global 2000* is not complacent about population growth.

Finally, as Kahn pointed out, there is a real difference in the way the two reports view nature. *Global 2000* is very much an environmentalist document; the subject is a core consideration in every topic discussed. Not so in the Simon-Kahn book: there is no mention of environmental considerations, for example, in the energy or agriculture articles. Kahn said the omission is appropriate: he took the Old Testament view that "everything that creeps or crawls exists for man's benefit," which, he said, is basically the attitude of traditional western culture and one that is shared by the authors (with an exception for Roger Revelle of the University of California at La Jolla who gave permission for a paper on land to be reprinted but who is "not very enthusiastic" about their approach). Kahn argued that *Global 2000* reflects a trend toward eastern thinking in which every living thing is believed to have an intrinsic right to exist.

Simon and Kahn's views correspond in many respects to those of the Reagan Administration. Some of the ideas are explicitly stated in a paper drafted in January, reportedly by presidential adviser Danny J. Boggs, for the Global Issues Working Group, which advises the Cabinet Council on Natural Resources and Environment.

After quoting from the somber introduction of *Global 2000*, the paper says "Rather, from our experience . . . if the economies and societies of much of the world remain reasonably free, if technological advance is permitted to continue, and if prices are permitted to bring changes in supply and demand into equilibrium, the world in the year 2000 will, in general, be a better place for most people than it is today. Although there will be more people in the world, each of them should have more individual living space. . . . There will very likely be greater material output for each person. . . . In many cases, technological and economic advance will be the key . . . to . . . environmental progress."



The paper goes on to discuss the value of global modeling and the improvement of government "foresight" capability, which has become of particular concern to the *Global 2000* groups. People like Train and Peterson are pushing hard to get some kind of legislation passed that would improve the government's ability to make comprehensive analyses and recommendations related to global population, resource, and environmental trends. Proposals vary, but basically the idea is not to have a monolithic global model—which all agree would be undesirable—but to facilitate interaction among various models, get the assumptions documented, the data more compatible and the inconsistencies made explicit. There are currently two bills pending: one introduced by Representatives Albert Gore, Jr. (D–Tenn.), and Newt Gingrich (R–Ga.) would establish an "office of critical trends analysis" in the White House to evaluate trends and the impact of government policies on them. The other by Senator Mark Hatfield (D–Ore.) would establish a "council on global resources, the environment and population" to improve projections. It also calls for a national policy of population stabilization.

Innocuous as the legislation may seem, people have definitely political reasons for supporting or opposing it. Boggs, in the White House document, says the "tendancy of such a centralization in an office would be to promote its capture and use by those who advocate a higher degree of governmental direction." He also notes that the "celebrated alarmist reports of the past . . . have underestimated the adjustive capacity and technological innovation of people" and have been "determinedly anti-market and anti-improvement by nongovernmental means."

Speaking at the conference held by the Global Tomorrow Coalition, Boggs pointed out that foresight can be wrong, as illustrated by such analyses as *Famine 75* by William and Paul Paddock and Paul Ehrlich's *Population Bomb*. "Would you want such an office run by Julian Simon or Herman Kahn?" he asked. Boggs later told *Science* he didn't think the problems of coordinating models was as great as *Global 2000* made out. He said "The notion that there is this commodity called foresight and if you will only buy a tube of it you'll come out with the right answers seems to me disingenuous." Supporters of the legislation, he felt, were saying "the world is going to hell in a handbasket and by passing this law we really want you to confess it and say so."

Boggs is on target in the last remark. An environmentalist told *Science* the Administration disliked the idea because it knew improved foresight would present facts that did not fit its dogma.

The Global Issues Working Group is currently preparing a report on the appropriate governmental role in global issues—presumably an expanded version of the January document. In view of the political and methodological poles represented by *Global 2000* and *Global 2000 Revised*, it will be interesting to see what they come up with. Alan Hill, chairman of the Council on Environmental Quality (CEQ), who initiated the study, says it will be a general document on resource, environment, and population issues to come out around the first of the year. "We've avoided saying this will be our answer to *Global 2000*," he says, explaining that the data will be better, thanks to the spadework done in the course of preparing that report. CEQ has also commissioned the World Wildlife Fund to look at the global data needs of the private sector.

It appears then that *Global 2000*, while it has not had the intended impact on government policy, is serving as the basis for ever-widening circles of dialogue about global issues. Rather than scaring the public out of its wits, as critics have claimed, the growing coalition spawned by the report is provoking others to reexamine common assertions about the world situation. So market forces are at work on the commodity, knowledge.

CHAPTER 4

# Bright Global Future



JULIAN L. SIMON

A FAIR SUMMARY of "Major Findings and Conclusions" from *The Global 2000 Report to the President* is found in this quotation from the first page:

> If present trends continue, the world in 2000 will be more crowded, more polluted, less stable ecologically, and more vulnerable to disruption than the world we live in now. Serious stresses involving population, resources, and environment are clearly visible ahead. Despite greater material output, the world's people will be poorer in many ways than they are today. For hundreds of millions of the desperately poor, the outlook for food and other necessities of life will be no better, for many it will be worse. Barring revolutionary advances in technology, life for most people on earth will be more precarious in 2000 than it is now—unless nations of the world act decisively to alter current trends.

In *The Resourceful Earth: A Response to Global 2000*, which I co-edited with the late Herman Kahn, we rephrased that summary to emphasize our contrasting viewpoint:

> If present trends continue, the world in 2000 will be *less* crowded (though more populated), *less* polluted, *more* stable ecologically, and *less* vulnerable to resource-supply disruption than the world we live in now. Stresses involving population, resources, and environment will be *less in the future than now*. . . . The world's people will be *richer* in most ways than they are today. . . . The outlook for food and other necessities of life will be *better* [and] life for most people on earth will be *less* precarious economically than it is now.

Here are the high points of our findings:

• Life expectancy has been rising rapidly throughout the world, a sign of demographic, scientific and economic success. This fact—at least as dramatic and heartening as any other in human history—is fundamental in any informed discussion of pollution and nutrition.

• The birthrate in less-developed countries has been falling substantially during the past two decades. The rate of growth of the world's population has dropped from 2.2 percent yearly in 1964–1965 to 1.75 percent in 1982–1983, probably a result of modernization and of decreasing child mortality, and a sign of increased control by people over their family lives.

• Many people are still hungry but the food supply has been improving since at least World War II, as measured by grain prices, production per consumer and the famine deathrate.

• Trends in world forests are not worrisome, though in some places deforestation is troubling.

• There is no statistical evidence for rapid loss of species in the next two decades. An increased rate of extinction cannot be ruled out if tropical deforestation is severe, but no evidence of linkage has been demonstrated.

• The fish catch, after a pause, has resumed its long upward trend.

---



Julian L. Simon is professor of business and social science at the University of Maryland and a senior fellow at the Heritage Foundation, Washington, DC.

- Land availability will not increasingly constrain world agriculture in coming decades.
- In the United States, the trend is toward higher-quality cropland suffering less from erosion than in the past.
- The widely published report of increasingly rapid urbanization of U.S. farmland was based on faulty data.
- Water does not pose a problem of physical scarcity or disappearance, although the world and U.S. situations do call for better institutional management through more rational systems of property rights.
- The climate does not show signs of unusual and threatening changes.
- Mineral resources are becoming less rather than more scarce, affront to common sense though that may be.
- There is no persuasive reason to believe that the world oil price will rise in coming decades; it may fall well below what it has been.
- Compared to coal, nuclear power is no more expensive, and is probably much cheaper, under most circumstances. It is also much cheaper than oil.
- Nuclear power gives every evidence of costing fewer lives per unit of energy produced than does coal or oil.
- Solar energy sources (including wind and wave power) are too dilute to compete economically for much of humankind's energy needs, though for specialized uses and certain climates they can make a valuable contribution.
- Threats of air and water pollution have been vastly overblown; these processes were not well analyzed in *Global 2000.*

LET US CONSIDER how benign developments in resource availability come about in human history, using energy as an example. England was alarmed in the 1600s at an impending shortage of energy due to deforestation for firewood. People feared a scarcity of fuel for both heating and the iron industry. This impending scarcity led to the development of coal.

It was not the English government that developed coal. Rather, individual entrepreneurs sensed the need, saw the opportunity and used all kinds of available information and ideas. They made lots of false starts which were very costly to many of those individuals but not to others in society and eventually arrived at coal as a viable fuel. This happened in the context of a competitive enterprise system that worked to produce what was needed by the public. And the entire process left us better off than if the shortage problem had never arisen.

In the mid-1800s Englishmen came to worry about an impending coal crisis. The great English economist, Jevons, calculated that a coal shortage would bring that nation's industry to a standstill by 1900; he also concluded that oil could never make a decisive difference. But ingenious profit-minded people developed oil into a more desirable fuel than coal ever was. And today we find England exporting both coal and oil.

Another chapter in the story bears telling. Because of greater world demand due to population growth and increased income, the price of whale oil for lamps jumped in the 1840s, and the American Civil War pushed it even higher, leading to a whale-oil "crisis." This provided an incentive for entrepreneurs to discover and produce substitutes. First came oil from rapeseed, olives and linseed, and camphene oil from pine trees. Then inventors learned how to manufacture coal oil, a flourishing industry by 1859. Other ingenious persons produced kerosene from the rock oil that seeped to the surface, a product so desirable that its price then rose from $.75 a gallon to $2.00. This was a stimulus to increase the supply of oil, and finally Edwin L. Drake drilled the famous well in Titusville, Pennsylvania. Learning how to refine the oil took a while. But in a few years there were hundreds of small U.S. refiners, and soon the bottom fell out of the whale-oil market.

This story is prototypical of the intertwined history of resources, population and civilization. Resource problems become opportunities and turn into the occasions for the advances of knowledge that support and spur economic development. We need more and bigger problems, rather than just having them solved, as conventional economics would have it.

And the story goes beyond energy resources. There is a direct connection between the deforestation crisis in England in the 1600s and modern transport. Development of coal created the new problem of water in coal mines. That problem led to the invention of steam engines for driving the pumps. Then someone had the bright idea to put wheels under the engine and run it on rails. From that came the railroad, the steam automobile, factories with central steam power and belts to transmit the power, and then steam-driven farm machines. Steam power in autos was replaced with the internal combustion engine, and that meant greater mobility—for better or for worse.

These ever-ramifying developments continue, leading to new resources that are cheaper than the old, besides being the founts of additional developments. For example, because of diminishing deposits elsewhere, enterprises are developing methods for making the frozen ground of Alaska and Canada yield its riches. Harbors and docks had to be built to handle the ships transporting the materials, but hauling concrete far north is costly. Instead, fresh water is poured on top of piled-up gravel. The water hardens into ice, docks are built cheaply, and we benefit from technology that we can use elsewhere, too.

ALL IS NOT WELL everywhere. Children are hungry and sick; people live out lives of physical or intellectual poverty and lack of opportunity. We do not guarantee a rosy future; war or some new pollution may do us in. But our study shows that for most relevant matters, aggregate global and U.S. trends are improving.

Nor do we say that a better future happens automatically. It will happen because people—as individuals, as enterprises working for profit, as voluntary groups, as governmental agencies—will address problems and will probably overcome, as they have throughout history. There are many who assume that if the government is not doing something, nothing is being done. If you suggest that the market will provide natural resources satisfactorily, many say: "You mean that the situation will take care of itself."

Yes and no. If you include the efforts of people as part of your concept of a market—and that is what underlies every market—and if you assume that that happens "automatically," then yes, the situation will take care of itself. But it is people who are taking care of the situation, and it does not seem automatic to them. If by "automatic" you mean that fine results descend from heaven, then the answer is no. The alternatives are not government action versus heaven-sent benefits but rather government action versus private action.

Of course, there are always local shortages and pollutions due to climate, increased population and income, or mismanagement. These should not be glossed over in any global assessment. Sometimes temporary large-scale problems arise. But the nature of the world's physical conditions and the resilience of a well-functioning economic and social system enable us to overcome such problems, with solutions that usually leave us better off. That is the great lesson of human history.

We are less optimistic about the constraints imposed upon material progress by political and institutional forces, such as those urged upon us by *Global 2000* and other frightening reports. These constraints include the views that resource and environmental trends point toward deterioration, that there are physical limits which will increasingly act as a brake upon progress, and that nuclear energy is more dangerous than energy from other sources. Thus the calls for subsidies and price controls, as well as for government ownership, management and allocation of resources.

The world already suffers from such policies—for example, agriculture in Africa—and continuation and intensification of government control could seriously damage resource production and choke economic progress. Refusal to use nuclear power could hamper the United States in its economic competition with other nations, as well as cause unnecessary deaths in coal mining and other types of conventional energy production.

Reports that things are getting worse, when they are really getting better, are being used to convince people that government action is warranted; yet in many cases those suggested actions would be counterproductive and perhaps disastrous. That is why it is so important to know that the trends of material life are generally improving.

When they see that historic trends have been toward things getting better, those on the other side of the fence often reply: "But history is not a good guide in this connection, because we are at a turning point in history." My response is that all throughout history people have felt that they are at a turning point, and it has not turned out to be so. More generally, if we cannot base our judgments about the future largely upon past experience, in conjunction with reasonable theoretical explanations of that experience, then all of our experience and all of our science are without value.

We are not suggesting an economy without rules. Sound rules are crucial to the working of a free and efficient enterprise society. We say, with Frederick Hayek, that the making of such rules is one of the most difficult and valuable services that one can perform. But the rules should be as automatic as possible, with as little arbitrary bureaucratic action as possible, to enhance decision-making.

The environment is one of the trickiest subjects for rule-making. Whenever there is a commonly owned good, such as the atmosphere or the rivers, there must be rules enforced by government to make the costs of using the resource function as a set of signals that will get people to use the resource in a way that best benefits the public. But too often the answer to environmental problems is simply more rules of the "Thou shalt not use this common property" or "Thou shalt not produce this or that pollution." Instead, we need rules that will provide maximum freedom for people to try new and better ways to take advantage of the opportunity to serve the public and to make a profit than an environmental constraint imposes. Making property private, and requiring polluters to pay the full social cost, are well-established economic principles.

PEOPLE FREQUENTLY ask: "If so much bad news about resources and the environment is false, why do we hear so much of it, and why do so many believe it?" The answer to this difficult question is complex and multifaceted.

The bad-news reports and the consequent public belief in them have a bootstrap effect. "How would it be possible to hear so much bad news if it weren't true?" people ask, in justification of their belief in the truth of that news. This is a variation of the "Where there's smoke there's fire" argument, which has its effect even though it is logically unsound. And ultimately the issue is important because of the effect of bad news on personal and national morale, and on enterprise.

There are several reasons why the standard compendium of bad news has seized popular thinking:

- personal and professional self-interest;
- exaggeration and other shortcuts of the truth out of a belief that urgency requires and justifies untruths;
- the inflammatory rhetoric that fuels fears about these matters;
- misunderstanding of the nature of resources, and of the economic theory and facts about environment and population;
- misunderstanding of, and affection for, a social order which is personal rather than impersonal, actuated by caring rather than profit-seeking, and directed by a centralized planning mechanism rather than by unplanned, decentralized, market decision-making.

In addition, there is a trait of mind, common among safety engineers, that, when applied to consideration of natural resources and the environment, paralyzes the social will and causes rejection of new technical possibilities. This application of "worst case" analysis, as game theorists term it, causes one to reject as unattractive many possibilities that on average are desirable. Much of the thinking in the environmental movement seems to be worst-case analysis. But, as Einstein said about nature, God may be tricky but is not malignly trying to do us in. Therefore, we need not take the same stance as safety engineers.

Nuclear power debates provide many instances of what we might call the "play-it-safe" syndrome. Those who are against nuclear power point to scenarios conceivably leading to, say, 50,000 deaths. Proponents point out that the risk of such a scenario occurring is minuscule, and the expected number of deaths—using "expected" in the statistical sense—is very small. The opponents are not impressed by such a probabilistic argument, saying that the worst case has a meaning to us that cannot be treated as part of any set of averages. Nor are they impressed by other examples of similarly large worst-case risks that we routinely accept, such as those of power-producing dams whose breach might kill hundreds of thousands, or airplanes falling into stadiums seating 70,000 people. Yet those risks are probably greater than those from nuclear energy. There seems to be a value judgment at the base of the argument, which cannot be rebutted logically any more than other values can be rebutted logically. But it is possible to point out costs of such policies that are neglected in the discussion.

It is appropriate for a safety engineer not to be concerned with the costs of avoiding a dangerous activity, because the cost/benefit calculation will be made at higher levels of management. But in discussion of such matters as nuclear power, all discussants should take a balanced view and not just focus on one side. After all, there is no arbiter in a court of public opinion who will take into account all views, as higher levels of management are responsible for doing in an industrial setting. Also, it seems appropriate to point out in such discussions that if we routinely adopt the "play-it-safe" approach, lives will be shorter and poorer, and fewer people will be able to enjoy life, because of the life-shortening effects of air pollution from coal and the industrial accidents that kill so many people in coal-mining and petroleum operations.

THE LONGRUN OUTLOOK is a more abundant material life rather than increased scarcity, in the United States as in the world. Adding people causes problems, but people are also the means to solve these problems. The main fuel to speed the world's progress is our stock of knowledge, and the brake is our lack of trained minds and of imagination. Such progress does not come about automatically, and we do not preach complacency. We agree with the doomsayers that our world needs the best efforts of all humanity to improve our lot. We part company with them in that they expect us to come to a bad end despite our efforts, whereas we expect a continuation of our history of successful efforts.

Their message is self-fulfilling, because if you expect your efforts to fail because of inexorable natural limits, you are likely to feel resigned, and therefore literally to resign from the task. But if you recognize the possibility—indeed the probability—of success, you can tap large reservoirs of energy and enthusiasm. Here lie hope, the excitement of creation, spiritual fulfillment, the betterment of our lives and inevitably the betterment of the lives of others in the community as well.

# SECTION II
# PRINCIPLES AND TRENDS IN GLOBAL ECOLOGY

CHAPTER 5

# Interacting with the Elements: Man and the Biogeochemical Cycles



ROBERT B. COOK

Human activity significantly influences the major biogeochemical cycles—those interactive processes by which carbon, nitrogen, sulfur, and phosphorus circulate through the global oceanic, terrestrial, and atmospheric systems. Without exception, man's use of natural resources is accompanied by the production of unwanted wastes. For example, burning coal and petroleum releases carbon, nitrogen, and sulfur compounds into the atmosphere. Increasing atmospheric carbon dioxide may cause a significant warming of global climate: nitrogen and sulfur emissions have been linked to the formation of acidic rain and snow, which can have deleterious consequences for some sensitive ecosystems. Additionally, the use of nitrogen and phosphorus fertilizers to promote increased yields from agricultural land may result in significant nutrient pollution of rivers, lakes, and coastal waters.

In order to analyze the long-term impact of man's waste emissions on the atmosphere, oceans, and freshwater and terrestrial ecosystems, the natural cycles of carbon, nitrogen, phosphorus, and sulfur must be understood. Only after this has been done can we determine the ways in which man modifies these cycles. Although we speak of natural resources as being renewable, only through careful management will they remain renewable under the increasing demands for improved conditions for a growing world population.

Since the mid-1970s, the natural biogeochemical cycles and their perturbation by man have been the focal point used by the Scientific Committee on Problems of the Environment (SCOPE) in its study of the global environment. The ultimate goal of SCOPE, which was established in 1969 by the International Council of Scientific Unions (ICSU), is to provide an integrated view of the global ecosystem. Such a view could serve as background for assessing the long-term consequences of the increasing use of the Earth's natural resources. This article is based on some of SCOPE's findings.

## Natural Biogeochemical Cycles

The availability, circulation, and interaction of carbon, nitrogen, sulfur, and phosphorus have been essential for the development of life on Earth. However, the development of living organisms has in turn modified the primordial distributions and circulations of these elements, and thus has largely determined the main features of their cycles.

The biogeochemical cycling of carbon, nitrogen, phosphorus, and sulfur originating under natural conditions can be illustrated in the establishment and development of soil and vegetation on a previously unoccupied site. If a new land surface is deposited following a volcanic eruption or exposed by the melting of a glacier,

---

Reprinted by permission from *Environment 26(7)*: 11–15, 38–40 (1984). 

Robert B. Cook is an assistant professor of geology at Indiana University, Indiana, PA.

the new surface is at first devoid of carbon and nitrogen. Phosphorus and sulfur are commonly present in the newly exposed material.

The development of plants on such a site initially requires that seeds be carried there by the wind. Germination and growth of the seed into a plant requires uptake of atmospheric carbon dixoide ($CO_2$) by photosynthesis and intake of atmospheric nitrogen, which is converted into a usable form by nitrogen-fixing microorganisms in the soil or roots. Phosphorus is derived from the soil; sulfur can be derived from the soil or can also be supplied from the atmosphere.

As the plants die and decompose, organic matter containing carbon and nitrogen accumulates and increases the fertility of the soil. Soil microorganisms convert the organic matter into compounds that are readily used by growing plants. Once the availability of nitrogen in the soils is high, plants without nitrogen-fixing capabilities can grow; thus, nitrogen-fixation peaks relatively early in soil development.

Terrestrial ecosystems are remarkable for their ability to economize on nutrients by efficient recycling. However, some nutrients—especially phosphorus—can be converted into unavailable mineral and organic forms or removed from the ecosystem through leaching by running water. This slow loss is irreversible under natural conditions because there is relatively little replenishment of phosphorus from the atmosphere. With time, as more and more of the nutrients are lost from the soil or become unavailable, the fertility decreases.

Besides supplying compounds to hydrologic systems, terrestrial ecosystems also release many gases into the atmosphere. Respiration and decay of organic matter are two major natural sources of $CO_2$, methane ($CH_4$), carbon monoxide (CO), and nitrogen oxides ($NO_X$). Gaseous sulfur compounds are produced in environments devoid of oxygen, such as swamps and bogs (which release hydrogen sulfide); by actively photosynthesizing aquatic plants (which produce dimethyl sulfide); as well as by volcanoes (which give off sulfur dioxide). The atmosphere transports gases emitted by terrestrial and aquatic ecosystems. Nutrient elements are returned to terrestrial ecosystems and the ocean by rain, snow, and dry deposition (direct gas-phase and particulate-matter deposition at the Earth's surface).

Atmosphere plays a decisive role in regulating the Earth's surface temperature. Without the presence of the atmosphere, average surface temperatures would be about -18°C. However, due to the trapping of out-going infrared surface radiation by $CO_2$, $H_2O$ and ozone ($O_3$) in the atmosphere, the mean surface temperature is about 17°C.

Trace gases play an important part in maintaining suitable growing conditions for organisms. For example, the composition of the atmosphere is strongly influenced by the chemistry of ozone. Its breakdown by ultraviolet solar radiation leads to the production of the hydroxyl radical ($OH-$), which reacts with many atmospheric compounds that would otherwise be much more abundant. The hydroxyl radical serves as a catalyst in the oxidation of both methane and carbon monoxide to yield $CO_2$. Similarly, reduced sulfur compounds and $SO_2$ are oxidized by the hydroxyl radical to sulfuric acid, and nitrogen oxides are converted to nitric acid.

The global circulation of carbon, nitrogen, phosphorus, and sulfur is also due to the motions of water. As shown in Figure 1, the hydrologic cycle begins with the evaporation of water (caused by solar radiation) and continues with the subsequent dispersion of water vapor about the atmosphere. Part of the water is deposited upon the continents and enters the oceans via rivers, coastal run-off, and subterranean discharge; part enters the ocean directly as rain or snow.

Atmospheric and terrestrial waters undergo continuous chemical alterations by numerous processes along their way to the oceans. Gas dissolution in the water droplets of the atmosphere, biological activity in the rivers, and continental erosion are examples of such processes. The chemical composition of rivers is determined by a number of diverse factors, including the geology of the watershed; the character of vegetation in the drainage basin; and hydrographic conditions, such as the occurrence of floods.

River waters mix with the sea in the estuarine zone. The extent to which chemical and biological processes may change the transfer of elements to the adjacent coastal zone depends to a large degree on the hydrodynamic properties of the estuary. Where fresh water meets salt water, flocculation (formation of loose clumps or clouds of material) and sedimentation remove part of the riverborne organic matter and some dissolved inorganic chemical species.

The decrease in turbidity associated with these processes and the consequent increase in sunlight penetration promotes high algal productivity when nutrients are in high supply. Such nutrients are then transferred from solution to particulate organic matter, such as wastes, and can be

Figure 1. The Hydrologic Cycle

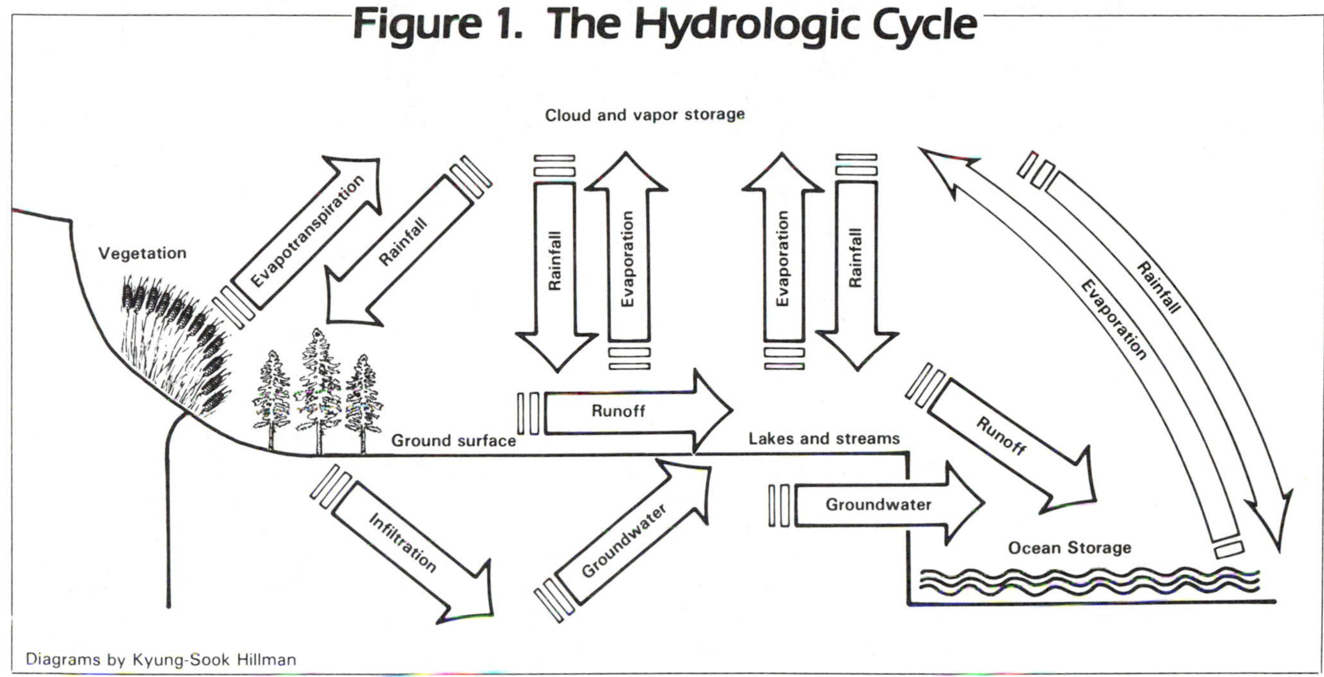


Diagrams by Kyung-Sook Hillman

trapped by sedimentation to deeper waters or to bottom sediments. Microbial activity in the sediments decomposes the organic matter to its component nutrients, which may then be released to the overlying waters.

The coastal zone, which is bounded by the 200-meter isobath (depth contour) and the continents, receives the material transported by the rivers and estuaries. This area is the source of about 25 percent of the primary productivity of the oceans; it receives about 90 percent of the organic matter that reaches the ocean floor and is buried.[1]

The open sea is by far the largest reservoir for carbon, nitrogen, phosphorus, and sulfur. Due to its size, the ocean responds slowly to imposed disturbances. Photosynthesis in the sea is limited to the upper 100 meters. The chemistry of some elements in the sea, most notably carbon, nitrogen, oxygen, phosphorus, and perhaps silicon and aluminum, is strongly mediated and linked by the biota. Algae incorporate these elements in fixed proportion, thus controlling their abundance.

Photosynthetic organisms are very efficient in utilizing available nutrients and recirculate much of the nutrient material at a rapid rate. Because algal growth is limited by the supply of phosphorus, most of this element is found in organic matter and little inorganic phosphorus exists in solution. A small fraction of the organic matter does settle out, mostly in fecal material and in dead organisms that are large enough to sink to the ocean floor.

Up to this point, the transport of carbon, nitrogen, phosphorus, and sulfur from the atmosphere to the oceans has taken from a few years to a few decades. Once the nutrients have been incorporated into ocean sediments, many millions of years will pass before the elements are brought back to the Earth's surface as, for example, sedimentary rocks or fossil fuel. The length of time it takes nutrients to move through the cycle is an important factor for predicting how quickly human activity will affect the various cycles.

After life developed on Earth, the global distribution of carbon, nitrogen, phosphorus, and sulfur was established by the interplay of biological, chemical, and geological processes and transport by wind and water. With the accelerated use of natural resources during the industrial period, human activity has also become a factor in the biogeochemical cycles of these elements.

## The $CO_2$ Cycle

The main sources of carbon dioxide to the atmosphere are the decomposition of organic matter by microorganisms, gas exchange in the ocean, deforestation, respiration by animals, and the burning of coal and petroleum (Figure 2). Photosynthesis converts $CO_2$ into the structural material for plants. When plants lose their leaves or fall to the ground, the structural material is decomposed by microorganisms and is then either released as $CO_2$ or incorporated into soil organic matter. The organic matter may remain in the soil or it may be leached into rivers and eventually carried to the sea.

The annual release of carbon dioxide to the atmosphere from the burning of coal and petroleum is equal to about 10 percent of the carbon dioxide used during natural photosynthesis by terrestrial plants.[2] This release has increased the amount of carbon dioxide in the global atmosphere from early nineteenth century values of approximately $265 \pm 30$ parts per million by volume (ppmv) to 335 ppmv in 1981. Estimates of the future combustion of fossil fuels based on an energy growth of between 2 and 3 percent each year (the current rate is 2.2 percent) imply that the atmospheric $CO_2$ concentration will be double our present one sometime in the next century.[3]

Because carbon dioxide is one of the atmospheric gases important in determining the Earth's climate, a doubling of $CO_2$, if maintained indefinitely, would cause a global surface air warming of between 1.5°C and 4.5°C.[4] These increases are predicted to vary considerably over latitude and over the seasons. The warming of the polar regions could be two to three times as great as that in the equatorial regions, and the warming of solar winters will probably be greater than of polar summers. Changes in climate—rainfall and temperature—may severely reduce the quantity of water resources in semi-arid areas such as the western United States.

Furthermore, summer soil moisture may decrease in the mid-latitudes. In addition to $CO_2$, the atmospheric concentration of several other "greenhouse" gases produced by human activity also appears to be increasing. These gases are nitrous oxide ($N_2O$), methane, and ozone. If increases of all these gases are forecast, their synergistic effect will probably cause climate changes to occur significantly earlier than if $CO_2$ were acting alone.[5]

The direct effects of elevated atmospheric $CO_2$ on terrestrial ecosys-

Figure 2. The Carbon Cycle

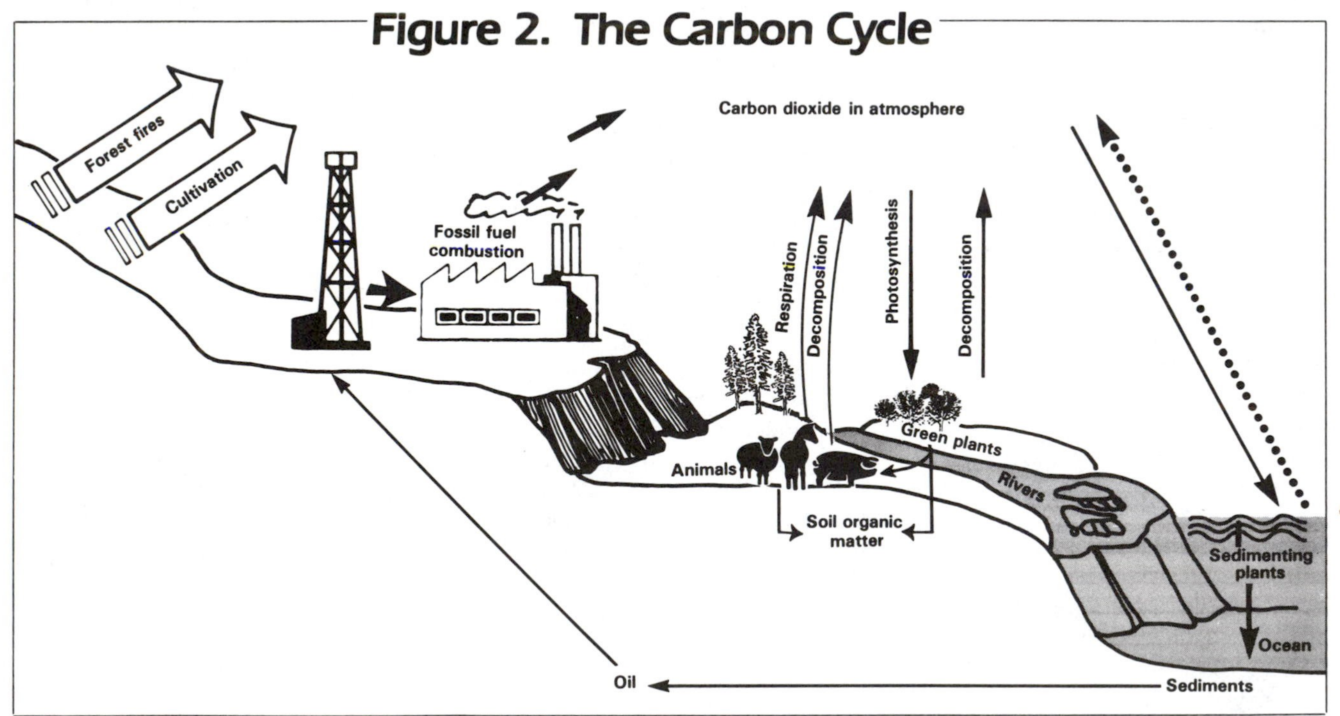

tems are not well known, beyond the fact that both photosynthesis and growth rates will increase in individual plants that are not strongly light-, water-, or nutrient-limited. The water-use efficiency of plants—the ratio of net primary production to evapotranspiration (the amount of water given off) is also increased because the $CO_2$ concentration gradient between plants and leaves is increased, and thus plant production is increased. It is not clear, however, that either natural or cultivated plants in the field are $CO_2$-limited often enough so that plant growth can be substantially increased in this way.[6] Some increase in terrestrial productivity (due to water-use efficiency) and some, but quite limited, increase in total biomass is the most likely outcome of elevated atmospheric $CO_2$ levels.

## The Sulfur Cycle

Atmospheric sulfur is derived from a number of sources, including volcanoes, the combustion of fossil fuels, and microorganism activity in tidal flats and the water-logged soils of swamps and bogs (see Figure 3). The sulfur cycle perhaps best exemplifies the difficulties that impede our understanding of how man's waste emissions disturb the natural biogeochemical cycles. Sulfur is emitted into the atmosphere from natural and anthropogenic (manmade) sources, but many of the effects of both types of emissions are identical. Only through careful measurements can the relative importance of the separate sources be determined.

Manmade sulfur emissions—primarily from the burning of coal and petroleum—probably exceed those from natural sources, especially in industrial regions. In the atmosphere, sulfur gases that have been converted to sulfuric acid are conveyed to the Earth's surface by precipitation or dry deposition. The rapid removal of this element from the atmosphere restricts the transport of anthropogenic emissions to short distances. Unlike fossil-fuel-produced $CO_2$, which has a global distribution, sulfur originating from coal and petroleum only disperses to a maximum of a few thousand kilometers.



Sulfur dioxide and other gases (e.g., ozone) have been shown to re-

**Figure 3. The Sulfur Cycle**

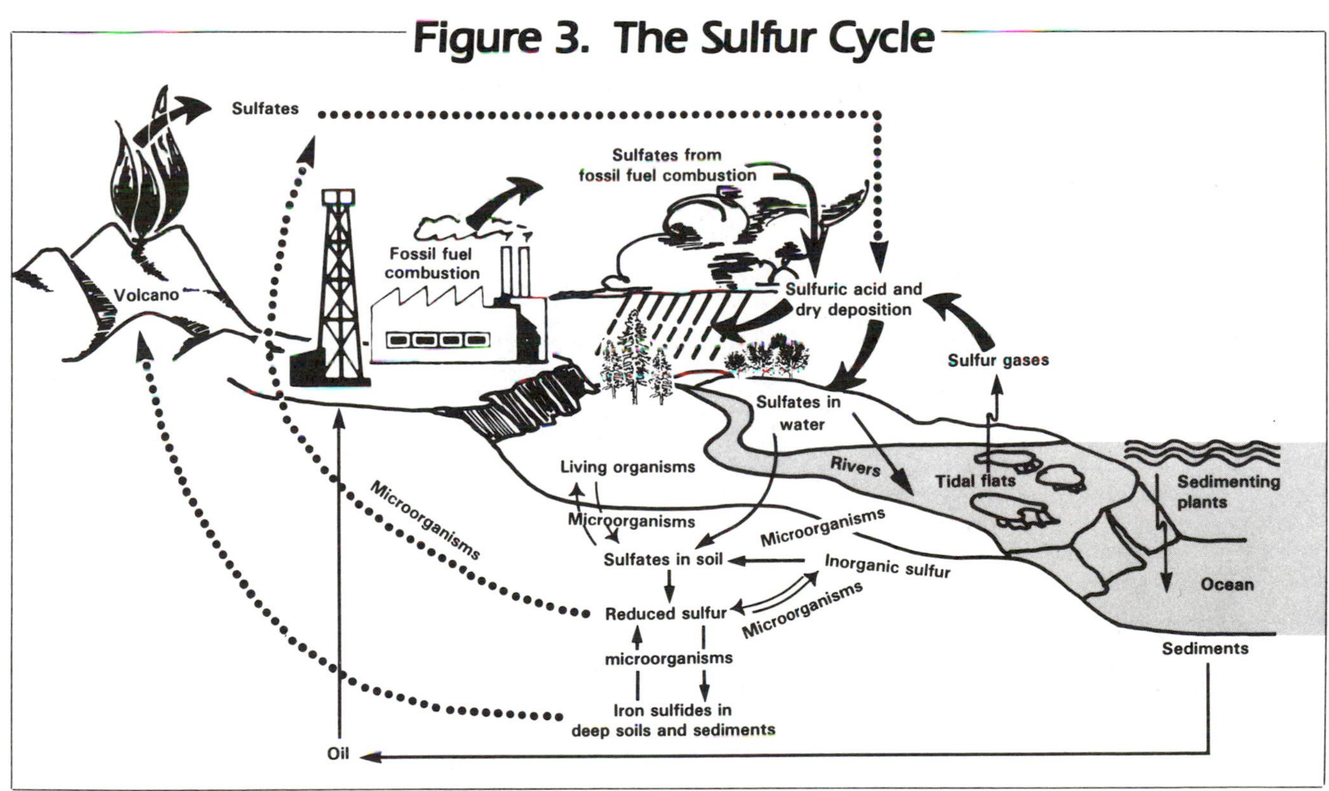

duce plant growth near major sites of fossil fuel combustion.[7] Most terrestrial ecosystems are resistant to sulfuric acid, however, and direct toxic effects of increased acidity have only been conclusively established in a few areas. For example, the toxicity of acidic precipitation has been implicated in an area of Vermont in which 50 percent of the trees died between 1965 and 1982.[8] Acid precipitation is also believed to be the cause of forest decline in Central Europe and in the Appalachian Mountains of the eastern United States.

Acidic precipitation may also cause increased leaching of sulfate and accompanying cations and trace metals to stream water and groundwater in many soils. The increased acidity and metal content of runoff and streams is causing the acidification of lakes in Scandinavia, Canada, and the northeastern United States, with damaging effects on fish and other aquatic organisms.

### The Nitrogen Cycle

The nitrogen cycle is intimately coupled to the carbon cycle. Like carbon, nitrogen is an essential constituent of living biomass, primarily as proteins and amino acids, and changes in carbon cycling are usually accompanied by changes in nitrogen cycling. Nitrogen compounds in the atmosphere originate from natural biological activity (denitrification by bacteria, for example), volcanoes, lightning, combustion of fossil fuels, and waste products of domestic animals, as shown in Figure 4.

The amounts of both nitrogen oxides and ammonia in the atmosphere are significantly influenced by combustion of fossil fuels and by wastes from domestic animal feed lots, respectively. Like the sulfur gases, nitrogen oxides react in the atmosphere to form acidic compounds that contribute to acidic wet and dry deposition. Nitrogen is also removed relatively quickly from the atmosphere, resulting in impacts only in industrial areas of the world.

The amount of nitrogen (as nitrogen oxides and nitrate) formed by the burning of coal and petroleum and also by fertilizer manufacturing is equal to about one-half of that produced naturally by the biosphere. In industrial regions of the world where terrestrial productivity is nitrogen-limited, the nitrogen in precipitation and dry deposition could fertilize ter-

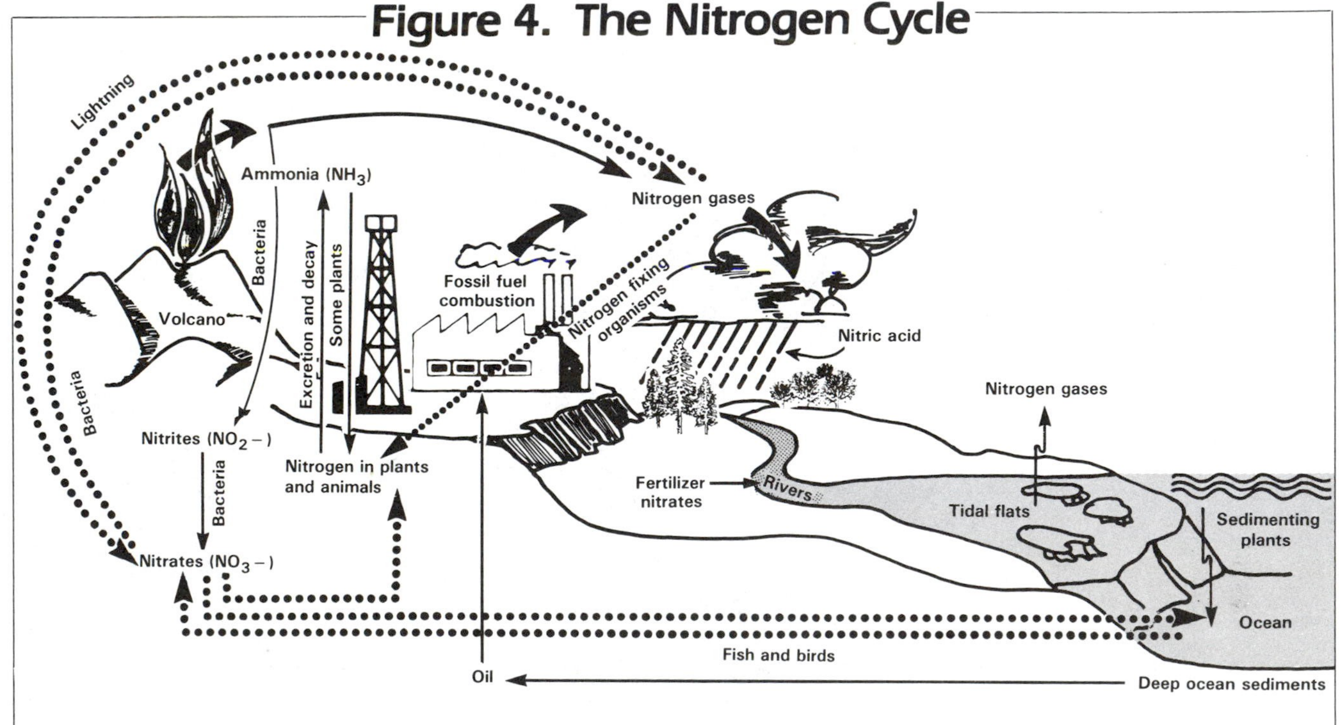

restrial biota, increasing net primary production. The effect could be significant over considerable areas, but there is no conclusive evidence for any enhancement of forest productivity attributable to increased atmospheric deposition of nitrogen. If forest biomass is increased by these nitrogen additions, some of the carbon from $CO_2$ released by fossil fuel emissions could accumulate in the increased biomass. One estimate of the maximum potential accumulation of carbon in this pool is 5 to 10 percent of the carbon released by combustion.[9]

### The Phosphorus Cycle

Figure 5 shows that unlike carbon, nitrogen, and sulfur, phosphorus does not have a significant gaseous phase in the atmosphere. Atmospheric phosphorus is, however, largely in the form of an aerosol (fine suspended particles that originates from crustal weathering, the oceans, and such human activities as detergent use and sewage processing.

The major impact of man on the phosphorus cycle is through the use of phosphorus fertilizers on agricultural land. Fertilizer use is necessary when the phosphorus in soil is naturally low or when it has been depleted due to intensive farming. Phosphorus is eventually removed via air or water transport to the ocean, where it is taken up by organisms and eventually incorporated into bottom sediments. After several million years, phosphorus in plant and fish remains is converted into phosphate minerals.

### Human Influences

Man's influence on the carbon, nitrogen, phosphorus, and sulfur cycles of terrestrial ecosystems is felt in most parts of the world. The natural forests, semi-deserts, and grasslands of the temperate regions have long been disruptively exploited by humans for food and fiber. During the last few centuries, about 10 percent of the land surface—primarily forests and grasslands—has been transformed into agricultural land. The increased intensity of land use takes many forms, but historically almost all have the long-term consequence of decreasing the reserves of carbon, nitrogen, phosphorus, and sulfur in soils.

During the 50 to 100 years of continuous cultivation in the North American Great Plains, approximately 50 percent of the carbon and nitrogen, 30 to 40 percent of the sulfur, and 10 to 30 percent of the phosphorus in the soil has been lost.[10] A portion of the carbon, nitrogen, and sulfur, and most of the phosphorus, is lost to harvest and erosion, but a larger fraction of the carbon, nitrogen, and probably sulfur is lost to the atmosphere in gaseous form or to aquatic ecosystems in dissolved or suspended form.

**Figure 5. The Phosphorus Cycle**

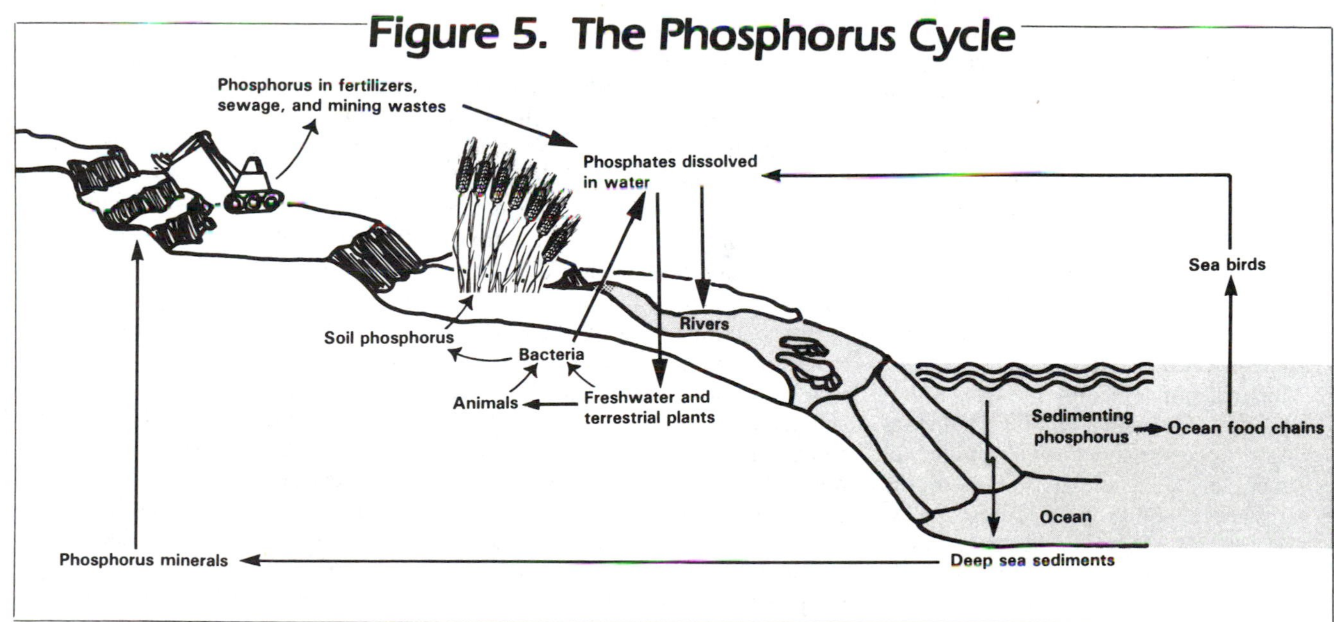

These losses of carbon, nitrogen, and sulfur occur because decomposition rates are increased in cultivated soils due to physical alterations and because less organic material is returned to the soil than in unaffected natural systems. When the addition of organic carbon to the soil is reduced, microbial activity declines and the ecosystem has a reduced capacity to immobilize the increased levels of nitrogen and sulfur released during decomposition. The nitrogen and sulfur released exceeds the amount that plants can take up, for at least part of the year.

After one to several years of cultivation, these losses reduce the ability of the soil to supply adequate amounts of nitrogen, phosphorus, and sulfur to crops, and fertilization is necessary to maintain high levels of production. Cultivated ecosystems can be managed, however, so that loss of soil fertility is minimized and even so that organic matter, nutrient content, and productivity of once-degraded sites are increased. However, these practices currently require the input of great amounts of energy in order to remain effective for several decades.

Fire is nearly universally employed in the conversion of forest land to other uses. Such fires generally have the same effects as natural fires except that natural fires, like fires used in shifting agriculture, are followed by relatively rapid plant regrowth. The increase in the amount of organic material consumed by fire as a consequence of human activities is substantial. According to one estimate, an average of about 0.3 percent of the total area of tropical forest is expected to be burned and converted to other uses during each of the next 20 years.[11]

Burning of the biomass in the process of deforestation and also in shifting cultivation and dry grass fires in the savannas produces many important trace gases such as CO, $CH_4$, $N_2O$ and $NO_X$.[12] Emissions of particulate matter have resulted in a high aerosol content in the atmospheric boundary layer (two to three kilometers) above the tropical savanna. The extent to which these aerosols may influence the heat balance of the atmospheric boundary layer is not known. However, frequent savanna and forest fires have occurred for hundreds or thousands of years, and it is not known if the amount of material consumed has increased in recent years or if this burning has caused a change in the flux of carbon, nitrogen, and sulfur from terrestrial ecosystems into the atmosphere.

The use of fertilizers, detergents, and industrial chemicals containing nitrogen and phosphorus has increased dramatically in the past few decades. As a consequence of waste disposal and runoff, the transport of nitrogen and phosphorus in rivers has increased twofold over the past 30 years.[13] Many lakes receiving elevated levels of phosphorus and nitrogen have exhibited enhanced biological productivity and the attendent production of organic matter, which can result in increased areas of anoxia (oxygen starvation) within these water bodies.

Introduction of phosphorus and nitrogen into marine coastal waters has also resulted in increased biological activity and disruption of fragile food chains. The changes in the ratios of nutrients entering the coastal zone over the past decades have brought about major changes as well. The entry of large amounts of phosphorus relative to nitrogen and silicon have lead to domination of marine communities by blue-green and brown algae, replacing the silica-containing diatoms that form the basis of many valuable food webs. This situation has become established in those areas not subject to extensive mixing with and dilution by the open ocean.

The replacement of diatoms by algae may have a harmful effect on coastal fishing, which constitutes 50 percent of the fisheries that provide society with food. Our understanding of coastal ecosystems is by no means adequate to judge how man is modifying the coastal waters and the ocean floor along the shore.

Man's effect on total carbon, nitrogen, phosphorus, and sulfur levels in the global ocean system is still small. Measurements of the pressure of carbon dioxide in equilibrium with midlatitude North Atlantic Ocean waters made over the past three decades suggest an increase in $CO_2$ pressure of 0.5 percent each year.[14] Even though pollution of the ocean surface layer can be detected all over the world, the direct impact of man on the nitrogen, phosphorus, and sulfur cycles has not been established by direct measurement in the open sea.

For example, if 20 percent of all phosphorus used by man reaches the open sea—which is most unlikely—the photosynthesis could increase by only 1 to 2 percent if all of this additional supply were used. Due to the size of the oceans, cumulative effects take place only gradually. However, it is important to monitor the oceans carefully in this regard, because small changes in concentrations correspond to the transport of enormous amounts of material, which may be of importance for the overall budgets of the elements.

The consequences of burning fossil

fuels are apparent from the previous discussion. So far, only a small percent of the total recoverable fossil fuel reserves have been used. The carbon and sulfur content of the untapped reserves are 50 to 100 times that which has been released to date. It is clearly crucial to assess what effects future emissions from these resources will have on the natural ecosystem.

The ultimate impact of man's interventions cannot be adequately deduced until the biogeochemical cycles are well understood. These cycles cannot be examined in isolation, and thus they require the simultaneous attention of a diverse group of scientists. Full comprehension of the carbon, sulfur, nitrogen, and phosphorus cycles and of the effect man's activities have upon them is difficult, and at the present time we are far from an integrated understanding of the global ecosystem.

## NOTES

1. B.B. Jorgenson, "Processes at the Sediment-Water Interface," in B. Bolin and R.B. Cook, eds., *The Major Biogeochemical Cycles and Their Interactions*, SCOPE Report Number 21 (New York: John Wiley, 1983), pp. 477–515.
2. B. Bolin, "The Carbon Cycle," in Bolin and Cook, ibid., pp. 41–45.
3. National Research Council, *Changing Climate: Report of the Carbon Dioxide Assessment Committee*, National Academy of Sciences (Washington, D.C.: National Academy Press, 1983).
4. Ibid.
5. Ibid.
6. Ibid.
7. A.H. Johnson and T.G. Siccama, "Acid Deposition and Forest Decline," *Environmental Science and Technology* 17(1983): 294A–305A; and G.H. Tomlinson, "Air Pollutants and Forest Decline," ibid., pp. 246A–256A.
8. Ibid.
9. J.M. Melillo and J.R. Gosz, "Interactions of Biogeochemical Cycles in Forest Ecosystems," in Bolin and Cook, note 1 above, pp. 177–220; and R.A. Houghton and G.M. Woodwell, "Effects of Increased C, N, P, and S on the Global Storage of C," ibid., pp. 327–344.
10. J.W.B. Stewart, C.V. Cole, and D.G. Maynard, "Interactions of Biogeochemical Cycles in Grassland Ecosystems," in Bolin and Cook, note 1 above, pp. 247–270.
11. J.P. Lanly and J. Clement, "Present and Future Natural Forest and Plantation Areas in the Tropics," *Unasylva* 31(1979): 12–20.
12. P.J. Crutzen et al., "Biomass Burning as a Source of Atmospheric Gases CO, $H_2$, NO, $CH_3Cl$, and COS," *Nature* 282(1979): 253–256.
13. R. Wollast, "Interactions in Estuaries and Coastal Waters," in Bolin and Cook, note 1 above, pp. 385–410.
14. T. Takahashi, D. Chipman, and T. Volk, "Geographical, Seasonal, and Secular Variations in the Partial Pressure of $CO_2$ in Surface Waters of the North Atlantic Ocean," in *Proceedings: Carbon Dioxide Research Conference; Carbon Dioxide, Science, and Consensus*, U.S. Department of Energy Report CONF-820970, part II, 1983, pp. 123–145.

CHAPTER 6

# Chemistry and the Biosphere

L.A. PIRUZYAN, A.G. MALENKOV AND G.M. BARENBOYM
TRANSLATED BY NORMAN PRECODA

The notion that mankind was becoming a power of geological scale was, by the beginning of the 20th century, clearly expressed by A. P. Pavlov in Moscow and, independently, by C. Schuchert in New Haven. They interpreted in a new way long-known facts on the changes in the environment caused by human activities, coming to the conclusion that their manifestation characterized the beginning of a new geological era. Ideas on the new geological era—"psychozoic" according to Schuchert, "anthropogenic" according to Pavlov—were developed in detail by V. I. Vernadski.*

Permission granted by the translator. Originally published in Russian in *Priroda*, No. 3 (1980). This abridged translation is from *Environment 22(10):* 25–30 (1980).

Lev Aramovich Piruzyan, Andrey Georgiyevich Malenkov, and Grigoriy Matveyevich Barenboym are all affiliated with the Interdepartmental Center on Biological Testing of Chemical Compounds of the U.S.S.R. Academy of Sciences.

Dr. Piruzhan is Director of the Center; Dr. Malenkov is Chief of the Center's Theoretical and Experimental Biophysics Section; and Dr. Barenboym is Chief of the Center's Molecular and Cellular Pharmacology Section.

Norman Precoda is a consultant in engineering science, Santa Barbara, CA.

According to such a viewpoint, the effects of man's activities should and must be commensurate in scale with the results of the action of tectonic [earth-deforming] forces. Indeed, man is remaking the earth's interior on an enormous scale, bringing immense quantities of petroleum and combustible gases to the surface, increasing the concentration of carbon dioxide gas in the atmosphere, redistributing water resources over great areas, and so forth.

Man's activities in the material sphere have qualitatively different directions. Investigating this phenomenon as a natural event, it is convenient to distinguish two sides of man's activities: physical (among these, thermodynamic, mechanical, electromagnetic), and chemical and biological.

It should be noted that, for the biosphere, different aspects of man's activities are, apparently, dangerous at various stages of the development of civilization. Until recent times the *biological* aspect was the most threatening: man annihilated the large animals and herded sheep and goats which literally "consumed" whole regions; he felled forests and in this way brought on degradation of the top soil, promoting the advance of the desert. This factor exists even today. However, among man's disruptive activities there arose a new and now dominating aspect, the *chemical*.

We will term as chemical activities of man those of his activities which directly affect the chemical form of the movement of material in the biosphere. This aspect of man's activities now gives rise to the greatest anxiety and danger. For it is the chemical activity that most clearly shows up in the disruptive effects on the biosphere of 20th century civilization. It is possible that in the future this role will come to belong to the *physical* factor (primarily, to the thermodynamic, i.e., disruption of the total thermal balance of the planet).

At the present time man's chemical activities have reached such rates that they surpass nature's capacity to neutralize them with the aid of natural feedback systems (that is, without any intervention by man). Nature ought not have to and can no longer wait for favors from mankind; we are obliged to help her.

***Alexi Petrovich Pavlov (1845-1929) taught physics and mathematics at Moscow University and was the author of numerous books and papers. Charles Schuchert (1858-1942), a paleontologist specializing in the historical geology of North America, worked for many years for the U.S. Geological Survey and then taught at Yale University. Vladimir Ivanovich Vernadski (1863-1945) was a Russian mineralogist and the founder of the science of biogeochemistry.**

### Increase in Man-made Chemicals

Man's chemical activities affect nature in many directions:

- The consumption of chemical substances from nature;
- The mass introduction of substances from the depths of the earth into the environment;
- Contamination of nature by the wastes from man's activities;
- The appearance in nature and, in particular, in the biosphere of new, highly active chemical compounds separated out from natural sources or synthesized by man.

More than 6 million chemical compounds with established chemical structures, obtained from natural sources and via synthesizing procedures, are already known. Two hundred thousand substances—mainly synthetic—are added to the list each year and of that number approximately 40,000 are our country's share.[1]

Man has created a whole universe of new chemical compounds and is rapidly expanding its boundaries. Each compound has its certificate-of-record or "passport" which contains a description of its basic physical, physical-chemical, and chemical properties. The form of this passport, containing a listing of its characteristics, is practically canonized by world science and practice. At the same time the substance is not, in general, characterized as to its potential effects on living matter and on man. This means that, in the universe of chemical compounds, products which are both beneficial and dangerous to the biosphere, and to man in particular, are concealed.

The data of contemporary science attest that all chemical compounds possess one or another type of biological activity—that is to say, introduced into a living organism, they have some effect on the organism's specific properties and functions. This circumstance is conditioned by the immense diversity of the material components and physical-chemical properties of living cells which allows them to interact with materials that are highly diverse in properties and structure. Even seemingly inert substances such as the noble gases (helium, neon, argon, and others) have a strong narcotic effect; having entered the human organism, insoluble, inert particles of asbestos can give rise to a cancerous tumor; fine particles of silicon are capable of effectively routing some insects.

Man has created some 10,000 drugs of synthetic origin, which together with food additives generate their own type of stress, continuously and concentratedly stressing man.[2] As a representative of a distinct biological species, what will man's reaction be to the effects of such substances not previously encountered along the paths of man's biological evolution? It is not possible to answer this question at the present time. It is obvious that chemistry has amassed a whole arsenal of means of affecting the biosphere without, in the overwhelming majority of cases, knowledge of their powers and potentialities. As an example it can be pointed out that, upon investigating Leningrad maternity homes in 1970-1972, it was found that the frequency of occurrence of inborn developmental defects had doubled in the past ten years.[3]

Eugene Odum's well-known book, *Fundamentals of Ecology*,[4] includes data which show the relationship of many respiratory ailments to contamination of the environment. It cites the in-the-literature concept which links the increase in deaths due to cancer and to lowered resistance to infections to increased pollution and, simultaneously, to the increased resistance of many pathogenic organisms to the processes by which water and food products are treated.

In general, man's activities directed toward changing the chemical potential of the planet have led to the emergence of a whole system of uncontrolled processes, the contents and interconnections of which are not clear in a majority of instances. Having initiated these processes, put our hopes on them, and promoted them, we condemn those consequences which turn out to be hazardous or harmful to us. However, so long as we do not understand the processes we are unable not only to regulate them but also to forecast their possible consequences.

### Causes of Our Quandry

What are the reasons for our helplessness?

The attainments of chemistry and industry have substantially exceeded the resources of biology and medicine to forecast scientifically the effects of chemical products on the biosphere. Mankind lacks adequate systems—in terms of compass, consistency, and informativeness—for tracking the chemical composition and state of the biosphere. Incomplete information about chemical changes in the biosphere on the one hand and about the biological activity of chemical compounds on the other leads to a situation in which man is not able to predict distant consequences of changes in the chemical composition of the environment, either on a planetary scale or on the scale of discrete biogeocenoses* and ecosystems.

It is only in recent times that man has begun to recognize the scale and

*"Biogeocenosis" is a term proposed by V. N. Sukhachev to indicate a complex consisting of a plant community (phytocoenosis) together with the animals dwelling in it (zoocoenosis) and the portion of the Earth's surface covered by it.

planet-wide importance of his industrial (in particular, chemical) activities. Until now man has not been ready, to the required degree, for a conscious, objectively oriented, collective effort toward optimization of this activity.

This gap between the potentialities of modern civilization and man's capacity to grasp them is conditioned not only by differences in the historical development of chemistry and biology but also by certain psychological barriers in the consciousness of people and of society as a whole.

The attainments of chemistry are obvious. But the genetic effects on man or the selective elimination of some group or other of insects or of soil microorganisms by the use of chemical agents are perceived by most people as something of a cold abstraction. The whole history of mankind's development is the history of a consumer attitude toward nature.

The problem posed by man's chemical activities and the need for nature preservation is vast and many-sided. Among a great number of aspects and questions are the obvious (such as the contamination of rivers by industrial wastes) and those which are complex in their nature (extensive and virtually uncontrolled use of chemical fertilizers and herbicides in agriculture). There are scientific problems which are far beyond the present state of man's knowledge (for example, the widespread use of antibiotics); there are problems which are obvious insofar as objectives but obscure insofar as method of implementation (for example, the creation of an adequately diversified and safe complex of contraceptive agents); there are also already-solved-by-science problems whose solutions have not yet been put into widespread practice (such as pharmacological means for adapting man to difficult and extreme conditions).

## Preserving Nature: Three Programs

In situations where the principles governing the effects of civilization on nature are obscure, the only feasible and intelligent choice is the *tactics of protection* (the protection of nature from society). The objective of this tactic is to reduce the pressure on nature during the period of our ignorance so that, by the time strong scientific concepts emerge, we will be able to utilize them instead of being helpless in situations where nature is dying. The maintenance of our tactics of protection must be the first program of practical work in the realm of nature preservation.

For developing the tactics of protection it is important to conduct *extensive scientific investigations* directed toward uncovering the system of principles involved in the interrelationships of man and nature. The objectives of these investigations and the manner of their organization must be contained in the second program of the work on nature preservation.

And, finally, it is necessary to set up a third program of activity in the realm of nature preservation which must determine the complex of scientifically grounded measures which permits building the man-nature relationship on the *principles of mutual benefits*.

### *The Tactics of Preservation*

At the present time man is morally ready to accept the tactic of preservation which is directed against the disruptive activity of industrial society. The principal elements of this tactic are:

- Development of a system for controlling the pollution of nature by wastes from the power engineering, industrial, and transport sectors;
- Development of a legislative code regulating the relationship between man and nature;
- Change-over of industrial processes to closed technological cycles, eliminating production's waste products;
- Preservation of the genetic stocks of life of the planet through creating a restricted-area system of "reservations" which provide for the conservation of a variety of living organisms, landscapes, biogeocenoses (preserves, national parks, hunting sectors, zoos, and others).

Such a tactic, together with proper governmental support and substantial participation by the public, will undoubtedly lead to positive results. We can even reduce the discharges into the atmosphere of harmful industrial wastes. This will not only to some degree protect the biosphere and man (as a part thereof) but also increase the effectiveness of industry itself. Reducing the air pollutants in the urban areas of the U.S.A. by 50 percent alone would save $2 billion annually thanks to the lowered costs of medical services and diminished illness-caused losses of worker time.[5]

For success along the line set by the preservation tactic the following measures are needed:

(1) The "departmental" approach to matters of nature protection must be systematically and relentlessly fought. This is one of the primary organizational obstacles hindering protection of nature. Wide publicity for all questions associated with protection of nature can assist here inasmuch as "departmentalism" fears publicity.

(2) Allocation of substantial material and labor resources is required. Intensive stimulation by the state of scientific efforts toward creating closed production

cycles, in particular, is required. Inasmuch as discoveries and inventions are the principal means for the development of any sphere of civilization, much depends on the efforts of inventors in the realm of nature preservation. It is especially essential, therefore, to encourage inventive activity in the realm. It seems advisable to us, the authors, to establish a special state prize fund; to simplify the procedure for issuing certificates of invention; and to guarantee automatic payment of reward for introduction, inasmuch as appraisal of the economic effects in the realm of nature in advance is not as yet possible.

(3) An orderly system of juridical and organizational measures must be developed which will ensure that institutes and establishments will design and work in those regimes which exclude contamination of the environment.

(4) Another aspect of the defensive tactic, the creation of reservations, has as its objective preservation of the diversity of animate nature until the advent of more propitious times. Toward this end it is necessary to work in three directions: first, expand the network of preserves; second, organize and finance the work of specialized scientific programs whose aim is the preservation of dying-out species in artificial or natural reserves; third, increase allocations for environmental protection inspections and raise the social and economic prestige of these professions.

One of the most important problems of the preservation tactic, without whose resolution transition to the next stage is not possible, is the creation of cadres of highly qualified, creative specialists who have received a sufficiently broad and at the same time mission-oriented education. Each such specialist must be sufficiently well-grounded in physics, chemistry, and biology, must have sufficiently penetrating ideas in the realms of ecology and pharmacology, and must be able to transform his ideas into workable engineering designs.

For the development of nuclear power in our country it was necessary, at one time, to create a Moscow physics and technical institute specifically structured to train cadres. The epoch of ecological technology likewise requires new structures and orientations for the preparation of cadres.

On the whole, the tactic of preservation must produce noticeable but limited results because it does not eliminate the fundamental reasons for the "man-nature" conflict. With the preservation tactic, as before, there continues to be uncontrolled growth in the number of people and the per capita consumption of resources. The last ones get either clearly finite or more slowly increasing consumption from existing resources. This factor has so general a character that its elimination or at least a decrease in its effects is possible only within the bounds of some postulates chosen *a priori*. It appears to us that the substance of these postulates is twofold:

- A consciously regulated demographic dynamic stipulated by the necessity for the growth of the economic and technological potentialities of society;
- Conscious limitation of per-capita consumption of material wealth on the basis of each person's reasonable, natural needs stipulated, in the last analysis, by limited resources, with priority given to the maximum possible satisfaction of spiritual needs and to the rational, collective needs for material and spiritual well-being.

### *Fundamental Scientific Investigations*

If one accepts these postulates, then resolution of the "man-nature" conflict is entirely dependent on the results of fundamental investigations directed to search for practical ways of optimizing man-and-nature interrelationships, that is, on realization of the second listed program.

At the present time it is difficult to describe the whole program of fundamental investigations. Clearly, however, it must eliminate the dangerous gap between scientific knowledge and the biological activity of chemical compounds and the scale of their use in the biosphere.

For solving this problem the most important means may be the organization of a system of biological testing for the entire body of chemical compounds synthesized in our country. The objective of this system would be formulation of an information bank of fundamental scientific knowledge on the biological activity of chemical compounds.

From our viewpoint the desired ratio of the quantity of the new chemical substances introduced and the quantity of new knowledge produced can be likened to the principle of the sand-clock (hourglass) wherein the sand grains symbolize the new chemical compounds. The upper part of the clock is the sphere where compounds are created; the lower, that of their use. The narrow neck section is where the biological activities of the chemical products are established. Only the grains of sand (compounds) move on which pass through the neck: there is no other route. The time of accumulation of the country's working chemical potential must be determined by the pass-through capacity of the "neck," that is to say, by the capacity of the biological testing system. Only those compounds would be introduced into practice which had been subjected to biological testing and, it goes without saying, in accordance with the tests' results.

The rate at which knowledge about biological activity is formed must be commensurate with the rate at which the body of new chemical compounds is generated–or even higher. On the one hand, then, it would be possible to look over the body of already existing ("old") chemical compounds and, on the other, to "design" compounds necessary for controlling specific processes in the biosphere.



From the practical viewpoint the system of biological testing must be oriented to two fundamental objectives: first, toward the search for materials useful to the human organism or able in a controlled manner to affect the productivity and biological equilibrium of natural and artificial ecosystems in accordance with the needs of society; second, toward uncovering substances which are harmful to the human organism or able to give rise to uncontrollable and dangerous disruption of the biological equilibrium of natural and artificial ecosystems. Substances not immediately found to be in one or the other of these two categories could be utilized by society in proportion to the development of science, which opens up the possibility of the use of compounds with still more varied properties.

Attainment of the first objective will help solve a number of urgent problems which confront mankind: the prevention and treatment of difficult or incurable chronic ailments; raising the resistance to acute infections and, particularly, to the viruses of illnesses, and treating them; optimization of individual development (including correction of hereditary defects) and expansion of man's physiological and intellectual possibilities; and regulation of the birth rate. Antitoxins for natural conditions may constitute a special class of substances, stabilizers, and regulators of natural biogeocenoses and the like.

Attainment of the second objective will permit the discovery, among the compounds being utilized in agriculture and industry, of materials dangerous to man and the environment. Moreover, the question is not simply about toxicity which even now is sufficiently reliably controlled but about the non-obvious types of biological dangers such as the mutagenic, carcinogenic, allergenic, and, also, threats to development of the fetus. In the case of substances widely disseminated by man in nature (on the strength of their functional purpose or in virtue of the imperfection of society's industrial activity), the ability to disrupt the natural biogeocenoses will be established.

As an example of the difficult to predict, sometimes clearly dangerous effects of man's chemical activities, there is the creation in nature of a background of chemical substances (including, also, radioactive) that give rise to mutagenic effects which, at the present time, have come to be one of the most important problems of preservation of the biosphere. Another example is the development and intensive use of antibiotics, which have changed the relationships between man and the bacterial world in a direction disadvantageous for man.

A third example is the intensive use of pharmacological substances which enable people with genetic defects at birth to survive. From one aspect this increases genetic diversity, opening up mankind's potential gene pool; but from another aspect, with the present imperfection of medicinal agents, it creates additional difficulties. Such difficulties can be overcome, through creation of a comprehensive system of remedies.

The enumerated examples illustrate negative consequences of man's activities and emphasize the necessity for their control, for forecasting the results.

## Beneficial versus Dangerous

We should mention that the technological principle of the simultaneous sorting of substances by a single "sieve" into beneficial and injurious is not the only important thing. Another principle is no less important: "beneficial *versus* dangerous." This principle means that, in the body of substances tested, chemical compounds may be found which reduce the unfavorable effects on the human organism caused by harmful physical and chemical factors. They can be divided into four groups:

(1) Reagents against chemical contamination in inanimate nature (precipitators, neutralizers and the like);

(2) Compounds against chemical contaminants on the biological level (antitoxicants, antimutagens, anticarcinogens, and the like);

(3) Substances such as adaptagens which increase the ability of biological matter to adapt to changed conditions;

(4) Compounds which promote the adaptation of natural biogeocenoses to new conditions or the generation of new natural or artificial ecosystems ("tools of ecological engineering").

A few examples are in order. World and Soviet science has found substances –antimutagens–which reduce by severalfold the probability of genetic changes induced by mutagenic factors in the environment.

Where substances toxic to living organisms come into the environment, other chemical compounds which are antitoxicants can be introduced. Thus in the Moscow Forestry Institute nearly 70 antidotes have been found for grasses and shrubs which die in an atmosphere containing ammonia, sulphur dioxide, and nitric oxides.[6]

Of great interest, in our view, are adaptagens of systemic action, both natural—obtained from plants or cultures of their tissues (for example, Eleutherococcus, ginseng)—and, for instance, derivatives of benzamidasols. They are able to increase the organism's resistance to the actions of the most diverse unfavorable environmental factors, both natural and artificial. They have the ability to reduce substantially the effects of stress actions.[7] These "medicines for the healthy" can be extensively utilized by people working under extreme conditions, raising their work productivity. Adaptagens are capable of easing the development of territories whose natural conditions are little suited for man. This is very important for the more uniform allocation of man's activities in nature.

Systemic adaptagens can reduce the consequences of stress in farm animals held in stalls. The modern commercial stock-raising complex concentrates hundreds and even thousands of animals in a limited area under conditions markedly different from natural ecological systems. This leads to over-load of the adaptive-protective mechanisms and to the emergence of "adaptation ailments."[8] It is evident that the importance of adaptagens for combating phenomena is extremely large. It must be noted that our country occupies a leading position in the sphere of basic research and practical operations on system adaptagens and it is for this reason that intensification of research is especially important.

The system of biological testing of chemical compounds is most advantageous in the case where both sides of the "coin" which marks the progress of chemistry can, with its aid, be checked. It is readily understandable that such a system will really be effective only when all chemical compounds synthesized in the country are tested.

## NOTES

1. L. A. Piruzyan, A. G. Malenkov, and G. M. Barenboym, "A System for Determining the Biological Activity of Chemical Compounds Comprising a Large Body," **Transactions of the U.S.S.R. Academy of Sciences**, Biological Series, No. 3, 1977.

2. L. A. Piruzyan, A. G. Malenkov, and G. M. Barenboym, "Investigations in the Chemistry of the Ocean," **Pravda**, November 17, 1976.

3. N. G. Veselov et al., in the collection: **Background Characteristics (Fonotipicheskiy) and Polymorphism and Diagnostics of Man's Ailments**, Moscow, 1976, pp. 14-15.

4. E. P. Odum, **Foundations of Ecology**, Moscow, 1975 [Eugene P. Odum, **Foundations of Ecology**, W. B. Saunders, Philadelphia, 1971].

5. Ibid.

6. For a detailed discussion on this see P. T. Obydennyy, **Forestry**, No. 6, 1977.

7. I. I. Brekhman, **Eleutherococcus**, Leningrad, 1969; I. I. Brekhman, **Man and Biologically Active Substances**, Leningrad, 1976; XIV Pacific Ocean Scientific Congress, Summaries of Papers, Symposium, Moscow - 3. M., 1979.

8. N. A. Urazayev, **Biogeocenosis and Animal Ailments**, Moscow, 1976, pp. 90-96.

CHAPTER 7

# The Role of Ocean Biota in Accelerated Ecological Cycles: A Temporal View



JOHN J. WALSH

Over the last century humans have accelerated both the return of carbon to the atmosphere via burning of wood, coal, and oil and the removal of nitrogen via use of fertilizers. Previous unbalanced global carbon and nitrogen budgets have not properly considered the role of marine biota in mediating such element fluxes among ocean, land, and atmosphere. Although anthropogenic carbon dioxide emission is tenfold more than carbon dioxide emission in glacial periods, methane emission and nitrogen fixation may have declined.

The times in which we live represent an anomaly, both climatologically and ecologically. For 90% of the past two million years, the earth's climate has been colder than today's Holocene interglacial period. During the 10,000 years since the last major glaciation, humanity has emerged as a runaway species, so far escaping environmental constraints on population abundance that the "normal" cycling of several elements has been accelerated.

Human abundance is projected to double by 2000, with continued consumption of oil and coal. Present release rates of $CO_2$ to the atmosphere from fossil fuel burning, cement production, and deforestation are perhaps high enough (3% annual increase) to double atmospheric carbon dioxide content by 2035. The increases of this and other gases—methane ($CH_4$), nitrous oxide ($N_2O$), and the freons ($CCl_3F$, $CCl_2F_2$)—may result in a concomitant 3°C rise in ocean temperature, almost ten times that over the last century and similar to the sea temperature increase since the last major glaciation (Hansen et al. 1981).

Three times less of the earth's area is now covered with ice sheets than was covered during the glacial periods, resulting in about a 3% increase in the ocean's volume (Milliman and Emery 1968). A melting of ice sheets, similar to the 0.1 cm $yr^{-1}$ sea level changes over the last century, would result in a sea level rise of only 0.3–0.6 m by 2050 (Gornitz et al. 1982). If the West Antarctic Ice Sheet were to slide off into the Southern Ocean in response to an accelerated anthropogenic temperature rise, however, sea level would quickly rise some 6 m, drowning most coastal cities.

Over the last century, human ability to extract nitrogen from the atmosphere has begun to rival $N_2$ fixation by plants. From 1950 to 1975, world production of agricultural fertilizers increased tenfold; increases in commercial crop yield during this time may be due to nitrogen application rather than to the use of hybrid strains (Boyer 1982). Nutrients from agrarian runoff, deforestation, and urban sewage have already affected primary production in streams, ponds, lakes, major rivers, and perhaps even

> It was the best of times, it was the worst of times, it was the age of wisdom, it was the age of foolishness. . . . For the streets were so full of dense brown smoke that scarcely anything was to be seen. . . . We drove slowly through the dirtiest and darkest streets that ever were seen in the world (I thought), and in such a distracting state of confusion that I wondered how the people kept their senses.
>
> C. Dickens (1853, 1859)

Reprinted by permission from *BioScience 34(8)*: 499–507 (1984). 

John J. Walsh is a graduate research professor with the Department of Marine Science, University of South Florida, St. Petersburg, FL.

continental shelf waters. But nutrients are routinely measured in only 25% of the world's 240 largest rivers, and few biological time series are available to document the extent of coastal eutrophication.

Man's perturbation of natural terrestrial nutrient cycles and adjacent aquatic ecosystems began over 2000 years ago. The rate of nitrogen and carbon deposition in Italian lake sediments increased tenfold from Roman road construction and deforestation in 171 B.C. (Hutchinson 1970). This was a small-scale phenomenon in these times, since European forests stretched so far in the drainage basins of the Rhine and Danube rivers that "no one in western Germany would claim to have reached the eastern edge of the forest even after traveling for 60 days, or to have discovered where it ends" (Caesar 53 B.C.).

After Roman times, around 1500 A.D., similar increases in aquatic sedimentation rates occurred on the upper slope of the Black Sea. A tenfold change took place between 800 B.C.–200 A.D. and Medieval times (Degens et al. 1976), presumably in response to conversion of previous Hercynian forests to steppes (Kempe 1979). Similar deforestation of eastern North America during its colonial period led to a tenfold increase in suspended solids in the rivers (Meade 1982) and perhaps in estuarine nitrogen.

With completion of colonial deforestation by 1850, subsequent farming, and today's addition of urban wastes, mean nitrogen in Lake Erie sediments increased sevenfold between 1850 and 1970 (Kemp et al. 1974). From 1930 to 1970 the nitrogen content of western Lake Erie increased by an order of magnitude to a winter–spring maximum of 30–60 $\mu$g-at N $l^{-1}$ (Burns 1976), similar to the nitrogen content of the Mississippi River in 1970 and ten times that of the pristine Amazon, Zaire, or Yukon rivers (Walsh et al. 1981). By this time, phytoplankton numbers had increased twentyfold in the western basin (Leach and Nepszy 1976), and the mean annual primary production of Lake Erie was about 250 g C $m^{-2}$ $yr^{-1}$—over two to four times that of the relatively oligotrophic Lakes Huron and Superior (Vollenweider et al. 1974) but similar to that of most continental shelves (Walsh 1983).

As much as one-third of the lake's annual carbon production in 1970–71 went unconsumed, accumulating instead on the bottom (Kemp et al. 1976). Estimates of the sedimentation rate within Lake Erie at that time were ten times that of the mean Holocene rate (Sly 1976). In the last decade, the amount of carbon and nitrogen sequestered within organic matter on the Lake Erie bottom may have doubled again (Fisher and Matisoff 1982).

With a tenfold increase in world population from 260,000,000 people in 14 A.D. (Botkin and Keller 1982) to 3.6 billion people by 1970, the magnitude of carbon and nitrogen perturbations has grown from local phenomena such as Lake Erie to impacts on a global scale. To place projected human activities in proper perspective, I have considered present fluxes of carbon, nitrogen, and phosphorus in relation to those during glacial periods.

## CARBON FLUXES

Recent studies of $CO_2$ concentration in Greenland (Neftel et al. 1982) and Antarctic (Delmas et al. 1980) ice cores provide information on glacial carbon fluxes among the atmosphere, ocean, and sediment reservoirs. The atmospheric $CO_2$ concentrations may have risen from 200 parts per million (ppm) during the coldest part of the Wisconsin glaciation, 15,000 to 20,000 years ago, to preindustrial Holocene values of 300 ppm. Based on the ice cores, this may have occurred rapidly over 1000 to 2000 years as a result of reduced photosynthesis in the open ocean at the beginning of the present interglacial period (Broecker 1982). Rapid nutrient removal at the continental margins by increased burial of organic matter with rising sea level and concomitant shelf inundation may have led to a reduction in oceanic photosynthesis and in $CO_2$ extraction from the atmosphere at this time (Broecker 1982).

More than 70% of world shelf sediments show no modern deposits since the Holocene transgression, or rise in sea level (Emery 1968). The organic carbon content of the slope sediments is ten times that of shelf sediments in the Mid-Atlantic Bight (Walsh et al. 1984). Some $1.2 \times 10^{12}$ tons of organic carbon (Broecker 1982) over the first 1000 years of the Holocene ($1.2 \times 10^9$ tons C $yr^{-1}$) may have been stored within continental slope or rise depocenters, where most of the long-term burial of carbon is thought to be located (Walker 1974). The actual sedimentation rate of organic matter produced by phytoplankton on continental rises, slopes, shelves, or within estuaries is unknown either for the glacial period or the present (Walsh et al. 1981).

From 1880 to 1980 atmospheric $CO_2$ values have increased from 280–300 ppm to 335–340 ppm at an exponential annual rate, which is now about 1 ppm $yr^{-1}$, or about $2.5 \times 10^9$ tons C $yr^{-1}$—at least ten times that of the so-called rapid early Holocene change. But our present annual carbon input to the atmosphere from fossil fuel burning, cement production, and tropical deforestation is estimated to be $6.2 \times 10^9$ to $7.2 \times 10^9$ tons C $yr^{-1}$, more than twice the actual atmospheric increment. The storage pools for the rest of today's anthropogenic $CO_2$ are not clear, but the ocean's dissolved inorganic and organic carbon reservoirs ($4 \times 10^{13}$ tons and $3 \times 10^{12}$ tons C, respectively) are 60 times that of the atmosphere ($6.9 \times 10^{11}$ tons C), presumably dampening part of the recent industrial input. Indeed, the potential of the sea's carbon cycle to yield fish and either store or release $CO_2$ with changes in primary production is a subject of continuing controversy.

Like the actual amount of $CO_2$ fixed annually during marine photosynthesis, the rates of phytoplankton carbon input to fish carbon or to sediment carbon are also unknown. Debates over potential fish harvest and $CO_2$ storage hinge on the amount and fate of both present and past marine primary production. Current estimates of annual total marine primary production range from 20 × $10^9$ to 55 × $10^9$ tons C $yr^{-1}$ (Walsh 1983, Table 1) and may be 3–10 times too low (Smith et al. 1984). The present range accounts for 25–50% of the estimated total net global carbon fixation (Woodwell et al. 1978), represents most aquatic primary production (Table 1), and is 400–1000% of either present fish yield or fossil fuel emissions.

Because of differences in physical habitat, the various ocean provinces have very different phytoplankton species assemblages (Walsh 1983). They also differ significantly in spatial and temporal variability of algal biomass as a function of nutrient input and grazing losses (Walsh 1976) and in the fate of fixed carbon (Eppley and Petersen 1980, Walsh et al. 1981). Most of the deep-ocean, rise, and slope primary production of calcite nanoplankton (coccolithophores) is consumed either by herbivores at the surface or by bacterioplankton in deeper water, such that less than one percent of the primary production is stored in bottom sediments of these regions (Table 1).

In contrast, as much as 50% of the siliceous netplankton (diatom) primary production in diverse shelf ecosystems off Peru; in the Mid-Atlantic Bight, the Bering Sea, and the Gulf of Mexico (Walsh and McRoy 1984); and off Sierra Leone (Longhurst 1984) is neither consumed by the shelf food web, nor deposited in shelf sediments. Instead, it appears to be stored in adjacent slope depocenters after partial decomposition of the 2.7 × $10^9$ tons C $yr^{-1}$ shelf export. Early diagenesis releases biogenic $CO_2$, which is stored in the sea's inorganic bicarbonate cycle (Broecker et al. 1979), and nutrients, which diffuse from pore water of the sediments to the overlying water column (Froelich et al. 1982). Between the Gulf of Mexico and Georges Bank, for example, this process puts the oxygen minimum layer near the slope bottom at 300 m, where low oxygen contents of 2.5–3.0 ml $O_2$ $l^{-1}$ are found; these values are similar to those for the oxygen minimum layer in the open Atlantic Ocean at depths of 800–900 m (Emery and Uchupi 1972, Richards and Redfield 1954).

The present upper estimate of slope marine carbon storage (Table 1) is derived from the organic carbon content, the porosity, and $^{210}$Pb measurements of sediment accumulation rates.[1] Similar to observed changes in sedimentation rate within Lago di Monterosi, Lake Erie, and the Black Sea, sediment accumulation rates over the last century may be tenfold that of either the whole Pleistocene Ice Age (Hathaway et al. 1979) or the last Wisconsin glaciation (Doyle et al. 1979, MacIlvane 1973). Modern slope removal constitutes most of the organic carbon storage of aquatic ecosystems (Table 1). The rate equals carbon isotope and atmospheric $CO_2$ changes at the end of the Wisconsin glaciation (Broecker 1982), but is ten times that of previous long-term estimates (Berner 1982, Holland 1978).

The area of inland waters is only 2 × $10^6$ $km^2$ (DeVooys 1979). Assuming an annual production of 200 g C $m^{-2}$ $yr^{-1}$, the same as continental shelves, the total net production is 0.40 × $10^9$ tons C $yr^{-1}$ (Table 1). The area of coastal salt marshes is 0.35 × $10^6$ $km^2$, that of open estuarine waters is 1.4 × $10^6$ $km^2$, and freshwater marshes extend over 1.6 × $10^6$ $km^2$ with respective average primary production of 1275, 675, and 945 g C $m^{-2}$ $yr^{-1}$. Ten to thirty-three percent of this primary production is assumed to accumulate in sediments. The area of coral reefs is estimated to be 0.11 × $10^6$ $km^2$ with an annual net production of 2725 g C $m^{-2}$ $yr^{-1}$, compared to the 20–60 g C $m^{-2}$ $yr^{-1}$ of surrounding oceanic water; organic carbon storage is estimated by carbon-to-carbonate accumulation in the reefs (Berner 1982). Finally, assuming production of 1500 g C $m^{-2}$ $yr^{-1}$ and a littoral area of 0.02 × $10^6$ $km^2$, the annual net production of seaweed macrophytes is 0.03 × $10^9$ tons C $yr^{-1}$; although the amount of carbon storage in this ecosystem is unknown (Smith 1981), it is considered negligible compared with marshes and coral reefs.

The total organic carbon sink since the beginning of the Holocene (Table 1), plus deep-sea inorganic carbon sinks of 2.0 × $10^9$ tons C $yr^{-1}$ (Broecker et al. 1979) as well as increases in the atmospheric $CO_2$ pool, would balance most of the modern anthropogenic carbon sources. What are the consequences of the associated nutrient losses and what is the required input of inorganic sediment?

If the sediment carbon content on all slopes is 1.5% of dry weight, then 60 × $10^9$ tons of *inorganic* detritus must be accumulating each year. The input of suspended solids from land runoff to the world estuaries is only 20 × $10^9$ tons $yr^{-1}$. All this is thought to be trapped within these nearshore systems in order to bury the 0.20 × $10^9$ tons C presently estimated to be accumulating in estuaries (Berner 1982). Are there other sources of inorganic, fine-grained sediments that reach the continental slope?

Similarly, with a C/N ratio of 9/1, and a C/P ratio of 150/1 by weight for slope sediments (Morse and Cook 1978, Walsh et al. 1984), about 10 × $10^7$ tons N $yr^{-1}$ and 6 × $10^6$ tons P $yr^{-1}$ must also be sequestered. How do these slope storage requirements affect present and past global nutrient and productivity cycles? Why did atmospheric concentrations of $CO_2$ not continue to double every thousand years during the Holocene, given such nutrient depletion rates?

## NITROGEN FLUXES

Unlike carbon, most of the earth's nitrogen is found as $N_2$ in the atmosphere (3.8 × $10^{15}$ tons). Of the small amount dissolved in the ocean (9 × $10^{11}$ tons), about 75% is in the form of nitrate. Any change in the sea's small storage pool must reflect changes in either input

[1] 0.1 – 0.3 cm $yr^{-1}$ in the Mid-Atlantic Bight and the southeastern Bering Sea (Walsh et al. 1984); off Peru (Koide and Goldberg 1982), California (Koide et al. 1972), Southwest Africa (De Master 1979), and Baja California (Koide et al. 1973); within the Gulf of Mexico (Booth 1979), the Gulf of Maine (Emery 1969), and the North Sea (Eisma 1973), assuming little sediment mixing.

**Table 1.** Aquatic photosynthesis and nitrogen fixation in relation to losses via carbon sediment storage, denitrification, and methane production in marine and freshwater ecosystems (after Capone and Carpenter 1982, DeVooys 1979, Sheppard et al. 1982, Walsh et al. 1981). Area in $km^2 \times 10^6$, production and losses in tons $yr^{-1}$.

| Region | Area | C input Net primary production ($\times 10^9$) | $N_2$ input Nitrogen fixation ($\times 10^7$) | C output Sediment organic carbon sink ($\times 10^9$) | $N_2$ output Denitrification loss ($\times 10^7$) | $CH_4$ output Methane emission ($\times 10^7$) |
|---|---|---|---|---|---|---|
| Open ocean | 310.00 | 18.60 | 0.43 | 0.19 | — | 0.36 |
| Continental shelf | 27.00 | 5.40 | 0.27 | — | 2.97 | 0.04 |
| Continental slope | 32.00 | 2.24 | 0.06 | 0.90 | 5.50 | 0.03 |
| Freshwater marshes | 1.60 | 1.51 | 2.21 | 0.15 | 6.40 | 3.10 |
| Estuaries/deltas | 1.40 | 0.92 | 0.06 | 0.20 | 1.04 | 0.60 |
| Salt marshes | 0.35 | 0.49 | 0.48 | 0.05 | 1.40 | 0.80 |
| Rivers/lakes | 2.00 | 0.40 | 1.88 | 0.13 | 0.26 | 5.10 |
| Coral reefs | 0.11 | 0.30 | 0.28 | 0.01 | — | 0.32 |
| Seaweed beds | 0.02 | 0.03 | — | — | — | 0.08 |
| Total aquatic area | 375.00 | 29.89 | 5.67 | 1.63 | 17.57 | 10.43 |

from rainfall, dry deposition, nitrogen fixation by phytoplankton, and land runoff or in losses from burial, formation of nitrous oxide, and release of $N_2$ to the atmosphere during denitrification.

Gaseous $N_2O$ is formed during the oxidation of reduced nitrogen compounds to yield both nitrite ($NO_2$) and nitrate ($NO_3$) (Elkins et al. 1978)—either by nitrification (Cohen and Gordon 1978) or by combustion of fossil fuels (Weiss 1981). During nitrification, the ocean may also lose nitrogen in the form of nitric oxide (NO) (Lipschultz et al. 1981). Present estimates of the ocean's $N_2O$ loss to the atmosphere vary by an order of magnitude (Hahn 1981), but a reasonable value appears to be $0.5 \times 10^7$ tons N $yr^{-1}$ (Liu 1979). Recent annual increases of 0.4% atmospheric $N_2O$ may have affected both ozone concentration (Crutzen 1981) and, together with increases of the other trace gases, methane (1.7%) and the freons (9–10%)—the greenhouse response to $CO_2$ changes between 1970 and 1980 (Lacis et al. 1981). An atmospheric $N_2O$ increase of 6 ppb over the last decade is equivalent to 10% of the greenhouse warming of $CO_2$; 200 ppm of added $CO_2$ plus all these trace gases, however, is equivalent to a 300 ppm $CO_2$ input. When bacterial decomposition of organic compounds exceeds the influx of oxygen to the water column and sediments, local anoxic habitats develop that show consumption of the oxides of nitrogen ($NO_3$, $NO_2$, $N_2O$, NO) and sulfur ($SO_4$) as well as liberation of other gases—$N_2$ (denitrification), $H_2S$ (sulfate reduction), and $CH_4$ (methanogenesis).

Previous nitrogen budgets (Emery et al. 1955, Holland 1973) had suggested that the input of combined nitrogen to the sea exceeded burial losses, with a steady state maintained by the loss of the excess nitrogen through biological denitrification. But more recent analyses (Fogg 1982, McElroy 1983) suggest that the $N_2$ losses to the atmosphere alone may exceed the total nitrogen input to the sea, implying that transient states of the marine nitrogen reservoir might occur every 10,000 years (McElroy 1983). Within anoxic slope waters now adjacent to productive coastal regions, for example, consumption of shelf carbon export by anaerobic denitrifying bacteria produces a total $N_2$ loss of $5.5 \times 10^7$ tons N $yr^{-1}$ from the eastern tropical North Pacific (Codispoti and Richards 1976), off the Peru–Chile coasts (Codispoti and Packard 1980), and in the Arabian Sea (Wajih et al. 1982).

Denitrification rates within the low carbon sediments on the Bering (Koike and Hattori 1979), Chukchi (Haines et al. 1981), and Washington shelves (Christensen 1981) suggest an annual loss of 1.1 tons N $km^{-2}$ $yr^{-1}$. In contrast, 7.4 tons N $km^{-2}$ $yr^{-1}$ may be lost in nearshore, carbon-rich sediments off the Belgian and Japanese coasts (Billen 1978, Koike and Hattori 1979), as well as within estuarine systems such as Danish fjords and Narragansett Bay (Seitzinger et al. 1984, Sorensen et al. 1979). Salt marshes (Kaplan et al. 1979) have even higher loss rates (40.0 tons N $km^{-2}$ $yr^{-1}$); lake denitrification losses are similar to shelf losses (Pheiffer-Madsen 1979). The annual loss of $N_2$ from the marine regions is thus $9.5 \times 10^7$ tons N $yr^{-1}$ (Tables 1 and 2). The gaseous loss of $10.0 \times 10^7$ tons N $yr^{-1}$ from the oceans is the same as the modern burial loss of nitrogen on continental slopes.

Recent estimates of $N_2$ extraction from the atmosphere by nitrogen-fixing algae *Oscillatoria* spp. in the open sea are only $0.5 \times 10^7$ to $1.0 \times 10^7$ tons N $yr^{-1}$ (Capone and Carpenter 1982, Fogg 1982), compared with $4.2 \times 10^7$ tons N $yr^{-1}$ fixed by man for agricultural fertilizer in 1975. Summing known $N_2$ fixation from all marine regions suggests an input of at most $1.10 \times 10^7$ tons $yr^{-1}$, compared with denitrification losses of 9.51 $\times 10^7$ tons N $yr^{-1}$ (Table 2). (If nitrogen

**Table 2.** A comparison of carbon and nitrogen fluxes in modern and glacial ecosystems. Net primary production and sediment organic carbon sink in $10^9$ tons $yr^{-1}$; nitrogen fixation, denitrification, and methane emission in $10^7$ tons $yr^{-1}$. Values based on areas given in Table 1.

| | Modern | | Glacial* | |
|---|---|---|---|---|
| | Marine | Marsh | Marine | Marsh |
| Net primary production | 27.49 | 2.00 | 21.17 | 28.40 |
| Nitrogen fixation | 1.10 | 2.69 | 0.77 | 38.20 |
| Sediment organic carbon sink | 1.30 | 0.20 | 0.20 | 2.84 |
| Denitrification loss | 9.51 | 7.80 | 0.00 | 110.80 |
| Methane emission | 1.43 | 3.90 | 0.78 | 55.38 |

*Assumes that total marsh area increased to $2.84 \times 10^7$ $km^2$ with disappearance of shelves and estuaries.

fixation by bacterial endosymbionts within floating diatom mats [Martinez et al. 1983] is globally significant, however, these estimates may represent serious underestimates.)

Salt and freshwater marsh ecosystems now fix twice as much total atmospheric $N_2$ each year as marine regions, despite two orders of magnitude less area (Tables 1 and 2). Nuisance blooms of nitrogen fixers, such as *Anabaena* spp., dominate both eutrophic and oligotropic lakes, where as much as 30% of the nitrogen demanded by annual primary production is met by nitrogen fixation. Terrestrial ecosystems fix an additional $10 \times 10^7$ tons N $yr^{-1}$ (Capone and Carpenter 1982).

If all the nitrogen now fixed by humans and by natural lake, marsh, marine, and terrestrial ecosystems ($20 \times 10^7$ tons N $yr^{-1}$) were directly transferred and/or retained within those aquatic ecosystems each year (Table 1), about half the estimated total burial ($20 \times 10^7$ tons N $yr^{-1}$ with a C/N of 9/1) and denitrification ($18 \times 10^7$ tons N $yr^{-1}$) losses would be offset by input. Passive transfer of atmospheric nitrogen to the sea from rainfall and dry deposition is now thought to be small ($0.8 \times 10^7$ tons N $yr^{-1}$, McElroy 1983), although past estimates ranged as high as $8.2 \times 10^7$ tons N $yr^{-1}$ (Soderlund and Svensson 1976); land runoff is the most likely source of terrestrial nitrogen. The soil nitrogen leaching rate from British arable lands, subject to intense fertilization has, for example, increased fourfold between the 1930s and 1970s (Forster et al. 1982), with severe effects on potable water supplies. The nitrate content of British (Wilkinson and Greene 1982) and French (Meybeck 1982) freshwater streams at times exceeds both US Public Health Service and World Health Organization standards (715–1615 μg-at $NO_3$-N $l^{-1}$).

Between 1850 and 1950, the nitrogen content of the Rhine River had only increased twofold (to 100 μg-at N $l^{-1}$) with a second doubling by 1970; over the last decade, it has doubled again (Van Bennekom and Salomons 1981). That of the Seine increased fourfold to 400 μg-at N $l^{-1}$ from 1965 to 1975 (Meybeck 1982), and that of the Thames doubled to >500 μg-at N $l^{-1}$ between 1968 and 1978 (Wilkinson and Greene 1982). The mean winter nitrate content of the North Sea shelf in the 1960s was already at least two to four times higher off the Thames estuary, the Wash, and the Rhine estuary than at the edge of the shelf (Johnston 1973). The winter phosphate content off the Dutch coast also doubled from 1961 to 1978 (Van Bennekom and Salomons 1981). Indeed, as much as 350 μg-at $NO_3$ $l^{-1}$, ten times the concentration of $NO_3$ in deep slope water, is now found both in the Scheldt estuary and 10 km off the Belgian coast (Mommaerts et al. 1979).

The nitrate content of the Mississippi River has at least doubled to 150 μg-at $NO_3$ $l^{-1}$ during spring flood over the past 10 years, and mean concentrations have increased from 40 μg-at N $l^{-1}$ in 1905 (Gunter 1967), 1935 (Riley 1937), and 1965 to >80 μg-at N $l^{-1}$ in 1980 (Walsh et al. 1981). Data for other US rivers that drain heavily populated areas, such as the Ohio (Wolman 1971), Potomac and Susquehanna (Carpenter et al. 1969), Delaware (Kiry 1974), and Hudson (Deck 1981), suggest that their nitrogen content has also increased over the past 25–50 years from sewage, fertilizers, and nitrate release to groundwater after deforestation (Likens et al. 1978). Similarly, US rivers such as the Columbia (Park et al. 1972) that drain less populated areas in the south and west used to contain ten times less inorganic nitrogen (7–10 μg-at N $l^{-1}$) than the above rivers before 1970. But now the dissolved nitrogen content of the Altamaha (Walsh et al. 1981), Alabama, Brazos, Sacramento, Columbia (Meybeck 1982), and Pamlico (Hobbie et al. 1975) rivers is now ≥30 μg-at N $l^{-1}$.

The location and discharge of the world's 30 largest rivers suggest that the freshwater flux to the sea from underdeveloped areas ($286 \times 10^3$ $m^3$ $s^{-1}$) is similar to that from developed areas ($234 \times 10^3$ $m^3$ $s^{-1}$) having six times the population and far more anthropogenic nitrogen. If we assume that half the world's total river discharge ($4 \times 10^{16}$ l $yr^{-1}$) has not yet been affected by deforestation, sewage, and fertilizers in these underdeveloped countries and that the low nitrate content (≤10 μg-at $NO_3$-N $l^{-1}$) of the Amazon (Ryther et al. 1967), Congo and Niger (Van Bennekom et al. 1978), Orinoco, Negro, Mackenzie, and Yukon (Meybeck 1982) rivers is typical, nitrate input to the shelves from such drainage systems is $0.3 \times 10^7$ tons $NO_3$-N $yr^{-1}$.

If we also assume that the other half of the world's river discharge now has at least 60 μg-at $NO_3$ $l^{-1}$ due to population distribution and the Industrial Revolution,[2] the annual nitrate input from developed terrestrial ecosystems would be $1.7 \times 10^7$ tons $NO_3$-N $yr^{-1}$.

The total present input of $2.0 \times 10^7$ tons $NO_3$-N $yr^{-1}$ from major river runoff would be at least three times that estimated more than 25 years ago (Emery et al. 1955) from data taken more than 50 years ago (Clarke 1924). Other estimates of present river nitrogen range as high as $3.5 \times 10^7$ tons N $yr^{-1}$ (Delwiche and Likens 1977). Using an estimate of $0.8 \times 10^7$ tons $NO_3$-N $yr^{-1}$ in river runoff, Garrels et al. (1973) have suggested that only 10% of the nitrogen fertilizers previously applied to fields had been released from the soil to streams by 1970. The nutrient time series over the past 25 years, however, suggest that $NO_3$ concentrations in major rivers has at least doubled within the past decade. If these data reflect increased mobilization of nitrogen from soil pools, we would expect future increases of primary production in the coastal zone as has apparently occurred off the Netherlands (Postma 1978) and within Lake Erie.

Rivers are not the only source of anthropogenic nitrogen to coastal waters. Analyzing present nitrogen loadings from coastal sewage outfalls, waste dumping, and river input to the shelves of the New York Bight and the Mediterranean, Japan, Baltic, North, and Irish seas, Segar (1984) suggests a range of 10–30 kg N $yr^{-1}$ $person^{-1}$ for each of these areas; the total urban nitrogen loading to Lake Erie was estimated to be 15 kg N $yr^{-1}$ $person^{-1}$ (Sly 1976). Extrapolating an estimate of 20 kg N $yr^{-1}$ for the agricultural, sewage, and other waste demands needed to support each of the three billion persons in North America, Asia, and Europe leads to an estimate of $6 \times 10^7$ tons anthropogenic N $yr^{-1}$. Taking into account the nitrogen fluxes from domestic wastes, as well as from feed lots, food-processing wastes, and a 20% soil leaching rate from fertilizers, Van Bennekom and Salomons (1981) have made a similar estimate ($6.4 \times 10^7$ tons N $yr^{-1}$). Such anthropogenic nitrogen input is ten times the preindustrial input (Table 3).

If part of the loss to continental slope ecosystems of nearshore primary production has increased in those coastal zones where anthropogenic inorganic nutrient supplies have been consistently increasing since the Industrial Revolution, then burial and diagenesis of this material in slope depocenters could represent the "missing carbon" in present global $CO_2$ budgets. Moreover, as much as $6 \times 10^7$ tons N $yr^{-1}$ of the $10 \times 10^7$ tons N $yr^{-1}$ estimated to be sequestered *with* carbon on the slopes may be anthropogenic, ultimately derived from nitrogen fixation in nonmarine ecosystems.

Because other human activities, e.g., overfishing off Peru (Walsh 1981), have added organic matter to the continental slope, the rates of marine denitrification may also have recently increased by 5–10% (Codispoti and Packard 1980). If we assume, however, that the rate of denitrification has remained the same since the reappearance of Holocene shelves and that the rate of anthropogenic nutrient input (Walsh et al. 1981) and burial (Walsh et al. 1984) has increased tenfold, the ocean's nitrogen depletion time would be halved (Table 3). Without significant burial, a rapid nitrogen removal time of even 10,000 years implies that the sea's nitrogen balance is not steady either on interglacial time scales (Fogg 1982, McElroy 1983) or over longer geological periods (Piper and Codispoti 1975).



## PHOSPHORUS FLUXES

Denitrification and burial losses may exceed nitrogen fixation in the sea whenever continental shelves are present during periods of warm temperature and high sea level. A warming trend of 2°C since the Wisconsin glaciation would decrease dissolved $O_2$ solubility in the sea by 5 percent, thereby increasing significantly both the volume of oxygen-deficient sea water and the rate of marine denitrification. Formation of extensive phosphorite deposits may have occurred after times of nitrogen depletion, leading to a net nitrogen loss, concomitant lower primary production, and precipitation of excess phosphorus as carbonate fluroapatite deposits (Piper and Codispoti 1975).

A number of large inorganic phospho-

**Table 3.** Pristine (5000 B.P.) and postindustrial (1980 A.D.) annual budgets of marine nitrogen and phosphorus fluxes and depletion times of their sea water reservoirs ($9 \times 10^{11}$ tons for nitrogen, $10^{11}$ tons for phosphorus). Input, losses, and net loss rate in $10^7$ tons $yr^{-1}$, depletion time in years.

| | Nitrogen | | Phosphorus | |
|---|---|---|---|---|
| | 5000 B.P. | 1980 A.D. | 5000 B.P. | 1980 A.D. |
| Input | | | | |
| Biological fixation | 1.1 | 1.1 | — | — |
| Passive rainfall | 0.8 | 0.8 | — | — |
| Land runoff | 0.6 | 6.0 | 0.04 | 0.40 |
| Losses | | | | |
| Surface evasion | 10.0 | 10.0 | — | — |
| Organic burial | 1.4 | 14.4 | 0.10 | 1.09 |
| Inorganic burial | — | — | 0.01 | 0.01 |
| Net loss rate | 8.9 | 16.5 | 0.07 | 0.70 |
| Depletion time | 10,000 | 5500 | 140,000 | 14,000 |

[2]More than 70 μg-at $NO_3$ $l^{-1}$ is now found in the Hudson, Potomac, Susquehanna, Delaware, Po, Vistula (Van Bennekom and Salomons 1981), Loire, Rhone, Garonne (Kempe 1982), Huang-He (Meybeck 1982), and Yangtze rivers (John Edmond, MIT, personal communication, 1981) in addition to the 150–300 μg-at $NO_3$ $l^{-1}$ in the Mississippi, Seine, Thames, Rhine, and Scheldt rivers.

rite deposits were formed in Cambrian, Ordovician, Devonian, Permian, and Cretaceous periods in relation to major marine transgressions, further adding to the organic removal rate of phosphorus during interglacial periods (Broecker 1982). The Permian Phosphoria deposit alone contains $7 \times 10^{11}$ tons P within an area of $3.5 \times 10^5$ km$^2$. This is equivalent to about seven times the total amount of dissolved phosphorus in the sea today contained in only 0.1% of its present surface area, i.e., the area of coastal upwelling ecosystems (Ryther 1969) or salt marshes (Table 1).

Unlike the carbon and nitrogen cycles, phosphorus fluxes between the atmosphere and the dissolved $PO_4$ ocean reservoir ($1 \times 10^{11}$ tons P) are negligible. Other than removal of organic phosphorus by anchovy-feeding seabirds to form island guano deposits (Hutchinson 1950), the transfer of phosphorus between land and ocean has depended on weathering and uplifting on very long geological time scales. By 1970, however, about $1.3 \times 10^7$ tons P yr$^{-1}$ were being mined for fertilizer (Garrels et al. 1973) in an N/P weight ratio of 3/1, comparable to human nitrogen fixation from the atmosphere. As indicated by the dissolved N/P ratio of 12/1 within rivers (Kempe 1982, Meybeck 1982), the leaching rate of the more insoluble phosphorus compounds from arable lands is less than that of nitrogen.

The phosphorus input to the sea from land run-off has nevertheless increased tenfold since the Industrial Revolution. Pristine rivers such as the Amazon, Zaire, Yukon, MacKenzie, and Orange contain 0.3 μg-at $PO_4$ l$^{-1}$, whereas most developed rivers have at least 3.0 μg-at $PO_4$ l$^{-1}$, with as much as 40 μg-at P l$^{-1}$ found in the Scheldt and 80 μg-at P l$^{-1}$ in the Thames (Meybeck 1982). At an annual river discharge of $4 \times 10^{16}$ l yr$^{-1}$, the preindustrial input of phosphorus was probably $0.04 \times 10^7$ tons P yr$^{-1}$ (Table 3). Assuming that half this discharge contains 3.0 μg-at $PO_4$ l$^{-1}$, the present agrarian flux is estimated at $0.22 \times 10^7$ tons P yr$^{-1}$.

In coastal ecosystems dominated by the discharge of domestic wastes, the N/P ratio of urban effluent is only 2.5/1 from sewage treatment and phosphorus detergents. As much as 25–50% of land-derived phosphorus may be from this source (Ryther and Dunstan 1971). Applying a human excretion rate of only 0.54 kg P yr$^{-1}$ person $^{-1}$ (Vollenweider 1968) to the three billion people in developed countries gives a sewage flux of $0.16 \times 10^7$ tons P yr$^{-1}$, suggesting a total contemporary input of at least $0.4 \times 10^7$ tons P yr$^{-1}$ (Table 3).

At the preindustrial rate of phosphorus input to the sea and a slope sedimentation rate of 0.03 cm yr$^{-1}$, a burial flux of $0.11 \times 10^7$ tons P yr$^{-1}$ would take 140,000 years to deplete the ocean's phosphate reservoir (Table 3). Considering phosphorus lost in the marine burial of fish scales, phosphorite, calcium carbonate–phosphorus tests of open-ocean coccolithophores and foraminifera, and organic phosphorus based on input of terrestrial carbon, Froelich et al. (1982) arrived at a similar residence time of 80,000 years.

If the shelf export of organic phosphorus were buried on the slope as fast as $1.10 \times 10^7$ tons P yr$^{-1}$, however, the ocean would be depleted of phosphorus in 14,000 years, a depletion time similar to that of nitrogen when denitrification and burial losses dominate over nitrogen fixation and runoff souces (Table 3). Comparison of the calculated phosphorus content of the deep glacial ocean (3.2 μg-at $PO_4$ l$^{-1}$) with the present interglacial concentration (2.2 μg-at $PO_4$ l$^{-1}$) suggests that 50% of the present ocean reservoir was buried as organic matter within 10,000 years, i.e., $0.50 \times 10^7$ tons P yr$^{-1}$. If the glacial nitrate content of deep water were also 48 μg-at $NO_3$ l$^{-1}$ rather than today's 32 μg-at $NO_3$ l$^{-1}$, the initial nitrogen reservoir would have been 150% larger, allowing a greater nitrogen depletion time of 8200 years with anthropogenic runoff of $6 \times 10^7$ tons N yr$^{-1}$.

So much anthropogenic phosphorus and nitrogen has only persisted for the last century, however. Without such runoff but with continuous substantial slope burial of nitrogen for most of the last 10,000 years, the nitrogen depletion time would still be 6000 years even at higher glacial $NO_3$ concentrations. At the Holocene carbon burial rate, the oxygen content of the atmosphere ($1.2 \times 10^{15}$ tons $O_2$) would have increased by 3% over 10,000 years, with a doubling in 400,000 years. Halfway through the Holocene, all nitrate would have been stripped from the sea, about half of the phosphorus pool would be left, and sulfate would have become the dominant oxidant in widespread anaerobic processes.

## CONTINENTAL SHELF OSCILLATION

Three major processes account for carbon consumption in ocean sediments: sulfate reduction, nitrate reduction, and oxygen metabolism.

Accounting for each mole of $H_2S$ produced in the sea, basically 2 moles of organic carbon are oxidized to $CO_2$ during sulfate reduction: about $67.5 \times 10^7$ tons C yr$^{-1}$ of organic carbon may be consumed for the $90 \times 10^7$ tons S yr$^{-1}$ thought to be now produced by marine sulfur bacteria in slope and other sediments (Ivanov 1977). Similarly, nitrate reduction involves oxidizing essentially 2 moles of C for each mole of $N_2$ produced, or $9.4 \times 10^7$ tons C yr$^{-1}$ consumed for the $5.5 \times 10^7$ tons $N_2$ yr$^{-1}$ now thought to be released from slope ecosystems (Table 1). Within oxygenated waters, however, only 1.3 moles of $O_2$ are required to turn 1 mole of plankton carbon into $CO_2$. This means that $20 \times 10^7$ tons C yr$^{-1}$ may now be oxidized by sediment oxygen uptake rates of 16 mg $O_2$ m$^{-2}$ day $^{-1}$ within most slope regions (Murray and Grundmanis 1980, Smith 1978, Smith and Teal 1973).

With these assumptions, $1 \times 10^9$ tons C yr$^{-1}$ might now be consumed in slope depocenters, compared with the $2.7 \times 10^9$ tons C yr$^{-1}$ presently imported from adjacent shelves (Walsh et al. 1981). Of the remaining $1.7 \times 10^9$ tons C yr$^{-1}$, about half has been estimated to accumulate today on continental slopes (Table 1); the rest is perhaps dispersed downslope by slumping events to the continental rise for either additional oxidation or burial.

Independent calculations (Broecker 1982, Walsh et al. 1981, Walsh et al. 1984) indicate that the deposition rates of organic matter at the continental margins may have been about $1 \times 10^9$ tons C $yr^{-1}$, both at the end of the Wisconsin glaciation and today. The rate of plankton burial must have changed by an order of magnitude. A low inorganic sediment burial rate would have allowed extensive bacterial destruction of the organic matter and subsequent biological return of most compounds to the water column. In contrast, a high burial rate would have removed carbon and nutrients from the sea, until sediment uplifting and subsequent weathering.

Because sea level has risen less and less during the last 10,000 years (from 1.0 to 0.1 cm $yr^{-1}$) on the mid-Atlantic Holocene shelf (Milliman and Emery 1968), the US Atlantic shoreline 10,000 years ago may have retreated 200 m $yr^{-1}$, about 5 m $yr^{-1}$ 5000 years ago, and only 1.5 m $yr^{-1}$ over the last 180 years (Belknap and Kraft 1981, Kraft et al. 1979). Ten times less nearshore fine-grained sediment would have been delivered to the slope 10,000 years ago than now because initial sea levels would have risen too fast to allow much erosion of shore face deposits (Swift 1970). Increased erosion of today's shore could have changed inorganic sediment accumulation rates on the slope by tenfold, from 0.03 cm $yr^{-1}$ at the beginning of the Holocene to 0.30 cm $yr^{-1}$ today, thus changing the carbon burial and decay rates as well (Walsh et al. 1984). The tenfold less carbon in early Holocene and late Wisconsin sediments on the mid-Atlantic slope as compared with today may indicate such changes (Doyle et al. 1979).

Deposition of $1 \times 10^9$ tons C $yr^{-1}$ 10,000 years ago, with eventual nutrient return to the ocean by remineralization, is thought to have increased atmospheric $CO_2$ concentrations (Broecker 1982); deposition and burial of the same amount of carbon today, with anthropogenic replacement of the buried nutrients may be *reducing* atmospheric $CO_2$ concentrations (Walsh et al. 1981). In addition to the suspected changes in shelf production, export, and adjacent slope storage of carbon, nitrogen, and phosphorus during the Holocene transgression of present shelves, the transition from interglacial to glacial habitats would have had great impact on global ecology. With shelf exposure during glaciations, most shelf carbon production, denitrification, and export would have stopped. Rivers would have discharged their nutrients and suspended solids to the open ocean rather than to estuaries or shelves, and the marshes might have covered more than 14 times their present area (Table 2).

Ancient freshwater peats, oysters, and even mastodon teeth are found on the outer part of the present continental shelf (Whitmore et al. 1967). Here productive macrophytes might have doubled the glacial primary production and carbon storage of ancient marshes, with possibly even greater changes in the nitrogen cycle. During glaciations, if a nitrogen outwelling rate of 37 g N $m^{-2}$ $yr^{-1}$ (derived from present South Carolina marshes, Kjerfve and McKellar 1980) is applied to an area of $2.84 \times 10^{13}$ $m^2$ (Table 2), as much as $1 \times 10^9$ tons N $yr^{-1}$ might have been released to the sea from glacial marshes, more than ten times present estimated anthropogenic inputs. During interglacial periods, the nitrogen fixed and buried by glacial marshes might also then be returned to the sea, after shore erosion by the slowly rising sea.

Atmospheric methane concentrations ($4.8 \times 10^9$ tons) have increased by 150 ppb over the last decade, or 1.7% annually (Rasmusson and Khalil 1981), contributing as much as 23% of the $CO_2$ greenhouse effect (Lacis et al. 1981). Changes in atmospheric $CH_4$ input are thought to be mainly biogenic, with only 10% attributed to fossil fuel production and consumption (Sheppard et al. 1982). Methane emission from aquatic ecosystems is also less than 10% of terrestrial input. At present rates per unit area (Cicerone and Shetter 1981, Harris and Sebacher 1981), extensive glacial marshes might have emitted almost 50% of today's total $CH_4$ fluxes. Such an emission rate would act as a negative feedback mechanism, returning global element cycles to their interglacial state by enhancing the greenhouse effect of the glacial atmosphere.



Admittedly, these budgets for the present and past transfer of carbon, nitrogen, phosphorus, and inorganic solids to and from the sea are still speculative, despite the enormous amount of recent data embodied in the above assumptions. But it is clear that since the Industrial Revolution, humans have altered carbon input to the atmosphere by ten times the "rapid changes" at the beginning of this interglacial period, as well as the input of nitrogen and phosphorus to the ocean and the seaward input of terrestrial detritus to rivers and estuaries.

If the carbon slope burial rate has also increased tenfold, then it might absorb some of the anthropogenic fluxes; if not, global carbon and nitrogen budgets would still be unbalanced. Continued burial of organic matter on the continental slope at these rates, however, *does* require anthropogenic *re*supply of nutrients to avert another decline in marine primary production and a further increase in atmospheric $CO_2$ concentrations.

Despite enormous changes in the burial of particulate carbon in the Cretaceous or of phosphorite in the Permian, atmospheric $CO_2$ and $O_2$ are thought to have changed little over the last several million years because of the effects of the relatively rapid biological processes of marine photosynthesis and respiration (Garrels and Perry 1974). If these processes were to be disrupted or overwhelmed by man's direct carbon and nitrogen interactions with the atmosphere, future global oscillations might be greater, perhaps leading to faster extinction of marine and terrestrial species. Glaciations, in fact, may have been induced by "regular" changes in biological processes of the sea (Broecker 1982, McElroy 1983) which humans now alter by overfishing, wetlands eradication, deforestation, and anthropogenic acceleration of the carbon, nitrogen, and phosphorus cycles.



## ACKNOWLEDGMENTS

This research was mainly funded by the Department of Energy (DOE) under contract No. DE-AC02-76CH00016 as part of the CASTOR/POLLUX (Carbon Storage/Pollution Export) and SEEP (Shelf Edge Exchange Processes) Programs at Brookhaven National Laboratory; additional support was provided by the National Science Foundation as part of the PROBES study (Processes and Resources of the Bering Sea). Manuscript preparation was completed during sabbatical leave at the Skidaway Institute of Oceanography, Savannah, Georgia. I am indebted to numerous discussions with Wally Broecker, Mike McElroy, Jim Walker, K. O. Emery, Lou Codispoti, Jeff Gaffney, Bill Reeburgh, Gene Premuzic, and Don Swift for the ideas presented in this paper. The composite "distracting state of confusion" within the text is my own.

## REFERENCES CITED

Belknap, D. F., and J. C. Kraft. 1981. Preservation potential of transgressive coastal lithosomes on the U.S. Atlantic shelf. *Mar. Geol.* 42: 429–442.

Berner, R. A. 1982. Burial of organic carbon and pyrite sulfur in the modern ocean: its geochemical and environmental significance. *Am. J. Sci.* 282: 451–473.

Billen, G. 1978. A budget of nitrogen recycling in North Sea sediments off the Belgian coast. *Estuarine Coastal Mar. Sci.* 7: 127–146.

Booth, J. S. 1979. Recent history of mass-wasting on the upper continental slope, northern Gulf of Mexico, as interpreted from the consolidation states of the sediment. Pages 153–164 in L. J. Doyle and O. H. Pilkey, eds. *Geology of Continental Slopes* SEPM Spec. Pub. 27.

Botkin, D. B., and E. A. Keller. 1982. *Environmental Studies*. C. E. Merrill, Columbus, OH.

Boyer, J. S. 1982. Plant productivity and environment. *Science* 218: 443–448.

Broecker, W. S. 1982. Glacial to interglacial changes in ocean chemistry. *Prog. Oceanogr.* 11: 151–197.

Broecker, W. S., T. Takahashi, H. J. Simpson, and T. H. Peng. 1979. Fate of fossil fuel carbon dioxide and the global carbon budget. *Science* 206: 409–418.

Burns, N. M. 1976. Temperature, oxygen, and nutrient distribution patterns in Lake Erie, 1970. *J. Fish. Res. Board Can.* 33: 485–511.

Caesar, G. J. [1980] 53 B.C. Page 226 in *De bello Gallico,* VI, 25 (translated by A. and P. Wiseman). D. R. Godine, Boston.

Capone, D. G., and E. J. Carpenter. 1982. Nitrogen fixation in the marine environment. *Science* 217: 1140–1142.

Carpenter, J. H., D. W. Pritchard, and R. C. Whaley. 1969. Observations of eutrophication and nutrient cycles in some coastal plain estuaries. Pages 210–221 in *Eutrophication: Causes, Consequences, Correctives*. National Academy of Sciences, Washington, DC.

Christensen, J. P. 1981. Oxygen consumption, denitrification, and sulfate reduction in coastal marine sediments. Ph.D. dissertation. University of Washington, Seattle.

Cicerone, R. J., and J. D. Shetter. 1981. Sources of atmospheric methane: measurements in rice paddies and a discussion. *J. Geophys. Res.* 86: 7203–7209.

Clarke, F. W. 1924. The data of geochemistry. *US Geol. Bull.* 770: 1–841.

Codispoti, L. A., and F. A. Richards. 1976. An analysis of the horizontal regime of denitrification in the eastern tropical North Pacific. *Limnol. Oceanogr.* 21: 379–388.

Codispoti, L. A., and T. T. Packard. 1980. On the denitrification rate in the eastern tropical South Pacific. *J. Mar. Res.* 39: 453–477.

Cohen, Y., and L. I. Gordon. 1978. Nitrous oxide in the oxygen minimum of the eastern tropical North Pacific. *Deep-Sea Res.* 25: 509–524.

Crutzen, P. J. 1981. Atmospheric chemical processes of the oxides of nitrogen, including nitrous oxide. Pages 17–44 in C. C. Delwiche, ed. *Denitrification, Nitrification, and Atmospheric Nitrous Oxide*. John Wiley & Sons, New York.

Deck, B. L. 1981. Nutrient-element distribution in the Hudson estuary. Ph.D. dissertation. Columbia University, New York.

Degens, E. T., A. Paluska, and E. Eriksson. 1976. Rates of soil erosion. *Ecol. Bull.* 22: 185–191.

Delmas, R. J., J. M. Ascencio, and M. Legrand. 1980. Polar ice evidence that atmospheric $CO_2$ 20,000 yr BP was 50% of present. *Nature* 284: 155–157.

Delwiche, C. C., and G. E. Likens. 1977. Biological response to fossil fuel combustion products. Pages 89–98 in W. Stumm, ed. *Global Chemical Cycles and Their Alterations By Man*. Dahlem Konferenzen, Berlin.

DeMaster, D. J. 1979. The marine budgets of silica and $^{32}$silicon. Ph.D. dissertation, Yale University, New Haven, CT.

DeVooys, C. G. N. 1979. Primary production in aquatic environments. Pages 259–292 in B. Bolin, E. T. Degens, S. Kempe, and P. Ketner, eds. *The Global Carbon Cycle*. John Wiley & Sons, New York.

Doyle, L. J., O. H. Pilkey, and C. C. Woo. 1979. Sedimentation on the eastern United States continental slope. Pages 119–129 in L. J. Doyle and O. H. Pilkey, eds. *Geology Of Continental Slopes*. SEPM Spec. Pub. 27.

Eisma, D. 1973. Sediment distribution in the North Sea in relation to marine pollution. Pages 131–150 in E. D. Goldberg, ed. *North Sea Science*. MIT Press, Cambridge, MA.

Elkins, J. W., S. C. Wofsy, M. G. McElroy, C. E. Kolb, and W. A. Kaplan. 1978. Aquatic sources and sinks for nitrous oxide. *Nature* 275: 602–606.

Emery, K. O. 1968. Relict sediments on continental shelves of the world. *Am. Assoc. Pet. Geol. Bull.* 52: 445–464.

______. 1969. The continental shelves. *Sci. Am.* 221: 106–122.

Emery, K. O., and E. Uchupi. 1972. Western North Atlantic Ocean: topography, rocks, structure, water, life and sediments. *Am. Assoc. Pet. Geol. Mem.* 17: 1–532.

Emery, K. O., W. L. Orr, and S. C. Rittenberg. 1955. Nutrient budgets in the ocean. In *Essays in Honor of Captain Allan Hancock*. University Southern California Press, Los Angeles.

Eppley, R. W., and B. J. Peterson. 1980. Particulate organic matter flux and planktonic new production in the deep ocean. *Nature* 282: 677–680.

Fisher, J. B., and G. Matisoff. 1982. Down-core variation in sediment organic nitrogen. *Nature* 296: 345–347.

Fogg, G. E. 1982. Nitrogen cycling in sea waters. *Philos. Trans. R. Soc. Lond.* B 296: 511–570.

Forster, S. S., A. C. Cripps, and A. Smith-Carington. 1982. Nitrate leaching to ground water. *Philos. Trans. R. Soc. Lond.* B 296: 477–489.

Froelich, P. N., M. Bender, N. A. Luedtke, G. R. Heath, and T. DeVries. 1982. The marine phosphorous cycle. *Am. J. Sci.* 282: 475–511.

Garrels, R. M., and E. A. Perry. 1974. Cycling of carbon, sulfur, and oxygen through geologic time. Pages 303–336 in E. D. Goldberg, ed. *The Sea,* Vol. 5. Wiley-Interscience, New York.

Garrels, R. M., F. T. MacKenzie, and C. Hunt. 1973. *Chemical Cycles And The Global Environment*. William Kaufmann, Inc., Los Altos, CA.

Gornitz, V., S. Lebedeff, and J. Hansen. 1982. Global sea level trend in the past century. *Science* 215: 611–614.

Gunter, G. 1967. Some relationships of estuaries to the fisheries of the Gulf of Mexico. Pages 621–638 in G. H. Lauff, ed. *Estuaries*. American Association for the Advancement of Science, Washington, DC.

Hahn, J. 1981. Nitrous oxide in the oceans. Pages 191–240 in C. C. Delwiche, ed. *Denitrification, Nitrification, and Atmospheric Nitrous Oxide*. John Wiley & Sons, New York.

Haines, J. R., R. M. Atlas, R. P. Griffiths, and R. Y. Morita. 1981. Denitrification and nitrogen fixation in Alaska continental shelf sediments. *Appl. Environ. Microbiol.* 41: 412–421.

Hansen, J., D. Johnson, A. Lacis, S. Lebedeff, P. Lee, D. Rind, and G. Russell. 1981. Climate impact of increasing atmospheric carbon dioxide. *Science* 213: 957–966.

Harriss, R. C., and D. I. Sebacher. 1981. Methane flux in forested freshwater swamps of the southeastern United States. *Geophs. Res. Lett.* 8: 1002–1004.

Hathaway, J. C., C. S. Poag, P. C. Valentine, R. W. Miller, D. M. Schultz, F. T. Manheim, F. A. Kohout, M. H. Bothner, and D. A. Sangrey. 1979. U.S. Geological Survey core drilling on the Atlantic Shelf. *Science* 206: 515–527.

Hobbie, J. E., B. J. Copeland, and W. G. Harrison. 1975. Nutrient cycling within the Pamlico Estuary. Pages 287–305 in L. E. Cronin, ed. *Estuarine Res.,* Vol. 1. Academic Press, New York.

Holland, H. D. 1973. Ocean water, nutrients, and atmospheric oxygen. Pages 68–81 in E. Ingerson, ed. *Proceedings of the Symposium on Hydrogeochem. Biogeochem.* Clarke Boardman Co., New York.

Holland, H. D. 1978. *The Chemistry of the Atmosphere and Oceans*. Wiley-Interscience, New York.

Hutchinson, G. E. 1950. The biogeochemistry of vertebrate excretion. *Bull. Am. Mus. Nat. Hist.* 96: 1–554.

______. 1970. Ianula: an account of the history and development of the Lago di Monterosi. Latium, Italy. *Trans. Am. Philos. Soc.* 60: 1–178.

Ivanov, M. V. 1977. Influence of microorganisms and microenvironment on the global sulfur cycle. Pages 47–61 in W. E. Krumbein, ed. *Environmental Biogeochemistry and Geomicrobiology. I. The Aquatic Environment*. Ann Arbor Sci., Ann Arbor.

Johnston, R. 1973. Nutrients and metals in the North Sea. Pages 293–307 in E. D. Goldberg, ed. *North Sea Science*. MIT Press, Cambridge, MA.

Kaplan, W., I. Valiela, and J. M. Teal. 1979. Denitrification in a salt marsh ecosystem. *Limnol. Oceanogr.* 24: 726–734.

Kemp, A. L., T. W. Anderson, R. L. Thomas, and A. Mudrochova. 1974. Sedimentation rates and recent sediment history of Lakes Ontario, Erie, and Huron. *J. Sediment. Petrol.* 44: 207–218.

Kemp, A. L., R. L. Thomas, C. I. Dell, and J. M. Jaquet. 1976. Cultural impact on the geochemistry of sediments in Lake Erie. *J. Fish. Res. Board Can.* 33: 440–462.

Kempe, S. 1979. Carbon in the rock cycle. Pages 343–378 in B. Bolin, E. T. Degens, S. Kempe, and P. Ketner, eds. *The Global Carbon Cycle*. John Wiley & Sons, New York.

______. 1982. Long-term records of the $CO_2$ pressure fluctuations in fresh water. *Mitt. Geol.-Palaont. Inst. Univ. Hamburg* 52: 91–332.

Kiry, P. R. 1974. An historical look at the water quality of the Delaware River estuary to 1973. *Contrib. Dep. Limnol., Acad. Nat. Sci. Phila.* 4: 1–76.

Kjerfve, B., and H. N. McKellar. 1980. Time series measurements of estuarine material fluxes. Pages 341–357 in V. S. Kennedy, ed. *Estuarine Perspectives*. Academic Press, New York.

Koide, M., and E. D. Goldberg. 1982. Transuranic nuclides in two coastal marine sediments off Peru. *Earth Planet. Sci. Lett.* 57: 263–277.

Koide, M., A. Soutar, and E. D. Goldberg. 1972. Marine geochronology with $^{210}Pb$. *Earth Planet. Sci. Lett.* 14: 442–446.

Koide, M., K. W. Bruland, and E. D. Goldberg. 1973. Th-228/Th-232 and $^{210}Pb$ geochronologies in marine and lake sediments. *Geochim. Cosmochim. Acta* 37: 1171–1187.



Koike, I., and A. Hattori. 1979. Estimates of denitrification in sediments of the Bering Sea shelf. *Deep-Sea Res.* 26: 409–415.

Kraft, J. C., E. A. Allen, D. F. Belknap, C. J. John, and E. M. Mauermeyer. 1979. Processes and morphological evolution of an estuarine and coastal barrier system. Pages 149–183 in S. P. Leatherman, ed. *Barrier Islands from the Gulf of St. Lawrence to the Gulf of Mexico*. Academic Press, New York.

Lacis, A., J. Hansen, P. Lee, T. Mitchell, and S. Lebedeff. 1981. Greenhouse effect of trace gases, 1970–1980. *Geophys. Res. Lett.* 8: 1035–1038.

Leach, J. H., and S. J. Nepszy. 1976. The fish community in Lake Erie. *J. Fish. Res. Board Can.* 33: 622–638.

Likens, G. E., F. H. Bormann, R. S. Pierce, and W. A. Reiners. 1978. Recovery of a deforested ecosystem. *Science* 199: 492–496.

Lipschultz, F., O. C. Zafiriou, S. C. Wofsky, M. B. McElroy, F. W. Valois, and S. W. Watson. 1981. Production of NO and $N_2O$ by soil nitrifying bacteria. *Nature* 294: 641–643.

Liu, K. 1979. Geochemistry of inorganic nitrogen compounds in two marine environments: the Santa Barbara Basin and the ocean off Peru. Ph.D. dissertation, University of California, Los Angeles.

Longhurst, A. 1984. Benthic-pelagic coupling and export of organic carbon from a tropical Atlantic continental shelf—Sierra Leone. *Estuar. Coast. Shelf Sci.* 17, in press.

MacIlvane, J. C. 1973. Sedimentary processes on the continental slope off New England. Ph.D. dissertation, MIT.

Martinez, L. A., M. W. Silver, J. M. King, and A. L. Alldredge. 1983. Nitrogen fixation by floating diatom mats: a source of new nitrogen to oligotrophic ocean waters. *Science* 221: 152–154.

McElroy, M. B. 1983. Marine biological controls on atmospheric $CO_2$ and climate. Nature 302: 328–329.

Meade, R. H. 1982. Sources, sinks, and storage of river sediment in the Atlantic drainage of the United States. *J. Geol.* 90: 235–252.

Meybeck, M. 1982. Carbon, nitrogen, and phosphorus transport by world rivers. *Am. J. Sci.* 282: 401–450.

Milliman, J. D., and K. O. Emery. 1968. Sea levels during the past 35,000 years. *Science* 162: 1121–1123.

Mommaerts, J. P., W. Baeyens, and G. Decadt. 1979. Synthesis of research on nutrients in the Southern Bight of the North Sea. Pages 215–234 in *Actions de Recherche Concertées*. Bruxelles, Belgium.

Morse, J. W., and N. Cook. 1978. The distribution and form of phosphorous in North Atlantic ocean deep sea and continental slope sediments. *Limnol. Oceanogr.* 23: 825–830.

Murray, J. W., and V. Grundmanis. 1980. Oxygen consumption in pelagic marine sediments. *Science* 209: 1527–1530.

Neftal, A., H. Oeschger, J. Schwander, B. Stauffer, and R. Zumbrunn. 1982. Ice core sample measurements give atmospheric $CO_2$ content during the past 40,000 yr. *Nature* 295: 220–223.

Park, K. H., C. L. Osterberg, and W. D. Forster. 1972. Chemical budget of the Columbia River. Pages 123–134 in A. T. Pruter and D. L. Alverson, eds. *The Columbia River Estuary and Adjacent Ocean Waters*. University of Washington Press, Seattle.

Pheiffer-Madsen, P. 1979. Seasonal variation of denitrification rate in sediment determined by use of $^{15}N$. *Wat. Res.* 13: 461–465.

Piper, D. Z., and L. A. Codispoti. 1975. Marine phosphorite deposits and the nitrogen cycle. *Science* 188: 15–18.

Postma, H. 1978. The nutrient contents of North Sea water: changes in recent years, particularly in the Southern Bight. *Rapp. R.-V. Reun. Cons. Int. Explor. Mer.* 172: 350–352.

Rasmusson, R. A., and M. A. Khalil. 1981. Atmospheric methane ($CH_4$): trends and seasonal cycles. *J. Geophys. Res.* 86: 9826–9832.

Richards, F. A., and A. C. Redfield. 1954. A correlation between the oxygen content of sea water and the organic content of marine sediments. *Deep-Sea Res.* 1: 279–281.

Riley, G. A. 1937. The significance of the Mississippi River drainage for biological conditions in the northern Gulf of Mexico. *J. Mar. Res.* 1: 60–74.

Ryther, J. H. 1969. Photosynthesis and fish production in the sea. *Science* 166: 72–76.

Ryther, J. H., and W. M. Dunstan. 1971. Nitrogen, phosphorus, and eutrophication in the coastal marine environment. *Science* 171: 1008–1013.

Ryther, J. H., D. W. Menzel, and N. Corwin. 1967. Influence of the Amazon River outflow on the ecology of the western Tropical Atlantic. I. Hydrography and nutrient chemistry. *J. Mar. Res.* 25: 69–83.

Segar, D. A. 1984. Contamination of polluted estuaries and adjacent coastal ocean—a global review. *MESA (Mar. Ecosyst. Anal.) N.Y. Bight Spec. Rep. Ser.*, in press.

Seitzinger, S., S. Nixon, M. E. Pilsen, and S. Burke. 1984. Denitrification and $N_2O$ production in nearshore marine sediments. *Geochim. Cosmochim. Acta,* in press.

Sheppard, J. C., H. Westburg, J. F. Hopper, K. Ganesan, and P. Zimmerman. 1982. Inventory of global methane sources and their production rates. *J. Geophys. Res.* 87: 1305–1312.

Sly, P. G. 1976. Lake Erie and its Basin. *J. Fish. Res. Board Can.* 33: 355–370.

Smith, K. L. 1978. Benthic community respiration in the N.W. Atlantic Ocean: *in situ* measurements from 40 to 5200 m. *Mar. Biol.* 47: 337–347.

Smith, K. L., and J. M. Teal. 1973. Deep-sea benthic community respiration: as *in situ* study at 1850 m. *Science* 179: 282–283.

Smith, R. C., J. S. Campbell, W. E. Esaias, and J. J. McCarthy. 1984. Primary productivity in the ocean. *Proceedings from a Workshop on Life from a Planetary Prospective: Fundamental Issues in Global Ecology,* in press.

Smith, S. V. 1981. Marine macrophytes as a global carbon sink. *Science* 211: 838–840.

Soderlund, R., and B. H. Svensson. 1976. The global nitrogen cycle. *Ecol. Bull. Stoch.* 22: 23–73.

Sorenson, J., B. B. Jorgensen, and N. P. Revsbech. 1979. A comparison of oxygen, nitrate, and sulfate respiration in coastal marine sediments. *Microb. Ecol.* 5: 105–115.

Swift, D. J. 1970. Quarternary shelves and the return to grade. *Mar. Geol.* 8: 5–30.

Van Bennekom, A. J., and W. Salomons. 1981. Pathways of nutrients and organic matter from land to ocean through rivers. Pages 33–51 in *River Inputs to Ocean Systems*. UNESCO, Switzerland.

Van Bennekom, A. J., G. W. Berger, W. Helder, and R. T. P. deVries. 1978. Nutrient distribution in the Zaire estuary and river plume. *Neth. J. Sea Res.* 12: 296–323.

Vollenweider, R. A. 1968. Scientific fundamentals of the eutrophications of lakes and flowing waters, with particular reference to nitrogen and phosphorus as factors in eutrophication. *OECD Tech. Rep.* DAS/CS1/68-27: 1–95.

Vollenweider, R. A., M. Munawar, and P. Stadlemann. 1974. A comparative review of phytoplankton and primary production in the Laurentian Great Lakes. *J. Fish. Res. Board Can.* 31: 739–762.

Wajih, S., S. Naqui, R. J. Noronha, and C. V. G. Reddy. 1982. Denitrification in the Arabian Sea. *Deep-Sea Res.* 29: 459–469.

Walker, J. C. G. 1974. Stability of atmospheric oxygen. *Am. J. Sci.* 274: 193–214.

Walsh, J. J. 1976. Herbivory as a factor in patterns of nutrient utilization in the sea. *Limnol. Oceanogr.* 21: 1–13.

______. 1981. A carbon budget for overfishing off Peru. *Nature* 290: 300–304.

______. 1983. Death in the sea: enigmatic phytoplankton losses. *Prog. Oceanogr.* 12: 1–86.

Walsh, J. J., and C. P. McRoy. 1984. Ecosystem analysis in the southeastern Bering Sea. *Cont. Shelf Res.*, in press.

Walsh, J. J., G. T. Rowe, R. L. Iverson, and C. P. McRoy. 1981. Biological export of shelf carbon is a neglected sink of the global $CO_2$ cycle. *Nature* 291: 196–201.

Walsh, J. J., E. T. Premuzic, J. S. Gaffney, G. T. Rowe, G. Harbottle, R. W. Stoenner, W. L. Balsam, and P. R. Betzer. 1984. Bomb-carbon and $^{210}Pb$ estimates of organic storage of $CO_2$ within the last century on the Mid-Atlantic slope. *Deep-Sea Res.*, in press.

Weiss, R. F. 1981. The temporal and spatial distribution of tropospheric nitrous oxide. *J. Geophys. Res.* 86: 7185–7195.

Whitmore, F. C., K. O. Emery, H. B. Cooke, and D. J. Swift. 1967. Elephant teeth from the Atlantic continental shelf. *Science* 156: 1477–1481.

Wilkinson, W. B., and L. A. Greene. 1982. The water industry and the nitrogen cycle. *Philos. Trans. R. Soc. Lond.* B 296: 459–475.

Wolman, M. G. 1971. The nation's rivers. *Science* 174: 905–918.

Woodwell, G. M., R. H. Whittaker, W. A. Reiners, G. E. Likens, C. C. Delwiche, and D. B. Botkin. 1978. The biota and the world carbon budget. *Science* 199: 141–146.

CHAPTER 8

# World Environmental Trends Between 1972 and 1982



MARTIN W. HOLDGATE, MOHAMED KASSAS AND GILBERT F. WHITE

## INTRODUCTION

The United Nations Conference on the Human Environment, which was held in Stockholm, Sweden, in June 1972, and is commonly referred to as 'the Stockholm Conference', focused the attention of the world's governments on environmental issues more sharply than ever before. The Conference's recommendations, ratified by the General Assembly of the United Nations in the following December, led directly to the establishment of the United Nations Environment Programme (UNEP), with its 58-nations' Governing Council, central secretariat based in Nairobi, Kenya, and voluntary Environment Fund (UNEP, 1978).

The Action Plan agreed at Stockholm had three major components: environmental assessment, environmental management, and supporting measures. The first area was designed to provide the essential foundation of information about trends in the world environment, the relative significance of new problems, and the rate at which old problems were responding to action. The Governments of the world needed this information if they were to make rational judgements about priorities for international action. Research, monitoring, and information exchanges, were all seen as essential for this process of evaluation and review, the information exchange activity being expected to be of direct practical benefit to the world community, because it would switch knowledge from country to country and hence avoid wasteful duplication of effort.

Over the past decade, the UNEP Secretariat has been developing this 'earthwatch' function through the

Reprinted from *Environmental Conservation*, the scientific journal devoted to global survival, founded and edited by Nicholas Polunin. Volume 9, No. 1, pp. 11–29 (Spring 1982). Published by Elsevier Sequoia S.A., P.O. Box 851, 1001 Lausanne 1, Switzerland, with kind permission of the Foundation for Environmental Conservation, 7 Chemin Taverney, 1218 Grand-Saconnex, Geneva, Switzerland.

Martin W. Holdgate is Chief Scientist and Deputy Secretary, Departments of the Environment and Transport, London.

Mohamed Kassas is head of the Department of Botany, Faculty of Science, Cairo University, Giza, Egypt, and President of the International Union for the Conservation of Nature and Natural Resources.

Gilbert F. White is emeritus professor of geography and director of the Institute of Behavioral Science, University of Colorado, Boulder, CO.

Global Environmental Monitoring Service (GEMS)* and INFOTERRA (as the International Environmental Information Network is popularly called).† It has also set up an international Registry of Potentially Toxic Chemicals (IRPTC)**. The evaluation and review process has, among other things, been reflected in annual reports on the State of the Environment, submitted to the Governing Council of UNEP. These Reports were required by the Council because of the task laid on it by the UN General Assembly in Resolution 2997 (XXVII) of 15 December 1972, namely to 'keep under review the world environmental situation in order to ensure that emerging environmental problems of wide international significance receive appropriate and adequate consideration'. Initially, the Reports were general and inevitably somewhat superficial, but in 1976 the Governing Council agreed that they would in most years deal only with four or five selected topics of contemporary importance. The subjects so treated in these Reports are listed in Table I.

The Governing Council having agreed that every five years a more comprehensive Report on the state of the global environment should be prepared, the first of these is being published and presented to the Council in 1982. It covers the decade since the Stockholm Conference (or, more precisely, the years between 1970 and 1979 or 1980, as the Report had to be sent for printing in the summer of 1981, and many statistical series run a year or two in arrears, so that little information was available for the most recent part of the period). The Authors of this paper are the Editors of that Report (Holdgate *et al.*, in press), upon which the present text draws heavily, the tables and diagrams being taken at least in substance from it.

The Report was prepared by an elaborate (and lengthy) process of consultation and compilation. In 1978 the Executive Director of UNEP, Dr M. K. Tolba, established a Senior Scientific Advisory Board (SSAB) to help the UNEP Secretariat to plan and compile the review. The SSAB, with the help of advisers, drew up the project plan, and Consultants from 16 different countries were then enlisted to write a series of reviews covering the main subject-areas. The SSAB, and a further series of expert reviewers, commented on these and sought additional information where gaps in coverage were apparent. We, as the Editors, together with some individual members of SSAB and the UNEP Secretariat, then produced a condensed and consolidated text which was circulated to United Nations agencies and Governments, and submitted for discussion by an international workshop that was held in Nairobi in March 1981. The comments received and supplementary data provided at that stage were then used by us to produce the final Report. In total, therefore, well over a hundred scientists and other specialists, from more than 50 countries, were closely involved in the production of the document while, through UN Agencies, Governments, and informal consultative processes, several times as many contributed indirectly. It is doubtful whether such a wide-ranging attempt to bring together the judgement of experts around the world has ever been made before, at least in the environmental field.

The Report is not a textbook or an encyclopaedia. It does not describe world environmental systems such as oceanic and atmospheric circulations, biogeochemical cycles, or biogeographical classifications—except where brief summaries are needed as a background to the pattern of change. It is selective, concentrating on global or regional phenomena and only citing some of the many examples of events in individual countries. It is also restricted in its time-span, deliberately excluding important environmental events that took place before 1970 unless these need description in order to place recent changes in a proper time-perspective.

The present paper begins with a summary of the main changes that have taken place during the past decade in the natural world and in the activities and processes by which human societies interact with and transform it. We then review some of the changes in attitudes towards the environment that, in our view, have been of overwhelming importance in determining the course of these events. Finally, we draw some conclusions about the present state of knowledge and the likely developments during the next decade. We should stress that the views which we express are our own, and not necessarily those of UNEP—or of the institutions of which we are members.

---

**See* the account by its Deputy Director, Dr Michael D. Gwynne, published on pp. 35–41 of this issue.—Ed.

†Formerly known as the 'International Referral System for Sources of Environmental Information'—*see* the report on its second management meeting ('INFOTERRA 2') published in *Environmental Conservation*, Vol. 7, No. 2, pp. 162–3, 1980.—Ed.

***See* the account by Dr Alexander I. Kucherenko & its Director, Dr Jan W. Huismans, published on pp. 59–63 of this issue.—Ed.

TABLE I. *Topics Treated in Annual State of the Environment Reports* (source: Holdgate *et al.*, in press).

| Subject Area | Topic | Year |
|---|---|---|
| The atmosphere | Climatic changes and their causes. | 1974*, 1976, 1980 |
| | Possible effects of ozone depletion. | 1977 |
| The marine environment | Oceans. | 1975* |
| Freshwater environment | Water resources and quality. | 1974*, 1976 |
| | Ground-water. | 1981 |
| Land environment | Land resources. | 1974* |
| | Raw materials. | 1975* |
| | Firewood. | 1977 |
| Food and agriculture | Food shortages, hunger, and losses of agricultural land. | 1974*, 1977, 1976 |
| | Use of agricultural and agro-industrial residues. | 1978 |
| | Resistance to pesticides. | 1979 |
| Environment and health | Toxic substances and effects. | 1974*, 1976* |
| | Heavy-metals and health. | 1980 |
| | Cancer. | 1977 |
| | Malaria. | 1978 |
| | Schistosomiasis. | 1979 |
| | Biological effects of ozone depletion. | 1977 |
| | Chemicals in food-chain. | 1981 |
| Energy | Energy conservation. | 1975*, 1978 |
| | Firewood. | 1977 |
| Environmental pollution | Toxic substances. | 1974* |
| | Chemicals and the environment: | |
| | – possible effects of ozone depletion; | 1977 |
| | – chemicals and the environment. | 1978 |
| | Noise pollution. | 1979 |
| Man and environment | Human stress and societal tension. | 1974* |
| | Outer limits. | 1975* |
| | Population. | 1975*, 1976* |
| | Tourism and environment. | 1979 |
| | Transport and the environment. | 1980 |
| | Environmental effects of military activity. | 1980 |
| | The child and the environment. | 1980 |
| Environmental management achievements | The approach to management. | 1974*, 1976 |
| | Protection and improvement of the environment. | 1977 |
| | Legal and institutional arrangement. | 1976* |
| | Environmental economies. | 1981 |

*Asterisks indicate brief treatment in early reports.

## CHANGES IN THE BIOSPHERE

One of the major advances in perception during the past decade concerned the linkage between the major biogeochemical cycles of carbon, nitrogen, oxygen, sulphur, phosphorus, and other elements—hence emphasizing the unsatisfactory nature of the traditional subdivision of The Biosphere into the media of atmosphere, hydrosphere, and lithosphere (Bolin *et al.*, 1979, and in press). The flux of elements between these sectors, and the interlinkages between ocean and atmosphere in determining climatic patterns, are of overwhelming importance. While it remains convenient to use the familiar sectoral units in the discussion of trends (and this is the approach adopted in the Report and in the paragraphs below), the essential interdependence of these sectors must therefore be borne in mind in considering the significance of the trends recorded.

*Changes in the Atmosphere*

Time-series measurements only allow us to make statements on a global basis about a few atmospheric constituents such as carbon dioxide, stratospheric ozone, and particulates. Other measurements permit general statements about regional trends in the acidity of rainfall and about sulphur dioxide emissions, while many developed countries have data on urban air pollution with $SO_2$, smoke, photochemically produced oxidants, airborne metals, and oxides of nitrogen. In contrast, climatic records are extensive and satellite imagery has provided good new information about the pattern of atmospheric circulation.

Carbon dioxide concentrations continued to increase during the decade at a rate of about 1 ppm (part per million) per year, reaching about 340 ppm by 1981 (Keeling & Bacastow, 1977, and *see* our Fig. 1). The burning of fossil fuels and the clearance of forests were the agreed main causes but their relative importance was in some dispute. Most national energy plans assumed an increasing use of carbonaceous fuel, making it almost inevitable that the trend in atmospheric $CO_2$ would continue. There was still, in 1981, uncertainty among scientists about the consequences of these continuing increases in $CO_2$ in the atmosphere, but a general agreement that the trend could lead to an increase in air and sea-surface temperatures, and a change in rainfall patterns—to the advantage of some areas and the disadvantage of others. There was agreement that these issues needed to be taken seriously, and an expanded programme of research (under the World Climate Programme) was in progress. The social consequences of possible changes in climate had not, however, been examined in any depth, and strategies for reducing emissions by curbing fossil fuel consumption were only beginning to be discussed.

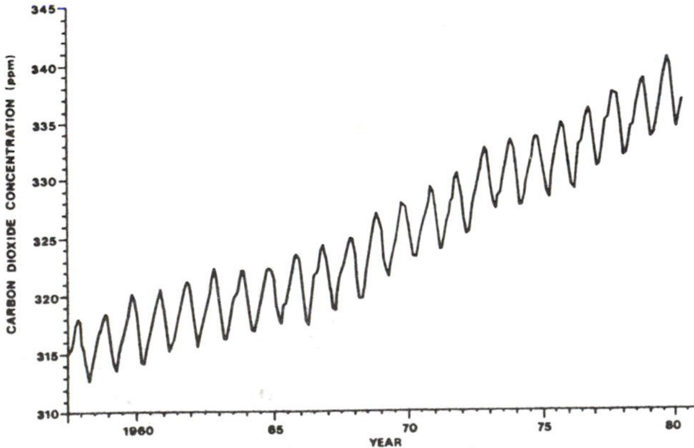


Fig. 1. *The rise in atmospheric carbon dioxide concentration since 1958, as measured in Hawaii. Data from Keeling & Bacastow (1977), extended by use of information from the National Oceanographic and Atmospheric Administration (NOAA), USA.*

Acid precipitation was established during the decade as a serious problem in northwestern Europe and northeastern North America (Wallén, 1980). Its principal cause was the deposition, as sulphuric and nitric acids, of sulphur and nitrogen oxides produced in fuel combustion. Over Europe, the sulphur content of precipitation rose by some 50% between the mid-1950s and the early 1970s, although there are some indications that it may not have changed much since. Nitrate concentrations in precipitation have continued to rise. The resulting impoverishment of freshwater ecosystems, especially in lakes over igneous rocks, was broadly established by 1981, although there was less certainty about the possible impact of acid rain and snow on forest soils and productivity. A regional Convention was negotiated in Europe covering both research and information exchange and possible action.

The depletion of stratospheric ozone, postulated early in the decade as a likely result of the use of chlorofluorocarbons as refrigerants and aerosol propellants, and because of the emission of exhaust gases from high-flying aircraft, had not been detected instrumentally by 1981. Estimates of the scale of the likely effect varied widely, and without demonstration by direct measurement, the degree to which community action was needed was not clear. During the 1970s, limits were imposed on the manufacture and use of the chlorofluorocarbons that were likely to have the most serious impact, and their production fell by 17% between 1974 and 1979.

Among other atmospheric trends noted in the past decade was a slight increase in stratospheric particulates, with uncertain consequences for climate, a probable continuing rise in the airborne dispersal of metals and toxic chemicals, a fall in photochemical oxidant smog and smoke and sulphur oxide concentrations in the cities of those developed countries that had adopted controls, but a rise in urban air pollution in the large conurbations of many developing countries, bringing a threat of damage to health. Although there were severe droughts, some disastrous floods, and other extreme climatic events, there was no evidence that the 1970s were worse in these respects than previous decades.

*Changes in the Marine Environment*

At global and regional levels there are reasonably complete records of fisheries landings (FAO, 1979*b*), and the International Whaling Commission has statistics of whale catches. World fishery landings rose during the 1960s to exceed 70 million tonnes in 1970–1, but there was a dip in the curve of growth in 1972–73 and some populations of fish were severely depleted as a result of over-exploitation. The World Conservation Strategy (IUCN/UNEP/WWF, 1980) argued, however, that landings in 1980 were 15 to 20 million tonnes less than they would have been had management been better. By the end of the decade some fish stocks showed signs of recovery as a result of tighter controls, and the contribution from mariculture was expanding rapidly—especially in south-east Asia. Whale catches were greatly reduced following the adoption of new procedures by the International Whaling Commission, but the total ban on commercial whaling called for at the Stockholm Conference had not been achieved by 1981.

Although the pollution of the oceans was not monitored on a global basis, there were good monitoring schemes in the north-east Atlantic (ICES, 1978, 1980), especially in the Baltic Sea and the North Sea, while during the decade UNEP's Regional Seas Programme stimulated action in the Mediterranean (UNEP, 1981*a*) and elsewhere (UNEP, 1981*b*). Such surveys showed that, in some coastal waters, pollutions by sewage, agricultural chemicals, oil, and metals, were causing concern, and there was evidence of local ecological changes. Table II illustrates the scale of contamination by reference to the North-east Atlantic. However, there was no evidence that marine productivity or fish stocks had been reduced by chemical contamination outside restricted coastal areas, or that floating oil had reduced the rate at which young individuals replaced adult members of fish and shellfish populations. Nonetheless, many marine biologists remained cautious about possible long-term changes resulting from the exposure of marine ecosystems to chemical contamination.

The minerals of the sea-bed—especially oil—were exploited on an increasing scale during the decade. In the tropics, development affected considerable areas of mangrove and coral-reef. The coastal zone everywhere came under greater and greater pressure, although marine nature reserves also began to be established. Oil pollution increased, and although tankers were the source of only about 5% of it, accidents to large vessels released large quantities in several confined areas (Table III). Oil

TABLE II. *Input of Contaminants to the North-east Atlantic (Oslo Convention Area). For Sources and Details see Holdgate, White & Kassas (*in press*).*

| | Domestic sewage | Industrial waste | Domestic + industrial wastes | Rivers | Dumping |
|---|---|---|---|---|---|
| Total flow (million $m^3$/yr) | 5,664 | 3,432 | 9,393 | 316,514 | – |
| *Contamination (tonnes/yr):* | | | | | |
| Nitrogen | 109,999 | 70,255 | 202,481 | 973,010 | 22,202 |
| Phosphorus | 29,759 | 25,042 | 56,249 | 94,794 | 13,048 |
| Suspended solids | 388,000 | 9,354,100 | 9,893,100 | 5,188,000 | – |
| BOD | 452,000 | 395,000 | 1,125,000 | 938,000 | – |
| Iron | 4,958 | 16,051 | 28,336 | 246,588 | – |
| Manganese | 310 | – | – | 30,207 | – |
| Cadmium | 43 | 38 | 80 | 421 | 89 |
| Copper | 598 | 891 | 1,492 | 2,786 | 2,426 |
| Chromium | 176 | 170 | 381 | 2,678 | 2,712 |
| Nickel | 219 | 173 | 391 | 2,417 | 527 |
| Lead | 246 | 785 | 1,726 | 3,831 | 4,248 |
| Zinc | 1,279 | 11,053 | 13,719 | 19,275 | 9,131 |
| Mercury | 17 | 6,3 | 23,3 | 36,4 | 35 |

pollution continued to cause great social concern because it killed many seabirds, fouled beaches, and affected tourism.

*Changes in Inland Waters*

From some standpoints, the condition of the world's freshwater resources deteriorated in the 1970s. The continued growth of human populations and their waste products, and an acceleration of water use in many areas, had by 1970 already begun to strain the water resources of some regions. During the 1970s these problems increased, aggravated by pollution and the continued prevalance of water-borne disease. Although the proportion of people in rural areas having access to safe water supplies doubled during the decade, it stood at only 29% in 1980, and only 13% had sanitary facilities. In urban areas, despite rapid population growth, access to water supplies improved, with 75% having safe sources in 1980 as against 67% in 1970: however, the proportion with sanitary facilities fell from 71% to 53%. Unless progress accelerated, it was clear that the targets of the Drinking Water and Sanitation Decade, 1980–90, would not be attained. The areas of ground-water depletion expanded during the past decade, and contamination of ground-water became more serious.

TABLE III. *Oil Spilled into Oceans Following Incidents Involving Tankers in which 5,000 Tonnes or More Were Shed*[(a)].

| Year | Incidents | Spillages |
|---|---|---|
| 1970 | 11 | 212,120 |
| 1971 | 4 | 121,250 |
| 1972 | 1 | 65,000 |
| 1973 | 3 | 58,000 |
| 1974 | 2 | 61,000 |
| 1975 | 6 | 123,000 |
| 1976 | 8 | 150,000 |
| 1977 | 6 | 134,500 |
| 1978 | 5 | 264,000[(b)] |

Source: IMCO (1979).

(a) These figures are certainly underestimates. Not only may some incidents have been missed, but many spillages of under 5,000 tonnes are omitted.

(b) The *Amoco Cadiz* contributed some 230,000 tonnes towards this total.

Great advances were, however, made during the 1970s in the assessment of the fixed stock of freshwater resources, and in methods for their management (UNWC, 1978). There was a great expansion in the use of water in irrigated agriculture, especially in arid regions (Table IV), and some of the problems arising from poor management of such systems began to be solved. Industrial use of water also increased, but in developed countries recycling of water became increasingly important. Pollution control also made headway in many developed countries, although the damage to aquatic ecosystems from acid precipitation was a growing concern in the 1970s. Spurred on by recurrent floods and droughts, engineers and scientists made considerable progress in flood management. Inland fisheries catches increased from about 5.2 million tonnes in the 1950s to about 10 million tonnes in the mid-1970s, and the creation of new fisheries in man-made lakes, together with expanding aquaculture, made a significant contribution. The overall balance of benefit from some man-made lakes was, however, questioned by some scientists.

*Changes in the Lithosphere*

During the decade, there were major scientific advances in our understanding of the basic structure of the Earth's crust, and of the processes by which mineral-rich zones and regions at high risk from earthquakes and volcanoes arose. Annual production of almost all major non-metallic minerals expanded during the 1970s, although there was considerable variation from year to

TABLE IV. *Increase of Water Withdrawal Over the World.*

| | Cubic kilometres per year | |
|---|---|---|
| Water Used | 1970 | 1975 |
| Domestic water-supply | 120 | 150 |
| Industry | 510 | 630 |
| Agriculture | 1,900 | 2,100 |
| Totals | 2,530 | 2,880 |

Source: USSR (1978).

year in response to economic and political factors. There was some advance in the prediction of earthquakes and also in designing buildings and urban facilities to reduce their impact. Although the decade was not unusual in the number of earthquakes and volcanic eruptions, one earthquake at Tangshang, China, in July 1976, had an intensity that has been reached only a few times in recorded history, killing at least 250,000 people.

Important improvements were made during the 1970s in the methods used to reduce the damaging impacts of mining, processing, and transporting, minerals. In the mid-1970s, it was estimated that some 40–60% of the land that had been disturbed by mining was either in the process of natural restoration or was being reclaimed. Recycling and substitution of scarce minerals increased, largely in response to economic forces. Some minerals remained an important factor in international diplomacy, because major nations remained heavily dependent on their import (the United States imported 94% of its bauxite, 97% of its manganese, and all of its titanium and columbium in 1976–9), while some other countries held a near-monopoly of the most economically-attractive sources of supply.

### *Changes in Terrestrial Biota*

Despite the fact that good satellite imagery has been available for many years, there is still a shortage of reliable statistics on the areas of the main vegetation formations of the world and the rates at which they are changing. However, it is clear that the natural ecosystems of the land were in retreat in many areas during the 1970s. Tropical forests were converted for agriculture and energy on a scale that was the subject of many conflicting estimates (figures ranged from 7 to 20 million hectares per year), but appeared greatest in Africa.

In the more arid regions, including those supporting savannas, steppes, and areas of tropical deciduous forest and mediterranean climate, desertification was a major problem (Table V), and its advances claimed large areas during the decade—despite the assembly of much scientific knowledge whose application would halt or even reverse it. One estimate put losses in the southern part of the Sahara alone at 650,000 square kilometres in the past 50 years.

Changes resulting from human impact were less severe in the temperate and polar zones than in the tropics, but even in the arctic tundra and in the Antarctic, mineral exploration and tourism caused some local damage to plant and animal life.

The need to conserve genetic resources, and to secure essential development while sustaining the stability and productivity of renewable natural resources and protecting wildlife, were major concerns during the 1970s and the focus of the *World Conservation Strategy* (IUCN/UNEP/WWF, 1980) at the end of the period.

### *Changes in Agriculture and Forestry*

According to the records kept by the Food and Agriculture Organization of the United Nations (FAO),

TABLE V. *Arid Lands Affected by Desertification* (in thousands of hectares).

| *Continent* | *Irrigated Land* | | *Rainfed Cropland* | | *Rangeland* | |
|---|---|---|---|---|---|---|
| | *Total* | *Area affected by desertification* | *Total* | *Area affected by desertification* | *Total* | *Area affected by desertification* |
| Africa | 7,756 | 1,366 | 48,048 | 39,633 | 1,182,212 | 1,026,758 |
| Asia + USSR | 89, 587 | 20,572 | 112,590 | 91,235 | 1,273,759 | 1,088,965 |
| Australasia; | 1,600 | 160 | 2,000 | 1,500 | 550,000 | 330,000 |
| Europe (Spain) | 2,400 | 890 | 5,000 | 4,200 | 16,000 | 15,500 |
| North America | 19,550 | 2,835 | 42,500 | 24,700 | 345,000 | 291,000 |
| South America | 5,389 | 1,229 | 14,290 | 11,859 | 384,100 | 319,380 |
| | 126,282 | 27,052 | 224,428 | 173,127 | 3,751,071 | 3,071,603 |

Source: UN (1980*a*).

more than 450 million people were estimated to be either chronically hungry or malnourished during the decade in question. Although total food production increased nearly everywhere (Table VI and cf. FAO, 1980), it was unable to match population growth in many parts of Africa, Asia, and Latin America. One FAO projection indicated that a 60% increase in food production would be needed simply to maintain current *per caput* food-consumption levels over the 20 years to AD 2000 (FAO, 1979*c*). The improvement in production in the 1970s was secured partly through bringing new areas under cultivation or irrigation, and partly through the use of higher-yielding crop strains and better husbandry techniques. Several analyses suggested that in future more intensive use of the more fertile and easily managed land was likely to be more rewarding than creating new farms in the wilderness.

There were, however, major losses of productive land during the decade. In developed countries, urbanization and associated transport networks consumed some 30,000 km$^2$, while world-wide the losses to such uses totalled around 50,000–70,000 km$^2$. Soil degradation through erosion, salt accumulation, and desertification, affected much greater areas; but the evidence

TABLE VI. *World and Regional Indices of Food Production* (1961–5 = 100).

| *Region and Country* | *1961–5* | *1970* | *1971* | *1972* | *1973* | *1974* | *1975* | *1976* | *1977* | *1978* | *1979* |
|---|---|---|---|---|---|---|---|---|---|---|---|
| *Total production:* | | | | | | | | | | | |
| Developed countries— | | | | | | | | | | | |
| United States | 100 | 115 | 126 | 126 | 128 | 122 | 134 | 137 | 143 | 144 | 152 |
| Canada | 100 | 109 | 130 | 122 | 123 | 112 | 128 | 142 | 144 | 148 | 138 |
| Western Europe | 100 | 115 | 121 | 121 | 124 | 129 | 125 | 123 | 129 | 136 | 134 |
| Eastern Europe | 100 | 116 | 123 | 132 | 135 | 141 | 138 | 145 | 145 | 150 | 148 |
| USSR | 100 | 136 | 135 | 128 | 155 | 144 | 128 | 153 | 148 | 163 | 142 |
| Japan | 100 | 110 | 103 | 110 | 110 | 111 | 115 | 109 | 118 | 117 | 116 |
| Oceania | 100 | 121 | 128 | 123 | 127 | 127 | 135 | 138 | 133 | 152 | 142 |
| Developing countries— | | | | | | | | | | | |
| Latin America | 100 | 129 | 131 | 130 | 139 | 145 | 152 | 158 | 163 | 168 | 173 |
| Africa[(a)] | 100 | 117 | 120 | 122 | 119 | 125 | 120 | 132 | 129 | 131 | 134 |
| West Asia | 100 | 122 | 127 | 137 | 127 | 141 | 154 | 168 | 166 | 171 | 168 |
| South Asia | 100 | 128 | 126 | 119 | 130 | 124 | 140 | 137 | 150 | 157 | 146 |
| East Asia[(b)] | 100 | 129 | 132 | 130 | 141 | 147 | 155 | 164 | 167 | 170 | 175 |
| World | 100 | 123 | 127 | 125 | 133 | 132 | 135 | 141 | 144 | 151 | 147 |
| *Per caput production:* | | | | | | | | | | | |
| Developed countries— | | | | | | | | | | | |
| United States | 100 | 106 | 116 | 114 | 115 | 109 | 119 | 120 | 124 | 125 | 130 |
| Canada | 100 | 97 | 114 | 106 | 106 | 95 | 106 | 117 | 117 | 110 | 108 |
| Western Europe | 100 | 109 | 114 | 113 | 116 | 119 | 115 | 113 | 118 | 124 | 122 |
| Eastern Europe | 100 | 111 | 117 | 124 | 126 | 131 | 127 | 133 | 128 | 135 | 133 |
| USSR | 100 | 125 | 124 | 116 | 139 | 129 | 113 | 134 | 128 | 140 | 121 |
| Japan | 100 | 102 | 94 | 100 | 98 | 97 | 100 | 94 | 101 | 98 | 96 |
| Oceania | 100 | 107 | 111 | 105 | 107 | 105 | 110 | 111 | 106 | 119 | 110 |
| Developing countries— | | | | | | | | | | | |
| Latin America | 100 | 107 | 106 | 102 | 106 | 108 | 110 | 111 | 112 | 112 | 113 |
| Africa[(a)] | 100 | 98 | 98 | 97 | 93 | 95 | 96 | 95 | 92 | 91 | 88 |
| West Asia | 100 | 102 | 102 | 107 | 96 | 104 | 110 | 117 | 112 | 113 | 108 |
| South Asia | 100 | 110 | 106 | 97 | 104 | 97 | 107 | 103 | 110 | 113 | 103 |
| East Asia[(b)] | 100 | 108 | 108 | 104 | 110 | 112 | 116 | 120 | 119 | 119 | 120 |
| World | 100 | 107 | 109 | 106 | 111 | 108 | 108 | 111 | 112 | 115 | 110 |

Source: USDA (1980).
[(a)] Excluding South Africa.
[(b)] Excluding Japan.

was incomplete. In some countries, up to half of the irrigated area was estimated to have been endangered through salinization or waterlogging, while the deterioration of semi-arid and arid lands destroyed or damaged about 60,000 km$^2$ annually.

Consumption of nitrogenous fertilizers increased from *ca* 33 million tonnes world-wide in 1969–70 to 54 million tonnes in 1978–9 and was predicted to reach 84 million tonnes in 1985–6 (FAO, 1979*a* and Fig. 2). In areas of intensive fertilizer and pesticide use, concern about the ecological impact of such substances mounted; yet it was clear that both would need to be employed in greater and greater quantities if production goals were to be met, while new methods of pest control would also be needed to combat increasing resistance to the chemicals that are currently being employed (UN, 1979*a*, 1981).

Wastage of crops after harvest, especially through fungal attack, remained a severe problem in many developing countries, for example in south-east Asia taking up to 37% of the rice crop. New methods of reducing this waste, and of making better use of agricultural residues, were developed, while biotechnology offered some hope of developing nitrogen-fixing cereals and bringing about other improvements.

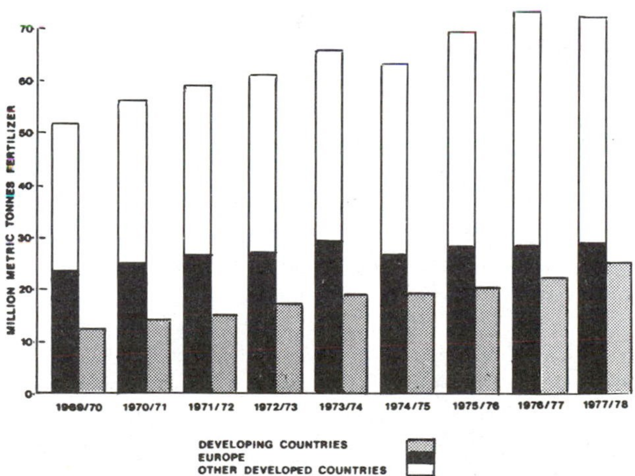


Fig. 2. *World Fertilizer Consumption (nitrogen, phosphate, and potash), 1969/70 to 1977/8. Data from* 1978 FAO Fertilizer Yearbook *(FAO, 1979*a*).*

CHANGES IN SOCIAL AND TECHNICAL SYSTEMS

One of the major lessons learned during the past decade has been that the crisis confronting humanity does not lie primarily in the natural world, despite the constraints which climate, soil, and water availability, place on human actions in many places. The major problems are those of mankind, because these impel people's interaction with the environment and underly actions which damage it and imperil its future productivity.

Millions of humans still endure squalor, inadequate housing, and a lack of the most basic services. The provision of health care and the improvement of settlements is barely keeping pace with population growth in many areas. Economic and technological changes are continually altering the nature of human impact in industrial areas. Conflicts for land—for example for food and fuel—are a serious problem elsewhere. These are the issues which are reviewed in the second and third sections of our Report (Holdgate *et al.*, in press), and will occupy much of the remainder of this paper.

### *Changes in Human Populations*

World, regional, and national, population trends are well documented by the United Nations (UN, 1979*a*, 1981), although the data for some developing countries are obviously not wholly reliable. These records are important because the implications of human population growth have been a controversial topic in recent years. At the beginning of the 1970s, the rapid growth in human numbers was seen by some as a major environmental problem while other groups argued that the problem lay much more with the quality of life and the resources available for development. The actual experience of the 1970s has been that, while world populations continued to increase—to pass 4,400 millions in 1980—the rate of growth has declined. In the first 5 years of the 1960s the annual population increase was 1.99%, whereas in the last 5 years of the 1970s it was 1.73% (Table VII). In many developed countries, birth- and death-rates became broadly equal, and generally speaking the developed world was certainly not experiencing a population explosion. Only in Africa did the annual rate of population increase fail to decline during the decade: the growth-rate climbed there to 2.9% per annum—more than in any other region.

TABLE VII. *Average Annual Rate of Population Growth* (per cent).

| | *1960–5* | *1970–5* | *1975–80* | *Percentage Change 1960–5/1975–80* |
|---|---|---|---|---|
| World | 1.99 | 1.95 | 1.73 | −13.1 |
| More developed | 1.19 | 0.84 | 0.71 | −40.3 |
| Less developed | 2.33 | 2.32 | 2.08 | −10.7 |
| Africa | 2.48 | 2.73 | 2.90 | +16.9 |
| Latin America | 2.80 | 2.54 | 2.45 | −12.5 |
| North America | 1.49 | 0.86 | 0.95 | −36.2 |
| East Asia | 1.94 | 2.02 | 1.38 | −28.9 |
| South Asia | 2.40 | 2.37 | 2.23 | −7.1 |
| Europe | 0.91 | 0.63 | 0.40 | −56.0 |
| Oceania | 2.08 | 1.85 | 1.47 | −29.3 |
| USSR | 1.49 | 0.95 | 0.93 | −37.6 |

Source: UN (1981).

TABLE VIII. *Population and Number of Cities in a Particular Size-class, 1975–2000* (population in millions).

| *Size-class (millions)* | *1975* | *1980* | *1990* | *2000* |
|---|---|---|---|---|
| | *More-developed Regions* | | | |
| Urban Population | 767 | 834 | 969 | 1,092 |
| 4 millions + | 121 | 142 | 171 | 207 |
| Number of Cities | 13 | 16 | 19 | 25 |
| 2–3.9 millions | 72 | 73 | 99 | 96 |
| Number of Cities | 26 | 27 | 36 | 36 |
| 1–1.9 millions | 77 | 99 | 119 | 131 |
| Number of Cities | 56 | 74 | 89 | 94 |
| | *Less-developed Regions* | | | |
| Urban Population | 794 | 792 | 1,453 | 2,116 |
| 4 millions + | 121 | 170 | 295 | 535 |
| Number of Cities | 17 | 22 | 33 | 61 |
| 2–3.9 millions | 61 | 82 | 144 | 183 |
| Number of Cities | 22 | 31 | 50 | 69 |
| 1–1.9 millions | 73 | 87 | 157 | 213 |
| Number of Cities | 51 | 65 | 115 | 154 |

Source: UN (1980*b*).

Both family planning and socio-economic advance were shown during the decade to be important and mutually reinforcing in reducing birth-rates, thus ending a controversy. Family planning programmes were strongly pursued in many developing countries, and contributed to the moderation of population growth there. However, in many developing nations, growth continued faster than education, health, sanitation, and other public services, could expand. Large-scale migration from rural to urban areas aggravated the problems of the latter. The pattern of international migration altered during the 1970s, with the predominant flows from developing countries to developed and Arab OPEC countries. Wars and political upheavals added to the stream of refugees, and in the last 3 years of the decade between 2,000 and 3,000 people throughout the world on average became refugees each day: by 1980 there were some 7.4 million such refugees, which is three times as many as in 1970.

### *Changes in Human Settlements*

United Nations statistics also give valuable documentation of major trends affecting human settlements (UN, 1980*b*). The expansion of the great cities of the developing world was a startling and disturbing feature of the decade (Table VIII). In 1950 there was only one city (Buenos Aires) in the developing world with a population of more than 4 millions: by 1975 there were 17 (compared with 13 in developed countries), and by 1980 the total had risen to 22 (compared with 16). By 2000, some 61 cities of over 4 million inhabitants seem likely in the developing countries, as against 25 in developed nations (Table VIII), and the total urban population in the developing world is predicted to double between 1980 and the end of the century. Because this growth is taking place against a background of low incomes, it has outstripped national capability to provide accommodation and services, so that squatter settlements have proliferated: in 1980 between 20% and 80% of the population of different new cities lived under these conditions. Even so, the majority of people in developing countries still lived in rural areas in 1980.

In rural and urban areas alike, large numbers of people in the Third World lacked safe water and waste disposal facilities, had overcrowded housing, and only meagre transport services. Nonetheless there were advances in the management of squatter settlements, with more flexible building design and an emphasis on self-help and community development.

In developed countries, in contrast, the decline of inner city areas became a major social problem, and the disposal of urban waste (which included increasing quantities of packaging and a widening range of potentially toxic chemicals) also caused concern.

*Environment and Health*

Data on world health trends are collected by the World Health Organization (WHO), which in 1980 published a global analysis of the health situation (WHO, 1980). But mortality statistics remain inadequate in many developing countries, where the registration of births and deaths is still not compulsory. Imperfect though the data are, they nevertheless reveal a striking contrast between developing and developed countries in the patterns of health and disease (Fig. 3). In the developing world, despite the successful eradication of smallpox, infectious diseases remained great killers, the 6 most serious together taking the lives of more than 5 million children every year.

Parasitic diseases were also rampant. Thus malaria remained the most serious disease in much of Africa, and one of the most important elsewhere in the tropics: an estimated 1 million children died from it every year. DDT-spraying reduced its incidence in the 1950s and 1960s, but in the 1970s there was a resurgence due partly to pesticide resistance in the mosquito vector,

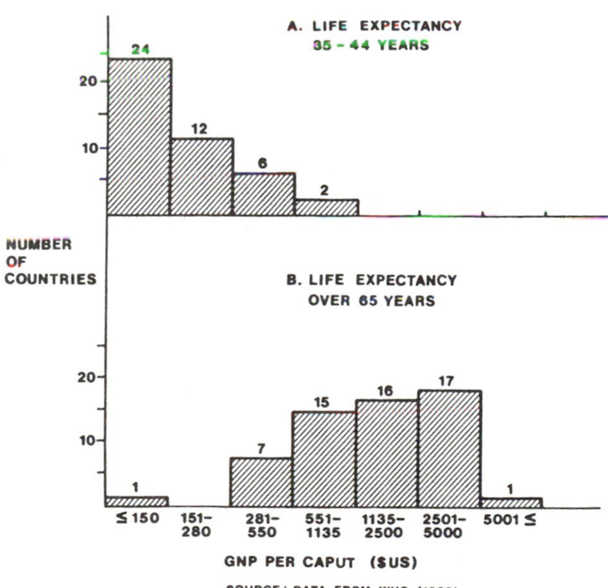


Fig. 3. *The relationship between national wealth and life expectancy. Data from WHO (1980).*

drug resistance in the parasite, and the increasing costs of control. Schistosomiasis, onchocerciasis, and viral diseases, remained widespread. A Special Programme for Research and Training in Tropical Diseases was sponsored by the World Health Organisation, the United Nations Development Programme, and the World Bank. In the less-developed regions, malnutrition remained a major contributory cause of death and illness. Altogether there was little improvement in the health of the vast rural and shanty-town populations, despite efforts to improve primary health care in such places.

In developed countries, coronary heart disease, hypertension, and cancer, were the main killers. There was concern over the role of nutrition and life-style in their causation. The significance of smoking as a major cause of lung cancer and a serious influence on other cancers and heart disease was confirmed: in the late 1970s there was a slight drop in tobacco consumption in some developed countries. In these regions there was also concern over the health-effects of chemicals in the workplace, and over such environmental questions as the increase in skin cancer that might follow the depletion of stratospheric ozone; also over effects of food additives, contaminants, pharmaceuticals, medical X-rays, and micropollutants in drinking water. Alcohol, drugs, and the role of environmental factors in mental health, received widespread attention.

*Industry and the Environment*

Industrial production in major sectors is documented in UN statistics, but records of expenditure to reduce environmental impact are much more scanty. The available records show, however, that the 1970s were a testing time for industry in most developed countries. The era of rapid growth, begun after the second world war, came to a close. Fuel prices rose sharply, and other costs also increased. At the same time, the demand for more stringent standards to protect the environment from pollutants discharged in industrial effluents imposed an added constraint in most countries.

Yet, by the end of the decade it was generally established that it was possible for industry to be efficient without creating damaging pollution, and that the margin of cost added by meeting acceptable standards was of the order of only 1% or 2% in those developed countries that had made estimates. Many individual industries reduced their effluents substantially without significant cost-increases. Some made a direct profit from waste-recovery (Table IX).

TABLE IX. *Costs and Benefits of Non-waste Technologies* (Examples from France).

| *Process* | *Company* | *Cost of operating conventional or destructive pollution control process (French Francs)* | | *Profit of alternative recovery process (French Francs)* | |
|---|---|---|---|---|---|
| Recovery of hydrocarbon in an oil refinery | Raffinerie Flf Feyzin (Rhône) | Investment:<br>Operating costs: | nil<br>2,438,000 | Investment:<br>Operating costs:<br>Sales of recovered product:<br>Gross operating profit: | 11,000,000<br>2,644,000<br>8,000,000<br>5,356,000 |
| Recovery of Methtonine mother liquor by evaporation | Société Alimentaire Equilibrée de Commentry (Allier) | Investment:<br>Operating costs: | 9,600,000<br>960,000 | Investment:<br>Operating cost:<br>Sales of recovered product: | 7,000,000<br>10,500,000<br>13,000,000 |
| Recovery of protein and potassium from a yeast factory | Société Industrielle de la Levure Fala (SILF), Usine de Strasbourg (Bas-Rhin) | Investment:<br>Operating costs: | 10,800,000<br>1,080,000 | Investment:<br>Operating costs:<br>Sales of recovered product:<br>Gross operating profit: | 5,200,000<br>860,000<br>1,015,500<br>155,500 |
| Recovery of lead and tin from furnace fumes | Société des Alliages d'Etain et Derives, Montreuil (Seine-Saint Denis) | Investment:<br>Operating costs:<br>Sales of recovered product:<br>Profit: | nil<br>nil<br>4,400<br>4,400 | Investment:<br>Operating costs:<br>Sales of recovered product:<br>Gross operating profit: | 300,000<br>200<br>8,930<br>8,730 |
| Conversion of phosphoric acid waste into plasterboard | Rhône Progil, Les Roches de Condrieu (Isère) Rouen (Seine-Maritime) | Investment:<br>Operating costs: | 9,000,000<br>5,000,000 | Investment:<br>Operating costs:<br>Sales of recovered product: | 35,000,000<br>73,000,000<br>73,500,000 |
| Water recycle in fiber-board plant | Isorel, Casteljaloux (Tarn-et-Garonne) | Investment:<br>Operating costs: | 5,000,000<br>500,000 | Investment:<br>Operating costs:<br>Sales of recovered product:<br>Gross operating profit: | 2,500,000<br>100,000<br>350,000<br>250,000 |
| Recycle of effluents in glue and gelatine manufacture | Société des Etablissements Georges Alquier, Bout-du-Pont-de-l'Ain, Mazamet (Tarn) | Investment:<br>Operating costs: | 534,000<br>53,000 | Investment:<br>Operating costs:<br>Reduced consumption of chemicals and sale of recovered product:<br>Gross operating profit: | 248,000<br>—<br>18,000<br>18,000 |
| Recovery of plum juice | Etablissements Laparre Castelnaud de Gratecombe (Lot-et-Garonne) | Investment:<br>Operating costs: | 768,000<br>77,000 | Investment:<br>Operating costs:<br>Sales of recovered product:<br>Gross operating profit: | 235,000<br>140,000<br>247,500<br>107,500 |
| Recovery of glycerine in a soap factory | Savonnerie de Lutterbach (Haut-Rhin) | Investment:<br>Operating costs: | 600,000<br>60,000 | Investment:<br>Operating costs:<br>Sales of recovered product:<br>Gross operating profit: | 400,000<br>101,700<br>280,000<br>178,300 |
| Recovery of quarry washings | Société d'Exploitation de l'Entreprise Mirsaint-Lary (Hautes-Pyrénées) | | | Investment:<br>Operating costs:<br>Sales of recovered product:<br>Gross operating profit: | 188,000<br>3,200<br>11,000<br>7,800 |

In the Third World, economic growth was given highest priority during the 1970s, but it was accepted that this must not be at the cost of environmental devastation and the loss of resources on which people would depend in the long term. Everywhere, as the decade progressed, the costs and benefits of alternative processes received increasing attention, and effort was being devoted to the reclamation of useful substances from what was previously regarded as waste.

*Energy and the Environment*

The 1970s saw a revolution in thinking about energy supplies. Oil ceased to be a cheap, secure, and universally available fuel. Developing countries whose plans depended on it found that sudden price-rises imposed a crippling economic burden. At the end of the decade, some were spending 25–65% of their 'hard-currency' earnings on oil imports. The 'crisis' of 1973 led to the realization that fossil fuels were finite, precious, and likely to be increasingly expensive.

The United Nations Yearbooks of World Energy Statistics (e.g. UN, 1979*b*) show that world energy consumption rose by about 30% between 1970 and 1978, about 80% of this growth being in developed countries, where *per caput* consumption was about 15 times as great as in developing countries (Fig. 4). In parallel with the pressure on fossil fuel supplies among developed countries, the Third World experienced a fuelwood crisis. Wood, charcoal, and agricultural and animal wastes, still provided up to 95% of the energy in Third World countries, and securing supplies from depleted woodlands consumed an increasing proportion of people's time, while also increasing the risk of erosion. Competition for land between food production, cash cropping, and energy production, increased in many areas.

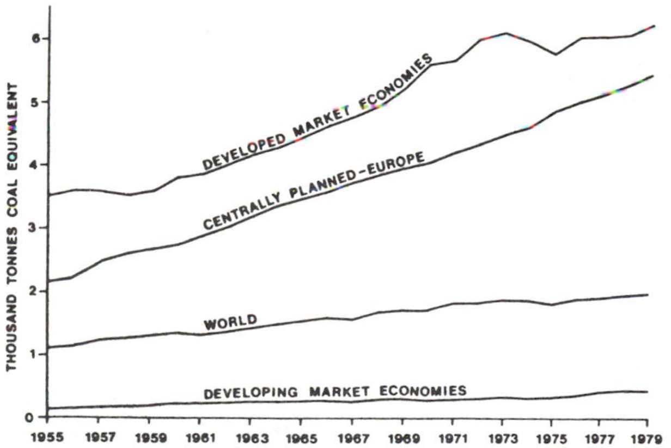


Fig. 4. Per caput *commercial energy consumption between 1955 and 1979 in different groups of countries. Data from UN Statistical Yearbooks (e.g. UN, 1979*a, *1979*b*).*

The changed perception of the 'energy future' led to new analyses of possible alternative supplies. By 1980 it was estimated that proven exploitable reserves of oil would be insufficient within 30 years, and that developed countries would need to make discriminating use of a wide range of available sources to sustain their industries. Many expected to use more coal, and interest in desulphurization and reduction in nitrogen oxide ($NO_x$) emissions mounted. The potential environmental impact of oil-shale and tar-sands exploitation also caused concern. The nuclear debate intensified. While nuclear power-generation was the source of only about 0.25% of the ionizing radiation dose received by the average person, reactor accidents and uncertainties over waste-disposal stimulated public unease, and the number of nuclear reactors on order around the world dropped sharply after 1973.

Renewable energy sources attracted increasing attention. Hydropower contributed 22% of world electricity in 1972, and although the percentage dropped slightly during the decade, the absolute amount rose, while great untapped resources remained in Africa and Asia. Geothermal, solar, tidal, wind-, and wave-power, were the subject of research, their attractiveness varying in part with environmental circumstances. Conservation was shown to have a great contribution to make, and several developed countries reduced their ratio of energy use to GNP by over 10%. Biological energy sources gained importance. Not only were measures to improve fuelwood supplies taken in many developing countries, but biogas production became prominent, especially in Asia (by late 1978 there were 7 million plants in China), while growth of crops for ethanol and methanol fuel production increased in countries such as Brazil, where land was abundant. All in all, the diversity of energy sources increased, and improved efficiency in the use of energy was clearly set as a goal for the 1980s.

*Changes in Transport*

In developing countries, animal power remains important and even as late as 1980 two-thirds of Indian rural transport depended on it. Pedestrian travel remained a dominant mode in urban areas even in developed countries, and cycling increased there during the 1970s.

United Nations statistics provide good records of the trends in the major forms of mechanical transport. However, the records confirm that in developed countries the dominant transport trend of the decade was the continuing growth in motor vehicles—especially passenger cars. In such countries the demands of private motoring for energy, and associated problems of pollution, noise, and traffic congestion, led some governments to attempt two counter-measures: (1) the encouragement of research and development of more economical, quieter, safer, and less polluting, vehicles, and (2) the positive stimulation of public transport, including unconventional para-transit systems.

Other modes of transport also grew in volume during the 1970s. Railway freight transport increased in both developed and developing countries, although unevenly from one to another. Passenger travel by rail followed trends that varied widely from country to country. Trains improved in speed and comfort. Goods transport by pipeline increased considerably (in the United States and Western Europe it rose by 27% between 1970 and 1977). Inland water traffic increased. There were 20% more sea-going ships at the end of the decade than at the beginning, and a 77% increase in tonnage. The size of oil tankers increased dramatically. Between 1970 and 1979, passenger travel by air (measured in passenger kilometres) more than doubled.

Between 1965 and 1975 the total length of motorway roads more than doubled in OECD (Organization for Economic Cooperation and Development) member countries, and the length of other roads increased by some 15%. Substantial areas of farmland were lost as a consequence during 1965–75, while energy consumption by transport doubled in some countries. Road accidents continued to be a serious consequence of transport growth, killing about 250,000 people every year during that decade. Transport remained the largest single source of marine oil-pollution.

All these trends inspired a range of counter-measures, including technical improvement, publicity, training, and measures to control the routing of vehicles, ships, and aircraft, so as to reduce risk and nuisance. One predicted impact that did not, however, create the problems expected at the time of the Stockholm Conference, was that of supersonic aircraft flying in the stratosphere on ozone concentrations—partly because such aircraft did not fly in the expected numbers.

*Changes in Tourism*

The number of international tourist arrivals increased by over 100 million during the past decade, from 180 millions in 1971 to over 280 millions in 1980 (Fig. 5). Over 72% of this travel was in Europe, 18% in the Americas, and 5% in Eastern Asia and the Pacific. Domestic tourism (i.e. within countries) was also of great importance, having a volume 4 times as great as international travel.

These trends were well documented for the earlier years by the World Tourism Organisation (WTO, 1977).

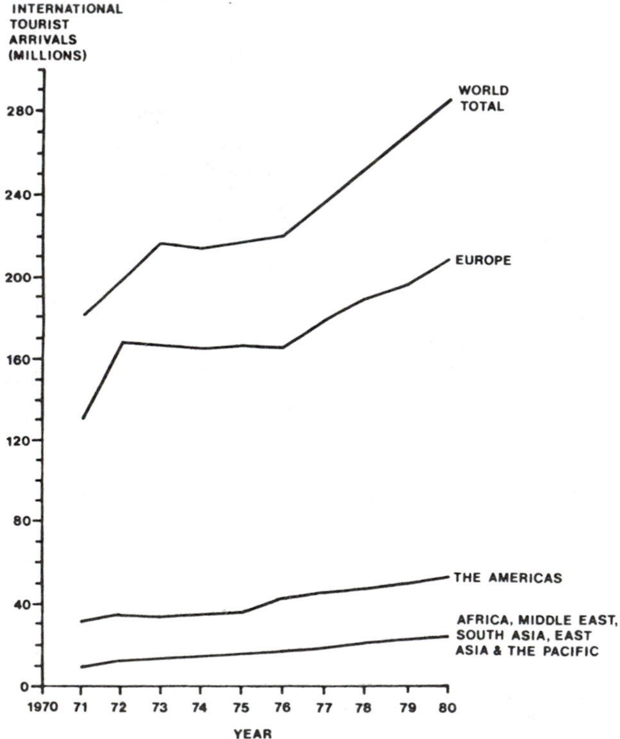


Fig. 5. *The growth of international tourism between 1971 and 1980. Data from World Tourism Organization.*

The social and environmental consequences are much less well recorded. It is clear that, by the end of the 1970s, international travel had become a major source of income, although part of the foreign exchange which it brought to developing countries went abroad again to pay for the goods and services used by tourists. Similarly, while tourism created jobs, the countries concerned paid heavily to support the hotels and other facilities involved, often diverting resources from the improvement of local settlements.

The social effects of tourism were mixed. On one hand it could re-vitalize local crafts but, on the other, it would stimulate mass-production of articles of low artistic merit and thus dislocate local culture. Similarly, the environmental effects ranged from a valuable stimulus in the establishment of natural parks and the protection of wild landscapes and historic buildings, to impacts that damaged fragile habitats on islands, coastlands, and mountains, through pollution, litter, erosion, and fires. Congestion with tourist traffic is also a problem in many old or historic towns and villages.

In many areas the limits of desirable tourist influx were clearly being reached during the 1970s, and the need for a balance between tourism and the environment was emphasized at the international level in the Manila Declaration (1980). This stressed the need for the type and scale of tourist development to be related to the carrying capacity of the environment.

### *Changes in Environmental Education and Public Understanding*

Environmental education was a major topic at the Stockholm Conference, and was taken up as a theme in the UNEP programme. Together with UNESCO, UNEP organized the First Intergovernmental Conference on Environmental Education, which was held at Tbilisi (USSR) in October 1977, where broad guidelines were laid down.

There are only limited data on trends in environmental education during the 1970s, but it is clear that the subject has grown and developed in many countries. By 1973 there were some 1,100 programmes under the broad heading of environmental education in *ca* 750 university and other post-secondary institutions in 70 countries. At the start of the decade such courses were not popular (attracting less than 2% of students in OECD countries). In the USA only 4 Universities offered a degree in environmental engineering before 1970; but by 1978–9 the number had risen to 27, and undergraduate enrolment climbed from under 300 students per year to more than 1,000 in 1976–7. In developing countries the trend was less marked—partly because of educational traditions, and partly because there were limited prospects for employment.

The past decade witnessed a dramatic growth in the number of scientific journals dealing with environmental and conservation topics. In 1969–70 there were 13 broadly environmental periodicals and 164 on conservation topics: in 1979–80 the totals were 686 and 486, respectively. Coverage by the popular media also expanded.

While in the 1950s 'environment' was a virtually unknown term, in the 1970s it took up hundreds of column-inches (Fig. 6). Between 1957–9 and 1967–9 the number of environmentally orientated articles appearing annually in periodicals in the United States rose from 68 to 226, while the *New York Times* published over 1,700 environmental articles in 1970 as against under 200 in 1960. Since then the coverage has fallen back somewhat, to below 700 in 1979,† although individual incidents (such as that at the Three Mile Island nuclear power-plant) provoked dramatic upsurges. Such dramatic events have triggered public concern throughout the period, but during the decade a trend in public interest was apparent. While concern over pollution remained, there was a mounting interest in the scarcity of natural resources, energy–environment relationships, and the relationship between environment, development, and conservation. Environmental groups led campaigns that affected the decision-making process in several developed countries—for example over the siting of nuclear power-stations, the chemical spraying of forests, and the addition of lead to gasoline.

### *Peace, Security, and the Environment*

Wars, and the threat of wars, are of concern in an environmental context for two main reasons. First, they cause environmental damage and disrupt human com-

†In answer to our query as to what has happened since, Professor White replied (*in litt.* postmarked 22 March 1982): 'An examination of the *New York Times* for 1981 shows that, by comparison with 1979, the number of articles on the environment had doubled, that the number on air pollution was about the same, and the number on water pollution had decreased to about two-thirds [of those of] the earlier year.'—Ed.

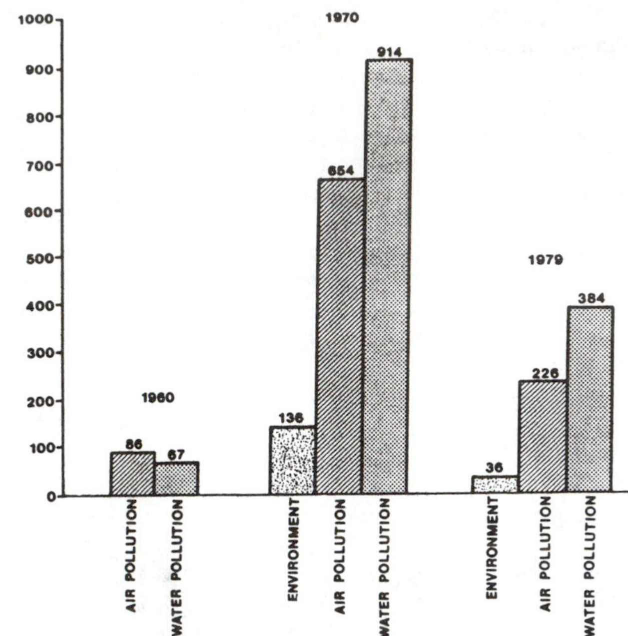


Fig. 6. *Number of* New York Times *articles on environmental topics in 1960, 1970, and 1979. Based on unpublished data collected by Gilbert F. White.* †

munities, leaving behind relics which are themselves capable of causing loss of life and interference with the environment. During the 1970s the cost, in human and material terms, of dealing with the relics of past wars received increasing attention. Surveys revealed that one nation alone had removed 58.5 million mines from 2.5 million square kilometres of its territory after World War II, while another suffered 3,800 fatalities and 8,000 injuries due to such relics. In South-east Asia, social organization, agriculture, and forest vegetation, were severely damaged by recent wars from which they are recovering only slowly.

Secondly, by 1980 global military expenditure was 30 times as great as at the turn of the century and 4 times as great, in constant money terms, as in 1964. The rate of increase was a little less during the 1970s than in previous decades, and it grew less rapidly than GNP; but the share attributable to the developing countries grew particularly rapidly, and the world armaments trade expanded by about 15% per annum between 1970 and 1975. The oceans, the atmosphere, and space, were all increasingly militarized. Despite widespread condemnation of nuclear weapons, 469 devices were exploded between 1970 and 1980, 41 of them in the atmosphere (SIPRI, 1981).

About 40% of the world's total expenditure on scientific research and development lay in the military sphere during this past decade, when the destructive capacity of weapons grew, and the possibility of using environmental modification as a weapon increased (although an international convention prohibiting such actions was agreed to). Despite a large and growing volume of international initiatives to secure disarmament, the harsh reality was that, during the decade, military investment was drawing away large volumes of scarce skills and finance which might have been employed for development purposes, and the arms race became a more and more severe threat to humanity and the environment.

## TRENDS IN ATTITUDE TO THE ENVIRONMENT

Four trends have been of major importance in the recent evolution of the environmental movement. The first is concerned with the process of development of environmental thinking, and the way in which what were originally separate scientific and conservation initiatives have come together and have become popularized. The second concerns the rapid growth in available data concerning the environment, but continuing difficulties in selection of those components that are most relevant to decision. The third is a rapid advance recently in our understanding of the nature of environmental systems, while the fourth is concerned with changes in our understanding of how the human social and political system responds to environmental issues. These four aspects will now be treated in turn:

### *1. Development of the Environmental Movement*

The international environmental movement that generated the Stockholm Conference has a long history and no sharp beginning. In part it can be traced to the great voyages of discovery and exploration that made people aware of the shape of the world and the diversity of its lands and waters, soils and rocks, floras and vegetation, faunas and cultures. This process of dis-

†*See also* footnote on preceding page.

covery merged imperceptibly into international scientific programmes, which gained strength during the 1960s and 1970s. The International Geophysical Year (IGY) of 1957–8 demonstrated that world-wide scientific programmes could be put into operation, and because they involve coordination of standardized observation on an agreed pattern of sampling in space and time, could yield information that was not previously available about the properties of the atmosphere, the magnetosphere, and the ionosphere. The IGY also showed that international cooperation was uniquely well equipped to provide information about remote areas, such as the Antarctic, where the research programmes to which it gave impetus have continued to evolve.

The IGY was moreover the direct inspiration of the International Biological Programme (IBP), which ran mainly between 1964 and 1974 and had as its basic theme 'biological productivity and human welfare'. These programmes in turn led on to others, coordinated under scientific and special committees of the International Council of Scientific Unions (ICSU), or under United Nations agencies such as UNESCO, whose Man and the Biosphere Programme (MAB) continued to develop some of the basic studies begun in IBP. Much of the understanding of world environmental systems which underlies our review of events during the 1970s, stems from these programmes or the somewhat similar operations of FAO, WHO, WMO, and other UN bodies.

The conservation movement also has a long history. Societies concerned with the efficient management of natural resources (and especially the improvement of agriculture) have existed in developed countries for at least 200 years. In the nineteenth century a parallel movement concerned with the preservation of natural habitats and human monuments sprang up, linked in many respects to popular societies for the study of natural history.

In the twentieth century the study of natural history and the concern for the conservation of places rich in wildlife and natural beauty grew together, especially under the guidance of professional ecologists. From the outset, therefore, environmental conservation developed with a scientific base, and strong links with the academic world. Moreover, because scientists and naturalists had always been concerned with the world-wide distribution of species and the assemblages of plants and animals that they composed, the conservation movement also early acquired a global dimension. During the twentieth century, appreciation of the environment grew in many countries outside Europe and North America. The establishment of the International Union for Conservation of Nature and Natural Resources (IUCN, formerly UIPN) in 1948 was one outcome of these developments.

These processes led, in recent decades, to the development of a more sophisticated approach in the environmental movement. Not only was this last concerned with literally all aspects of the natural environment—on land, in the oceans, in the lakes and rivers, and in the atmosphere and space—but it also turned towards the human situation. Thus it became concerned with the relationship between man-made and natural environments, and with the essential need to provide human communities with a quality of living which allowed individuals to realize their full potential. Whereas the earlier 'Protection of Nature' movement was concerned with safeguarding certain natural resources and particularly threatened species against over-use or destructive change, the newer environmental movement, while including this, went beyond it.* It became concerned with a much wider range of environmental phenomena on the grounds that the violation of ecological principles was reaching the point where, at best, the quality of life was threatened and, at worst, the long-term survival of humanity could be imperilled.

This process involved the public more directly in the 1960s than at any previous time. Their involvement was stimulated by certain widely publicized and disturbing environmental incidents such as the air pollution episodes in London between 1952 and 1966, the fatal outbreaks of mercury poisoning in Japan between 1953 and 1965, the decline in aquatic life and changes in freshwater ecology in some of the North American Great Lakes, the deaths of many birds as an unexpected side-effect of organochlorine pesticides, and the massive oil pollution from the wreck of the *Torrey Canyon* in 1966. These widely publicized events caused many people in developed countries to fear that pollution was indeed already jeopardizing the human future. The rise in coverage of environmental matters in the popular media, described in a preceding section, brought environmental matters to the attention of an ever-widening public, and directly stimulated the development of

---

*Hence the change of name after some years from the 'International Union for Protection of Nature' (IUPN) to the present style of 'International Union for Conservation of Nature and Natural Resources' (IUCN).—Ed.

political environmental movements including 'ecology parties'.

The popularization of the environmental movement did, however, bring with it a major change in philosophy. Scientific environmentalists have, like other scientists, been concerned to measure the scale and the rate of change of environmental phenomena. They have become aware of the imprecision of many measurements, and of the uncertainty that resides in the behaviour of many natural systems. They have made predictions with due caution, usually with some attempt to indicate the likely margins of error. The results, however imperfecly, have been that they have engaged in a process of risk estimation, namely an attempt to state objectively the probability of an event of given magnitude.

Meanwhile the popular environmental movement has been concerned with the impact of such possible or probable events on the community at large, and people's reactions to the stated problems have varied according to their own perceptions of their relative significance to them as individuals or groups. This perception of risk may vary to a considerable degree from the order of rank in which a scientist may place the phenomena he is measuring. The popularization of the environmental movement has led to a decline in the scientific precision with which some issues are debated by the community at large, and the scientific community has not yet been successful in interpreting environmental phenomena in a fashion which ensures that the wider community acquires a precise understanding of the phenomena of, and the risks attached to, environmental change.

## *2. Changes in the Information Base*

The 1960s and 1970s witnessed what has been called a data explosion (Brown, 1971). The number of scientific journals and other publications expanded rapidly in this period, as has already been explained. There were also rapid advances in the development of instruments that would record environmental parameters automatically and with a high frequency of sampling. The precision of analytical instruments made it possible to detect substances in the environment at extremely low concentrations, thereby demonstrating the world-wide distribution of many contaminants such as DDT and other organochlorine pesticides. Satellites brought the prospect of more efficient surveillance of changes in the atmosphere and on the ground. There were rapid advances in the capacity of computers, and rapid reductions in the costs of electronic data-storage and retrieval systems. More scientists came to work in areas of environmental research than at any time in the past. The result was an almost unlimited capacity to acquire data about the world environment.

Unfortunately this expansion was not accompanied by the steps which are essential to make much of that volume of data meaningful as information. Instead often different methods, with different resolutions, were adopted in different places, the sampling systems often not being designed in a fashion that allowed the results to be compared. Where samples were interchanged between analytical laboratories, variations in the precision of analysis were commonly revealed, and one consequence has been the rejection by some scientists of large quantities of earlier data: for example, analysis of trace substances in marine life acquired before the mid-1970s are now commonly regarded as insufficiently precise for use in making comparisons or defining trends.

One of the major conclusions of the 'State of the Environment' Report (Holdgate *et al.*, in press) is that, although large volumes of numerical information are available, there are still some startling gaps and an especial lack of reliable quantitative information about the environment in the developing world. This makes projections about many aspects of the changing state of the world extremely difficult if not impossible, because many of the statements in the literature are based only on the scantiest evidence about what, in fact, has been happening.

On the world scale, there are measurements of major meteorological elements and of carbon dioxide and ozone concentrations in the atmosphere. There are regional records of sulphur oxide emission and a large number of local measurements of the concentration of oxides of nitrogen, particulates (especially metals), and the oxidants derived from photochemical reactions in urban areas. There are some good regional monitoring programmes in the seas. There are reasonably complete records of the surface distribution and flow of fresh water. On the other hand it is not possible to say with certainty whether world-wide climatic warming has begun, or is likely to begin.

Some specialists believe that a depletion in stratospheric ozone may be about to become detectable, but confident judgements of possible human influence on this parameter remain difficult. The possible influence of rising carbon dioxide concentrations on precipitation patterns (which are likely to be more important than changes in temperature as such) remains a matter of speculation. It is not possible to make any general statement about the pollution of the oceans—except that, assuming that the most heavily polluted seas are those which are currently being monitored, and recognizing that their fish faunas and marine ecosystems do not show any signs of marked change, there are no grounds for suspecting that the world oceans as a whole are currently endangered.

On land, the extent and processes of soil transformation are imperfectly understood, and there is conflicting evidence over the scale and rate of conversion of forests even though in the public's mind this is one of the dominant world environmental crises. There is rather better documentation about some of the major human activities that affect the environment, because food production and fishery statistics are extensive, and so are many industrial records (including fertilizer production and use).

Demographic information about human populations, including birth- and death-rates, is fairly complete, although the lack of compulsory registration in some developing countries does reduce the quality of health statistics. There are many reasons for this situation: one is a lack of agreement on the parameters whose monitoring should receive priority, while another is the lack of resources—including trained personnel—especially in developing countries. The important point is that the task which the world community set itself in Stockholm in 1972, namely the development of a capability for assessing the state of the world environment and the rate of change in major and important components of it, is still not achieved, and may not be for some years to come.

### 3. *Changes in Understanding of Environmental Systems*

One major advance in the past decade has been in the understanding of environmental systems. Their essential unity has become apparent especially through the study of biogeochemical cycles, despite the great geographical and ecological diversity of the world. A major review by SCOPE (Bolin *et al.*, in press) shows how closely the global cycles of carbon, nitrogen, phosphorus, and sulphur, are interlinked, both on land and in the sea. The study has also revealed that human actions have been influencing all of them on an increasing scale. In 1980, the annual release of carbon dioxide into the atmosphere by human activities was about 10% of the amount that was being used by green plants in photosynthesis. In the past century about 10% of the land surface has been transformed into agricultural land, and this has caused a major movement of nitrogen compounds and other nutrients from the soils to the rivers and lakes, and ultimately into the sea. In 1980 the formation of nitric oxides and nitrate in the processes of combustion and fertilizer manufacture was about half of what the biosphere produced naturally. More sulphur dioxide was entering the atmosphere, mainly from fossil-fuel combustion, than was being exchanged naturally between the air, land, and oceans, as a result of the decay of dead organic matter.

All these human impacts were tending to grow in scale, and therefore to bulk larger and larger in proportion to the natural components of the cycles. It also became clear that the modification of any one cycle by Man would affect the others. On a more subtle scale, it further became apparent that there could be interlinkages between phenomena in parts of the world that appeared dramatically remote from one another. Changes for example in the sea-surface temperatures in areas of the northeastern Pacific were shown to affect the weather patterns in areas as remote as Western Europe. The acidification of lakes and rivers could be traced to the release of sulphate and nitrate through human activities at distances up to thousands of kilometres away. Such manifestations gave strength to the emphasis in the Declaration adopted by the Stockholm Conference (UNEP, 1978) on the need for all nations to be mindful of the impact of activities within their jurisdiction on the environment of other states, and on all areas beyond the limits of their national jurisdiction.

A further advance occurred in the modelling of environmental systems, and in appreciation of the uncertainty which is inherent in many of them. There are components in many complex ecosystems whose behaviour is best treated as random. The result is to introduce substantial uncertainty into the prediction of their likely future behaviour, and of their response to stress by Man. However, there is an indication that the so-called 'balance of Nature' is not a simple dynamic equilibrium, but that many systems have relatively stable states, and are resilient within certain outer limits of stress. But if they are perturbed beyond those limits, they may change rapidly to a quite different but possibly equally resilient condition: the reversal of the process may either be virtually impossible or take a very long time. Knowledge of the interactions between populations of different species, including predators and prey, parasites and hosts, or valued species and the so-called pests that compete with or prey upon them, made substantial advances and provided a logical basis for some forms of pest control.

*4. Changes in Human Responses to Environment*

The 1970s were characterized by an abundance of high-level meetings, colloquia, scientific symposia, and other gatherings, at which information about the environment was exchanged, problems were analysed, and action plans duly prepared. A number of these action plans were soundly based scientifically, to the limits of the data-base available at the time (the Plan of Action to Combat Desertification was one such case). Yet the past decade also demonstrated the width of the 'application gap' in many of these areas. The knowledge available to the scientific community and to Governments was only imperfectly applied in many areas by the operators changing the environment in the field. One reason for this gap was the lack of training and education within the countries, particularly in the Third World. Another reason was the inability of the political system to cope with social dislocations.

The so-called 'green revolution' provided many illustrations of the rapid introduction of higher-yielding crop varieties in areas where the farming community was organized and trained so as to understand the requirements of these species, fertilizers were available, and appropriate cultivation methods could be introduced. Where these prerequisites were not available, the mere provision of the cultivars themselves was less effective (Hutchinson, 1978), and in some areas the gain in production was accompanied by severe social disruption.

Rather similarly, the 1970s revealed that industrial processes which were not damaging to the environment could be evolved at a cost that was often not significantly greater than that of traditional but more polluting processes, but their rate of introduction and the distribution of their costs were very much dependent upon national administrative, regulatory, and economic, circumstances. Much the same applied in the public health field: the eradication of smallpox was a major success, and WHO has developed plans for the elimination or substantial reduction in six other major infectious diseases that kill millions of children annually in the Third World: there is little doubt that this programme is entirely feasible with today's scientific knowledge, if the necessary resources of money and manpower are committed to it.

The experience of the 1970s suggests that international cooperation is relatively easy where it is concerned with the acquisition of information and its exchange between scientists and Governments. It is more difficult where the need goes beyond this collection, analysis, and dissemination, of information, and involves the joint management of commercially important resources—especially where national interests conflict. It is also more difficult when effective action demands changes in traditional values and approaches. Many communities are understandably resistant to the importation of ideas that appear to them inseparable from alien value-systems, and there are many instances of where the uncritical transfer of technology to developing countries, and especially from temperate to tropical regions, have in themselves been disastrous. The critical analysis of all the relevant components of the socio-economic system is therefore crucial to the planning of action for the environment.

In a sense the world's traditional institutions were themselves under trial at the end of the decade. At the Stockholm Conference it was generally assumed that the world system of national governments, regional groupings, and international agencies, had the power to take effective action, and that the limiting factors were scientific and economic. But by the early 1980s there was less confidence in the working of national and international managerial systems, and in the effectiveness with which international debates lead to action to improve the well-being of people 'on the ground'. On a wider horizon, there were arguments over the adequacy of international economic arrangements and a demand for new approaches to the so-called north–south gap (Brandt Commission, 1980). The capacities of existing political and economic systems to bring about the necessary social and environmental developments were increasingly challenged, especially following the energy crises of the early 1970s. At the same time, by 1982 it seemed clear that speedy change in these international systems or their national counterparts was unlikely, and that the solutions would have to be found by progressive adjustment of the existing machinery if the many pressing problems of the 1980s were to be dealt with effectively.

## CONCLUSIONS

At the end of this review of the state of the environment in the decade up to 1982, we find ourselves with a small number of general conclusions, as follows.

The first is that the proportionate influence of Man

as against natural processes in the cycles of The Biosphere is increasing steadily. There is no evidence at the time of writing that this changing pattern threatens imminent collapse. But it is equally clear that the resilience of natural systems could be threatened, and at the very least (if human actions become predominant in any process), the need for human management to regulate those actions will become inescapable. In many developed regions human activities are now releasing as much sulphur to The Biosphere as natural cycles do, and this may be the first area where concerted action in the 1980s becomes necessary.

A second conclusion must be that, at the present time, the short-term productivity of The Biosphere has not seriously been reduced by human activities. Agricultural and fishery yields increased in general during the 1970s, and there are good reasons for believing that they can continue to do so. Although there have been substantial losses of soil through erosion and desertification in many areas, others have shown success in reversing these processes, and there is little doubt that the scientific knowledge for such reversal is now available. However, the extent to which long-term productivity of soil is being affected by modern agricultural practices is not clear. The pollution of the seas has not yet reduced their productivity except in a few localized areas, and there are encouraging signs, for example in regional conventions and action under the Regional Seas Programme of UNEP, that, in the long term, damaging marine pollution on any extensive scale could be avoided. The development of low-waste or non-waste technologies, and of industrial processes that reduce the generation of pollution, offer grounds for optimism.

Thirdly, scientific insights into the functioning of complex environmental systems will undoubtedly grow rapidly in the next ten years and offer real grounds for hope. It is necessary to move from an essentially responsive role to a predictive one, if socio-economic systems are to be adapted in order to avoid wasteful misuse of environmental resources. The scientific components essential for this change are beginning to become available.

Fourth and perhaps of overwhelming importance, education, training, and the development of social strategies for application of available scientific knowledge practically on the ground and especially in developing regions, is of paramount importance—and is indeed probably the major environmental need in the world at the present time. Many people are now living in squalid settlements, with poor facilities and a lack of essential health-care and nutrition, who would not need to suffer these disamenities if the socio-economic structure were better organized.

## ACKNOWLEDGEMENTS

We wish to express our thanks to Dr Mostafa K. Tolba, Executive Director of UNEP, who encouraged us to write this paper, which draws particularly on the materials assembled for the major UNEP Report on *The World Environment, 1972–1982* (Holdgate *et al.*, in press).

## SUMMARY

The United Nations General Assembly has instructed the Governing Council of the United Nations Environment Programme to keep the world environmental situation under review. In 1982, 10 years after the UN Conference on the Human Environment at Stockholm, the first comprehensive report on the state of the global environment is being published. The present paper, by the Editors of that Report, summarizes its main findings. It first reviews changes in the sectors of The Biosphere (while recognizing that the interlinkages between them have been stressed increasingly during the past decade), before turning to the human components of the total Man–environment system.

In the atmosphere, rising carbon dioxide concentrations, acidification of rain and snow in or by industrial regions, and stratospheric ozone depletion, remain the chief concerns, although the last has not yet been demonstrated instrumentally. In the oceans, pollution (including oil) has not been shown to have more than a local impact on ecosystems, and overall fishery yields have continued to rise slowly and erratically despite some overexploitation. The world's freshwater resources are better known than in 1970, and pollution control and the prevention of problems in irrigated agriculture have advanced; but the targets of the Drinking Water and Sanitation Decade appear less attainable as time passes. Mineral production rose without a concomitant increase in environmental damage. Changes in terrestrial life—especially loss of tropical forests—were the subject of widely varying estimates. Food production rose, but fell short of needs in many areas, while desertification, waterlogging, salinization, pest-resistance, post-harvest crop-losses, and the side-effects of agricultural chemicals, remained serious problems.

The dominance of the human element in the Man—environment system was increasingly recognized during the decade. Human population growth slowed somewhat, except in Africa, although the world total passed 4,400 millions in 1980. The cities of the developing world expanded rapidly, outstripping public services and threatening new problems. In the Third World, infectious and parasitic diseases remained major killers, whereas hypertension, coronary heart disease, and cancers—some due to self-inflicted influence—dominated the statistics in developed nations: environmental factors remained important in both. The 1970s showed that industrial growth could occur without environmental damage or unacceptable cost. The energy crisis of 1974 had a serious impact on developing countries with strategies based on cheap oil, and firewood shortages led to severe environmental problems there also: in contrast, many developed countries were able to adjust their energy plans with only moderate difficulty.

Transport and international tourism grew dramatically during the decade, consuming energy and land, and inspiring countermeasures to curb pollution, increase safety, and avoid social and environmental disturbances in areas that were frequented by many visitors. Environmental education schemes expanded—especially in developed countries, where the coverage of environmental issues in popular media grew dramatically between 1960 and 1970, falling back subsequently. The environmental impact of past wars and increasing military preparations caused concern, and the arms race continued to absorb resources that developing countries could ill afford.

Reviewing the decade, four dominant trends can be recognized. First, scientific and popular interest in environmental protection have come together to form a new kind of conservation movement. Second, there has been a data explosion in the environmental field, but much of the information is of limited value in assessing trends or as a foundation for decisions and actions. Third, new understanding of the structure and functioning of environmental systems offers a prospect of more reliable planning. Fourth and finally, it has become apparent that the lack of social organization, education, training, and political will, are commonly the limiting factors in environmental improvement, rather than a shortage of scientific knowledge.

## REFERENCES


BOLIN, B., DEGENS, E. T., KEMPE, S. & KETNER, P. (Eds) (1979). *The Global Carbon Cycle.* (SCOPE 13.) John Wiley & Sons, Chichester—New York—Brisbane—Toronto: xxxv + 491 pp., illustr.

BOLIN, B., CRUTZEN, A. P. J., VITOUSEK, P. M., WOODMANSEE, R. G., GOLDBERG, E. D. & COOK, R. B. (in press). *The Biogeochemical Cycles and Their Interactions.* (SCOPE Report No. 24.) John Wiley & Sons, Chichester, England, UK.

BRANDT COMMISSION (1980). *North:South:A Programme for Survival.* Pan Books, London & Sydney; 304 pp.

BROWN, Harrison S. (1971). Scientific information today—A scientist's view. Pp. 36–40 in *UNISIST: Intergovernmental Conference for the Establishment of a World Science Information System (Final Report).* UNESCO, Paris, France: 60 pp.

FAO (1979*a*). *1978 FAO Fertilizer Yearbook.* FAO Statistics Series No. 33). Food and Agricultural Organization of the United Nations, Rome, Italy: xix + 115 pp.

FAO (1979*b*). *1978 Yearbook of Fishery Statistics: Catches and Landings.* Food and Agriculture Organization of the United Nations, Rome, Italy: x + 372 pp.

FAO (1979*c*). *Agriculture Toward 2000.* Document C79/24, Food and Agriculture Organization of the United Nations, Rome, Italy: [not available for checking].

FAO (1980). *1979 FAO Production Yearbook, Vol. 33.* Food and Agriculture Organization of the United Nations, Rome, Italy: v + 309 pp.

FRANCE (1978). *Clean Factories.* Ministère de l'Environnement et du Cadre de Vie, Paris, France: [not available for checking].

HOLDGATE, M. W., KASSAS, M. & WHITE, G. F. (Ed.) (in press). *The World Environment, 1972–1982: A Report by the United Nations Environment Programme.* Tycooly International Publishing, Dublin, Ireland.

HUTCHINSON, J. B. (1978). The Indian achievement. In *Conservation and Agriculture* (Ed. J. G. Hawkes). Duckworth, London, England, UK: xi + 284 pp., illustr.

ICES (1978). *Input of Pollutants to the Oslo Commission Area.* (Co-operative Research Report No. 77.) International Council for the Exploration of the Seas, Copenhagen, Denmark: [not available for checking].

ICES (1980). *A Review of Past and Present Measurements of Selected Trace Metals in Sea Water in the Oslo Commission and ICNAF (NAFO) Areas.* (Co-operative Research Report No. 97.) International Council for the Exploration of the Seas, Copenhagen, Denmark: [not available for checking].

IMCO (1979). Where the oil was spilled: 1962–1978. *IMCO News* (London), 1 for 1979, pp. 12–3, map. Intergovernmental Maritime Consultative Organization [Renamed International Maritime Organization as from 22 May 1982.]

IUCN/UNEP/WWF (1980). *World Conservation Strategy: Living Resource Conservation for Sustainable Development.* International Union for Conservation of Nature and Natural Resources, United Nations Environment Programme, and World Wildlife Fund, 1196 Gland, Switzerland: Special pack of brochures, etc., totalling *ca* 50 pages.

KEELING, Charles D. & BACASTOW, Robert B. (1977). Impacts of industrial gases on climate. Pp. 72–95 in *Energy and Climate.* (Studies in Geophysics.) National Academy of Sciences, Washington, DC, USA: xiv + 158, illustr.

MANILA DECLARATION (1980). *World Tourism Conference, 1980.* World Tourism Organisation, Madrid, Spain: [not available for checking].

SIPRI (1981). *World Armaments and Disarmaments.* (SIPRI Yearbook, 1981.) Taylor & Francis, London, England, UK: xxvii + 518 pp., illustr.

UN (1979*a*). *World Population Trends and Prospects by Country 1950–2000: Summary Report of the 1978 Assessment.* United Nations, New York, NY, USA: v + 98, mimeogr.

UN (1979*b*). *An Overview of the World Energy Situation: Report of The Secretary-General.* United Nations Economic and Social Council, New York, NY, USA: 37 pp., mimeogr.

UN (1980*a*). *Study on Financing the UN Plan of Action to Combat Desertification.* (A/35/396.) United Nations, New York, NY, USA: [not available for checking].

UN (1980*b*). *Patterns of Urban and Rural Population Growth.* (Department of International Economic and Social Affairs, Population Studies, No. 68.) United Nations, New York, NY, USA: ix + 175 pp.

UN (1981). *World Population Prospects as Assessed in 1980.* United Nations Department of International Economic and Social Affairs, New York, NY, USA: [not available for checking].

UNEP (1978). *Compendium of Legislative Authority.* United Nations Environment Programme, Nairobi, Kenya, and Pergamon Press, Oxford–New York–Toronto–etc.: 287 pp.

UNEP (1981*a*). *Co-ordinated Mediterranean Pollution Monitoring Programme (MEDPOL), Part 1.* (Summary Scientific Report, February 1975–June 1980.) United Nations Environment Programme Document UNEP/3/WG.46/3, Geneva, Switzerland: [ii] + 166 pp., illustr., mimeogr.

UNEP (1981*b*). *The State of Marine Pollution in the Wider Caribbean Region.* United Nations Environment Programme, Paper UNEP/CEPAL/WG.48/INF5, Geneva, Switzerland: i + 31 pp., mimeogr.

UNWC (1978). *Proceedings of the United Nations Water Conference.* (Mar del Plato, Argentina, March 1977.) Pergamon Press Oxford–New York–Toronto, etc.: Part 1 pp. xxi + 369, Part 2 pp. xxi + 371–905, Part 3 pp. xxi + 907–1663, Part 4 pp. xxi + 1665–2646.

USDA (1980). *World Food Production.* US Department of Agriculture, Washington, DC, USA: [not available for checking].

USSR (1978). *World Water Balance and Water Resources of the Earth.* USSR Committee for the International Hydrological Decade, 1974. (English translation published by UNESCO, 7 Place de Fontenay, 75700 Paris, France: 663 pp., illustr., 1978.)

WALLÉN, C. C. (1980). *A Preliminary Evaluation of WMO–UNEP Precipitation Chemistry Data.* (MARC Report No. 22.) GEMS and MARC, Chelsea College, University of London, London, England, UK: [iii+] 17 pp., illustr.

WHO (1980). *Sixth Report on the World Health Situation, 1973–1977: Part 1, Global Analysis.* World Health Organization, Geneva, Switzerland: viii + 290 pp., illustr.

WTO (1977). *Tourism Compendium.* World Tourism Organisation, Madrid, Spain: vii + 192 pp.

CHAPTER 9

# An Inquiry into the State of the Earth



M. MITCHELL WALDROP

Quietly, but ever more forcefully, momentum is building for the largest cooperative endeavor in the history of science: a study of the earth and its environs as an integrated whole.

The International Geosphere-Biosphere Program (IGBP), as it is known, would encompass the global climate, the biosphere, and the biogeochemical cycles of all the major nutrients. It might well include the pulsations of the sun and the tectonic processes in the core of the earth. It would take data from satellites in orbit and instruments on the ground. It would involve a sharing of effort among scientists from every part of the world. And it would somehow have to be sustained for decades.

The obstacles, both economic and political, are clearly horrendous. But the enthusiasm within the scientific community is growing nonetheless. The International Council of Scientific Unions, meeting in Ottawa, has just endorsed a 2-year study to draw up a detailed plan for the IGBP. And in the United States, the National Academy of Sciences is also formulating a detailed plan in conjunction with the National Aeronautics and Space Administration (NASA), the National Science Foundation (NSF), the National Oceanic and Atmospheric Administration (NOAA), and a host of other agencies. By 1986, the IGBP could be ready to move.

There is nothing new about big, international programs, of course. The IGBP is very much in the tradition of the International Geophysical Year (IGY) of 1957–58, as well as such modern heirs of the IGY as the World Climate Program or the International Biological Program.

During the last decade, however, issues such as ozone depletion, carbon dioxide buildup, and now acid rain have dramatized the need for a truly global program. Humans are beginning to perturb the climate and the biosphere on a planetary scale, and yet there are enormous gaps in our knowledge of the system: governments have been faced with making expensive and controversial policy decisions on the basis of scientific guesswork.

"The picture now is full of programs that are competitive with each other and ad hoc," says Herbert Friedman, chairman of the National Academy's Commission on Physical Sciences, Mathematics, and Natural Resources, and one of the originators of the IGBP idea. The existing programs also tend to be of limited duration, even though many global processes take place on a time scale of decades or centuries. "You cannot address these questions without 10 to 20 years of rigorous research," he says.

So the IGBP would both complement the existing international programs and go beyond them, says Friedman. Instead of following the traditional division of the earth into atmosphere, lithosphere, and oceans, the IGBP will try to look at processes in a more holistic framework. In particular, it will lay a much greater stress on biology and chemistry, especially the biogeochemical cycles of such key nutrients as carbon, nitrogen, sulfur, and phosphorus.

As an example, consider the problem of methane, $CH_4$, a natural product of bacterial fermentation and the digestive processes of certain ruminants. Methane is a trace gas in the atmosphere, with a concentration of about two parts per million. During the last decade, however, its concentration has been rising at a comparatively enormous rate, roughly 1 to 2 percent per year. No one knows why. A larger population of cattle, perhaps? Increased cultivation of crops such as rice, which grow in waterlogged fields?

The rise has to be understood, however, because methane, like carbon dioxide, is a greenhouse gas and could thus have a significant effect on the climate. In the stratosphere, methane interacts

---

Reprinted from *Science 226:* 33–35 (1984). 

M. Mitchell Waldrop is a member of the editorial staff of *Science.*

strongly with chlorine radicals liberated from halocarbons and thus has an indirect effect on stratospheric ozone. In the troposphere, it is a controlling factor in the concentration of the hydroxyl radical, OH, which is itself a key to such things as smog and ozone formation.

A second reason for the sudden interest in an IGBP is technological: rapid advances in computers and the relative maturity of remote sensing have just begun to make the effort possible.

Only in the last decade, for example, have the sensors been available to give synoptic, large-scale views of the earth from space. NASA's Landsat series, begun in 1972, pioneered in geological surveys and the monitoring of crops and snow cover. The infrared instrument on NOAA's polar-orbiting weather satellites is being used to compile weekly maps of a "vegetation index," which dramatically illustrate the march of the seasons across the continents. In 1978 NASA's Seasat measured winds and wave heights over most of the world's oceans. An Upper Atmosphere Research Satellite, scheduled for the late 1980's, will monitor the chemistry and physics of the stratosphere and mesophere. And the list goes on.

Meanwhile, computers are critical for testing geophysical theories with ever more sophisticated and complex models, and also for handling the shear mass of data. The Landsat-5 satellite is already threatening to swamp the available computers with some 85 million bits of image data per second; the output of a global, multi-decade IGBP would be staggering. One of the earliest challenges in the program will be to set up an international data archive, using the most advanced computer techniques available.

A final reason that IGBP is gaining ground, and in some ways the most important reason, is that the idea has acquired some strong champions.

At the National Academy of Sciences, the inspiration came in 1983 from Friedman, who had been a participant in the International Geophysical Year of 1957–58. "I thought that, rather than simply celebrate the 25th anniversary of the IGY, why not see what we could do with new techniques?" he says. He was also concerned at the deepening chill in U.S.-Soviet relations: "All the international cooperation and communication that had come out of the IGY seemed to be fading. I thought we might be able to revive some of that."

An early convert was Thomas F. Malone of Resources for the Future, past foreign secretary of the academy and a veteran of the Global Atmospheric Research Program. Like Friedman, Malone was convinced that IGBP would not work as a genuine international effort unless it was coordinated by the International Council of Scientific Unions, ICSU, the same body that had been responsible for the IGY. Malone has thus spent a good part of the past year in airplanes.

"People ask me why I'm doing this at my age," says Malone, "and I tell them it's because I have ten grandchildren and two more on the way." The overseas response to IGBP has been remarkable, he says. "I was particularly heartened by discussions in Beijing and Moscow, where I found a high degree of interest, not only from the academicians, but from the scientists in the trenches."

Still, he says, things do have to move at their own pace. On 25 September, at ICSU's request, Malone and Juan G. Roederer of the University of Alaska conducted a symposium on the IGBP at the ICSU General Assembly in Ottawa. The following day ICSU approved the next phase: a 2-year sequence of workshops that, if successful, will result in a multinational plan for implementing IGBP. If that plan is approved in turn by the next ICSU general assembly in 1986, says Malone, an international coordinating committee will be formed, and the IGBP will move forward in much the same way as the IGY.

Meanwhile, in parallel with the ICSU effort, an academy committee under John A. Eddy of the National Center for Atmospheric Research has begun to design a set of sharply focused initiatives for the IGBP. In effect, Eddy's group is organizing a U.S. national plan. "We're not looking for IGBP to be a big, unwieldy umbrella program," says Friedman. "What we need is a limited set of high-priority thrusts around which a long-range program can grow." The committee is expected to report within the next few months.

About a year before the IGBP, NASA launched an independent, but very similar, "Global Habitability" program. It was originally conceived one Saturday in February 1982 in an all-day bull session between NASA associate administrator Hans Mark, now chancellor of the University of Texas, and Harvard University geophysicists Richard Goody and Michael McElroy. The idea was to take a good look at the factors affecting the earth's ability to support life, primarily the biogeochemical cycles and the climate; as Mark cheerfully admits, he needed a way to bring coherence and focus to the agency's Earth observations program—and to protect it against the Reagan Administration's budget cuts.

Unfortunately, NASA went public with Global Habitability long before it was ready. In August 1982 NASA administrator James M. Beggs presented the still-embryonic concept to UNISPACE, the United Nations space conference in Vienna (*Science*, 21 September 1982, p. 916). Somehow, the message came across as "Here's what NASA's going to do. Join us."

The reviews were scathing. In Vienna, third-world delegates were insulted at the implied condescension, and worse, members of the international science bureaucracy saw Global Habitability as undermining the existing global programs, which had gotten under way only after years of painful effort. Back in Washington, Global Habitability looked like an attempt to grab turf away from NSF, NOAA, and the other science agencies.

The upshot was that NASA hurriedly backed off, and by fall of 1982 Global Habitability had a very low profile.

It was far from dead, however. For one thing, the scientific merits of Global

Habitability were undeniable, and a lot of scientists were endorsing it. For another, there was something about it that had a way of turning skeptics into enthusiasts. "When Global Habitability first crossed our path we all said 'Terrible idea,' " says one convert. "But when you actually looked at it, it was a very good idea—just packaged terribly."

Finally, Global Habitability kept going because NASA officials at the working level kept fighting for it as an *inter*agency program. In particular, they welcomed the National Academy's IGBP initiative. "We're extremely supportive," says Robert Watson of NASA's earth observations division. There are still no formal ties, he adds, "But many of us view IGBP as *the* right vehicle for making Global Habitability an international program."

All this sounds very rosy. But, of course, IGBP still faces a host of unresolved issues. For example:

• Scientific scope. The processes affecting the earth fall naturally into three groups, with relatively weak couplings between them: solid earth geophysics, solar-terrestrial interactions, and climate/biosphere/chemical cycles. The latter group dominates the environment on the 10- to 100-year time scales and it practically defines the NASA Global Habitability program. Many scientists think that IGBP ought to concentrate its efforts in this area, if for no other reason than practicality.

But Friedman, for one, still thinks that the IGBP ought to look at the interactions as broadly as possible, especially since many of those "weak" couplings seem to have profound effects. On a million-year time scale, he points out, the chemistry of the ocean is dominated by hydrothermal action at the mid-ocean ridges. Even on a 100-year time scale, variations in the sunspot cycle are correlated with "little ice ages" on Earth.

• Credibility, especially in the Third World. It is one thing to trade weather data, which everyone does quite freely. But people get very sensitive when someone else starts looking at, say, their crop yields. If IGBP ever suffers the slightest taint of being a front for military intelligence or for economic exploitation, it will be in serious trouble.

On the other hand, says Friedman, one of the attractive things about IGBP is that it gives Third World nations a membership card into top-notch science. "Some awfully important things can be done very simply," he says. "Measuring sea level, for example. It doesn't cost much, but the records have to be accurate and they have to be kept over a very long time, so that we can tell if the sea level is changing."

• Institutional framework. Friedman, a man in his late 60's, talks calmly about a program that will not hit its stride until 1995, and that will not return some of its most interesting data for 20, 40, or 50 years after that. But the question is how to sustain such an effort.

"We see no reason why, if we set up an efficient structure, it can't continue," says Friedman. But what kind of structure? Will it suffice to set a small advisory council to coordinate otherwise independent national efforts, as happened in the IGY? Or will it be necessary to set up a new international body? And not incidentally, who will manage the global data archive?

No one should have any illusion about how long this will take. The Global Atmospheric Research Project was first proposed by President John F. Kennedy in 1961; the first experiment was begun in 1978.

• Government support. IGBP has not yet gained much visibility in policy circles, so it is hard to say how enthusiastic the various national governments will be. On the positive side there is an honest scientific rationale to the program, there is a potential prestige value in participating, and there is the painful fact that in many countries, problems like acid rain and deforestation are of real practical concern.

On the other hand, there is the matter of money. The cost of IGBP is still nebulous, although the proponents have tried to be reassuring on at least one point: IGBP will *not* come on top of the existing international programs. Nor will it come at their expense. "IGBP will be a focus for their interaction," says Francis P. Bretherton of the National Center for Atmospheric Research, chairman of NASA's new Earth Systems Science Committee. "It will complement them."

• The U.S. program. "A U.S. national program doesn't make sense unless it's embedded in a global program," adds Bretherton. "The problems are so big, both conceptually and observationally, that the United States simply cannot do it alone.

"But to get a strong world program you need effective leadership," he adds, "and the United States is the *only* country able to take strong leadership. We have such a large fraction of the world's scientists that if this country can't get its act together, then the rest of the world can't."

Most scientists would probably agree with that statement. In essence, the Eddy committee is trying to formulate such a U.S. program. Congress seems receptive to having the United States take a lead. The science adviser's office at the White House likes the idea. But then, no one has asked for any money yet, either.

In the last analysis, of course, the question is really one of political will. "Scientists are more than willing to join forces," said Friedman in his keynote address at the ICSU meeting. "Governments must be persuaded that it is in their interests to support international cooperation."

# SECTION III
# PROBLEMS OF POLLUTION AND ENVIRONMENTAL DETERIORATION

CHAPTER 10

# Air Pollution, Acid Rain and the Future of Forests



SANDRA POSTEL

Three hundred million years ago, when much of Europe and North America basked in a moist tropical climate, forests of fast-growing trees spread across vast areas of swampy lowlands. Giant "scale-trees" bearing little resemblance to the trees of today stood with stately cordaites, the forerunners of modern-day conifers. After these trees died, the brackish water in which they grew protected them from decay. Time and the increased pressure of sediment helped transform the trees and surrounding vegetation into solid masses of carbon, which now comprise the extensive coal fields of the British Isles, the U.S. Appalachians, the Ruhr Basin of West Germany and Belgium, the Saar-Lorraine Basin of West Germany and France, and the Donets Basin of the Soviet Union.[1]

In an odd twist of fate, humanity's use of the fossilized remains of these arboreal giants now threatens the health and productivity of their modern-day descendants. Over the past decade, scientists have amassed considerable evidence that air pollutants from the combustion of fossil fuels, both oil and coal, and the smelting of metallic ores are undermining sensitive forests and soils. Damage to trees from gaseous sulfur dioxide and ozone is well documented. Recently, acid deposition, more commonly called acid rain, has emerged as a growing threat to forests in sensitive regions. Acid deposition refers to sulfur and nitrogen oxides that are chemically transformed in the atmosphere and fall to earth as acids in rain, snow, or fog, or as dry acid-forming particles. Although acid deposition is now known to have killed fish and plants in hundreds of lakes in Scandinavia and eastern North America, its links to forest damage remain circumstantial. Yet studies of sick and dying trees in Europe and North America make the connection impossible to ignore.



Sandra Postel is Senior Researcher, Worldwatch Institute.

I sincerely thank Dieter Deumling, Andrea Fella, Georg Krause, Richard Rice, H. Steinlin, John Thorner, George Tomlinson, Gregory Wetstone, and Helen Whitney for their helpful comments on early drafts of this manuscript.

Temperate forests have a long history of stress and acidification, a history that offers a critical backdrop for considering new stresses from air pollution and acid rain. Since the end of the last continental glaciation 10,000-15,000 years ago, soils have slowly formed from the sterile layers of gravel, sand, and silt left behind by the retreating ice. Pioneering plants, animals, and microorganisms aided this soil development, helping form an intricate cycle of nutrient uptake and release. Death and decomposition of these inhabitants, then as now, generated acids in the soil. Where acids developed faster than other natural processes could neutralize them, the soils gradually acidified, a process that continues today.

Centuries of human use and abuse of forest ecosystems have added to this natural acidification. Many temperate forests in Europe and North America are now recovering from decades of intense burning, grazing, and timber cutting. The spruce-fir forests of New England and the Adirondacks, for example, had nearly all been clearcut for pulp by the early 1900s. Logging was often followed by burning that destroyed the forest floor. The soil formation processes taking place as these forests recover naturally increase the soil's acidity.[2]

Air pollutants and acids generated by industrial activities are now entering forests at an unprecedented scale and rate, greatly adding to these stresses carried over from the past. Many forests in Europe and North America now receive as much as 30 times more acidity than they would if rain and snow were falling through a pristine atmosphere. Ozone levels in many rural areas of Europe and North America are now regularly in the range known to damage trees. Despite air quality improvements made during the seventies, the average concentration of sulfur dioxide in many areas is high enough to diminish tree growth.

A comprehensive look at worldwide forest damage reveals multiple pollutants—including acid-forming sulfates and nitrates, gaseous sulfur dioxide, ozone, and heavy metals—that acting alone or together place forests under severe stress. Needles and leaves yellow and drop prematurely from branches, tree crowns progressively thin, and, ultimately, trees die. Even trees that show no visible sign of damage may be declining in growth and productivity. Moreover, acid rain's tendency to leach nutrients from sensitive soils may undermine the health and productivity of forests long into the future. Taken together, these direct and indirect effects threaten not only future wood supplies but the integrity of whole ecosystems on which society depends.

North Americans must travel to isolated mountain peaks in the eastern United States to see the kind of massive tree disease and death now spreading throughout central Europe. The loss of West Germany's woodlands is now a potent political and emotional issue among that nation's citizenry. "Waldsterben"—literally forest death—is now a household word. A survey in the summer of 1983 showed that West Germans were more concerned about the fate of their forests than about the Pershing missiles to be placed on their land later that year.[3] Environmental scientists in Poland and Czechoslo-

"'Waldsterben'—literally forest death—is now a household word in West Germany."



vakia warn that forests may become wastelands if plans for increased burning of their high-sulfur coal go unchecked.

Although scientists cannot yet fully explain how this forest destruction is occurring, air pollutants and acid rain are apparently stressing sensitive forests beyond their ability to cope. Weakened by air pollutants, acidic and impoverished soils, or toxic metals, trees lose their resistance to natural events such as drought, insect attacks, and frost. In some cases the pollutants alone cause injury or growth declines. The mechanisms are complex and may take decades of additional research to fully understand. But this growing body of circumstantial evidence is one more telling sign that fossil-fuel combustion has ecological limits, and that society will pay a price for overstepping them.

## Signs of Destruction Unfold

In just a few years, forest damage has spread with frightening rapidity through portions of central Europe. Trees covering between 3.5 and 4 million hectares—an area roughly half the size of Austria—now show signs of injury linked to air pollutants. (1 hectare = 2.47 acres.) No nation has better documented the destruction occurring within its borders than West Germany, where forests cover 7.4 million hectares—roughly a third of the nation's land area. Following an extensive survey in 1982, the Federal Minister of Food, Agriculture and Forestry estimated forest damage at 562,000 hectares—8 percent of West Germany's forests. Just a year later, in the fall of 1983, a second survey found damage on over 2.5 million hectares, 34 percent of the nation's forests. (See Table 1.) Some of this increase resulted from a more thorough investigation the second year, but nonetheless, the damage has spread markedly. Visible injury typically takes the form of yellowing and early loss of needles, deformed shoots, deteriorating roots, a progressive thinning of tree crowns, and, in its severest stages, tree death. The symptoms appear on trees of various ages and in forests of both single and mixed species.[4]

In the heavily wooded West German states of Bavaria and Baden Württemberg, home of the famed Black Forest, trees covering nearly half the forested area are damaged. Nationwide, three-quarters of the fir trees are affected, up from 60 percent a year ago. Damage to spruce, the most important species for the forest products industry, has risen from 9 percent to 41 percent, and a similar increase is evident with pine. These three conifer species, which together compose two-thirds of West Germany's forests, are the most severely struck. But damage has also been found among hardwood species

**Table 1: Changes in Forest Damage in West Germany, 1982 to 1983**

| Species | Area Showing Damage | | Portion of Forest Affected | |
|---|---|---|---|---|
| | 1982 | 1983 | 1982 | 1983 |
| | (thousands of hectares) | | (percent) | |
| Spruce | 270 | 1,194 | 9 | 41 |
| Fir | 100 | 134 | 60 | 76 |
| Pine | 90 | 636 | 5 | 43 |
| Beech | 50 | 332 | 4 | 26 |
| Oak | 20 | 91 | 4 | 15 |
| Others | 32 | 158 | 4 | 17 |
| Total | 562 | 2,545 | 8 | 34 |

**Source:** Der Bundesminister Für Ernährung, Landwirtschaft und Forsten, "Neuartige Waldschäden in der Bundesrepublik Deutschland," Bonn, October 1983.

such as beech and oak. Since most trees in the advanced stages of decline are removed from the forest, more have been affected than even these alarming survey results indicate. Dr. Georg Krause of the Land Institute for Pollution Control in Essen recently stated that "hardly anyone in Germany denies the great danger to forest ecosystems."[5]

In neighboring Czechoslovakia, forest damage covers an estimated half million hectares. Trees on some 200,000 hectares are believed to be severely damaged, and those on 40,000 hectares in the Erz Mountains reportedly have died. Dead and dying trees are plainly visible northeast of Prague in the Krokonose National Park, which has 34,000 hectares of forest, mostly populated with spruce. Not only are the spruce dying, they reportedly stopped regenerating in the park's mineral soils several years ago.[6] Further north in Poland, another half million hectares of forest are affected. Forest researchers in Katowice, near Krakow, say that fir trees are dead or dying on nearly 180,000 hectares and that spruce trees in areas around Rybnik and Czestochowa, also in the industrialized southern region, are completely gone. Environmental scientists warn that by 1990 as many as 3 million hectares of forest may be lost if Poland proceeds with its present industrialization plans calling for increased burning of the nation's high-sulfur brown coals.[7]

Accounts of forest damage in other European countries are not as well documented, but collectively they add to evidence of unprecedented forest devastation. Acute damage to pine trees was found in areas of the Netherlands in the spring of 1983, and pine and fir over a wide area in the eastern part of the country are now losing needles prematurely. Pine damage has also appeared in France and Italy. Some 12 percent of East Germany's forests are believed to be affected, and specialists in Romania have noted that 56,000 hectares of that nation's 6.3 million hectares of forests are damaged from industrial emissions.

**"Forests are suffering on both acidic and alkaline soils, and in areas where atmospheric concentrations of sulfur dioxide are low."**



In parts of Switzerland, 25 percent of the fir trees and 10 percent of the spruce reportedly have died within the past year, and many more trees show signs of injury. Various accounts also claim that trees are suffering from air pollution in the United Kingdom, Austria, and Yugoslavia.[8]

In the late autumn of 1983, early signs of tree injury began to emerge in northern Europe. Sweden, whose dying lakes first brought international attention to acid rain, now appears to have forest damage as well. Symptoms similar to those of the declining forests of central Europe have been reported by public and private foresters primarily in the southern and western portions of the country. Spruce and pine show the most injury. Although official estimates of damage are not yet available, early reports suggest that 10 percent of the timber stock in certain regions may be affected. In southern Norway, spruce reportedly are also showing injury.[9] A rare environmental report from the Soviet Union's Communist Party paper *Pravda* recently revealed that vast areas of forest are dying from air pollution near the automobile-manufacturing city of Togliatti, about 1,300 kilometers east of Moscow. According to the report, nearby forests along the Volga River may soon resemble a wasteland.[10]

Unable to attribute this massive destruction to natural events alone, scientists have turned their attention to air pollutants—an external stress that has increased greatly in recent decades. Sulfur dioxide and the acids into which it transforms have led the list of pollutant suspects. Scientists surveying West Germany's forests found that damage was greater on west-facing mountain slopes exposed to more rain and fog and thus probably to more acid deposition. The needles of ailing conifers in portions of Bavaria near the Czechoslovakian border contained more sulfur than those of healthy trees. Yet injured trees elsewhere have not shown this effect. Moreover, forests are suffering on both acidic and alkaline soils, and in areas where atmospheric concentrations of sulfur dioxide are low. Consequently, attention is increasingly being focused on the combined effects of gaseous sulfur and nitrogen oxides, heavy metals, and ozone. The debate is broadening with the apparent realization that different pollutants and mechanisms may be key factors in the damage on different sites.[11]

Although forest destruction of the magnitude occurring in central Europe is not visible in North America, trees are suffering from air pollutants there as well. In the United States, forest damage is most evident in the Appalachian mountain ranges of the east and in the Sierra Nevada of California. Field and laboratory studies have documented not only tree disease and death, but sustained declines in growth as well. From the Appalachians of Virginia and West Virginia, northward into the Green Mountains and White Mountains of New England, red spruce is undergoing a serious dieback, a progressive thinning from the outer tree crown inward. Damage is most severe in the high elevation forests of New York, Vermont, and New Hampshire, on peaks forested mainly with red spruce, balsam fir, and white birch. Because of high precipitation rates and the ability of conifers to intercept cloud moisture, these high mountain forests generally receive 3-4 times more acid deposition than those at lower elevations. In addition, the soils of these forests have shown a marked increase in lead concentration over the past two decades, believed to come almost entirely from the atmosphere.[12]

Detailed documentation of red spruce decline has come from research on Camels Hump in the Green Mountains of Vermont. There, with the benefit of two detailed tree inventories spanning the period 1965-79, researchers have found that seedling production, tree density, and basal area have declined by about half. In 1979, over half the spruce on Camels Hump were dead. A 1982 survey of spruce throughout the Appalachians has led researchers to conclude that spruce are declining over a wide area in a variety of forests. So far, no such decline is evident in commercially valuable spruce stands found at lower elevations in northern New England and Canada. Yet in light of the large wood volume declines on Camels Hump since 1965, botanist Hubert Vogelmann of the University of Vermont warns that "if such losses in only a few years are representative of a general decline in forest productivity, the economic consequences for the lumber industry will be staggering."[13]

As in central Europe, acid deposition has been linked to this dieback of spruce, although the evidence so far remains circumstantial. Studies on three varieties of pine in the New Jersey Pine Barrens provide the most convincing evidence to date for acid deposition's role. Analysis of tree rings shows that these pines have undergone a dramatic reduction in annual growth over the past 25 years, a pattern of decline not evident elsewhere in the 125-year tree ring record. Growth rates corresponded closely to the acidity measured in nearby streams, which in turn is a good index of the acidity of rain. With other factors such as drought, fire, insect pests, and ozone apparently not responsible, acid rain emerged as a likely cause. The researchers conclude that no other events in the trees' growth history are "as widespread, long-lasting, and severe in their effects."[14]

Although acid deposition's link to tree injury is still debated, scientists have firmly documented tree disease and death from ozone and other pollutants in the family of "photochemical oxidants." Ozone forms from nitrogen oxides reacting with hydrocarbons (produced mainly by automobile engines) in the presence of sunlight. Its formation and concentration is often closely tied to weather patterns and geography: A highly concentrated mass of pollutants mixing under sunny conditions is a ripe setting for ozone's creation.

Ozone has killed thousands of pine trees in the San Bernardino Mountains east of Los Angeles, California, a city now infamous for its yellow-brown photochemical smog. Tree injury was evident by mid-century as air pollutants from the growing urban area were carried

**"Air pollutants may be quietly undermining the productivity of large areas of temperate forests."**



east by marine winds. Over the past three decades, as pollution has worsened, losses of the stately ponderosa and Jeffrey pines have increased dramatically. Researchers discovered that 4-6 percent of these trees in higher elevations died over a six-year period. Losses have been greater in the western part of the mountain range, which receives higher pollutant doses. Moreover, the growth rings in ponderosa pine cores show that annual radial growth declined 38 percent over the period 1941-1971 compared with 1910-1940, a decline attributed to the rise in air pollutants. In areas receiving the highest ozone doses, the marketable volume of 30-year-old pines declined by 83 percent. Researchers conclude in their study that "this reduction in growth, along with air pollutant caused tree mortality, combine to limit production of timber in the San Bernardino Mountains."[15]

Similar damage appears to be occurring in the Appalachians. Estimates now are that 4-5 percent of the eastern white pines are dying in the southern Appalachian and Blue Ridge Mountains and north into Pennsylvania and Ohio. Ozone again is the leading cause, although, both here and in the San Bernardinos, damage is exacerbated by insects attacking trees that are weakened by air pollutants. Along with tree mortality, studies of Appalachian pines have also shown substantial growth declines.[16]

Yellowing and early loss of needles, dieback of tree crowns, and ultimately, tree death are obvious signs that forests are suffering. Measurements of tree rings on weakened trees have shown in many cases that these visible symptoms are accompanied by substantial and sustained declines in growth. But the most disturbing of air pollution's effects on forests is that growth and productivity can be declining in trees that show no visible symptoms at all. Having documented this "hidden injury" for white pine growing in the Appalachians, researchers at the Virginia Polytechnic Institute and State University conclude that it is "highly probable that growth loss in forests subjected to low-level and long-term exposures to air pollutants may be occurring unnoticed and/or unevaluated."[17]

Tree disease and death have unfolded in central Europe and limited areas of the United States at an alarming pace, and as yet show no signs of abating. The ultimate severity and extent of this damage is a looming question. As forests not yet showing injury remain exposed to acid deposition and high pollutant concentrations for longer periods of time, the damage may well spread. Moreover, if growth declines are occurring unnoticed, air pollutants may be quietly undermining the productivity of large areas of temperate forests.

---

## Tracing the Pathways of Pollution

Although a variety of pollutants are implicated in the forest damage and growth declines now occurring, most trace back to sulfur and nitrogen oxides emitted during the burning of fossil fuels and the smelting of metallic ores. Coal and oil contain sulfur and nitrogen that are released into the atmosphere as gaseous oxides during combustion. The quantity of pollutants emitted depends on the sulfur and nitrogen content of the fuel and, for nitrogen oxides, with the temperature and efficiency of combustion. The sulfur content of coal, for example, varies from less than 1 percent to as much as 6 percent. As a result, burning a metric ton* of coal may release 3-60 kilograms of sulfur. Smelting, a process of separating a metal from its ore, also releases large amounts of sulfur dioxide into the atmosphere when the ore contains sulfur. Common metals such as copper, nickel, lead, and zinc are smelted largely from sulfur-bearing rocks.[18]

Over the past century, fossil-fuel and smelting emissions have altered the chemistry of the atmosphere at an unprecedented pace. Today the atmosphere receives about as much sulfur from human activities as it does naturally from oceans, swamps, and volcanoes—on the order of 75-100 million tons per year. Yet most of the emissions from human activities occur on just 5 percent of the earth's surface, primarily the industrial regions of Europe, eastern North America, and East Asia. In these areas, energy combustion and smelting add 5-20 times more sulfur to the atmosphere than comes from nature. One smelter, for example, the International Nickel Company near Sudbury, Ontario, annually emits more than twice as much sulfur as Mt. St. Helens discharged during its recent most active year of volcanic eruptions. Emissions of nitrogen compounds are harder to estimate, but those from human sources also far exceed those from natural sources in many industrial areas. In the United States, human sources are thought to account for 75-90 percent of nitrogen oxides in the air.[19]

Fossil-fueled power plants, industrial boilers, and nonferrous smelters lead the list of sulfur dioxide emitters. The relative contribution of these sources to total emissions can vary substantially in different countries. (See Table 2.) Electric utilities account for two-thirds of sulfur dioxide ($SO_2$) emissions in the United States, for example, and in West Germany, they account for over half. In contrast, Canada's electric utilities contribute only 16 percent of $SO_2$ emissions, while about a dozen smelters emit nearly half. Motor vehicles add little to sulfur emissions, but their internal combustion engines are the biggest source of nitrogen oxides ($NO_x$) in most industrial countries. In the United States, Canada, and West Germany, motor vehicles account for roughly half of total $NO_x$ emissions, while utilities generate a third or less, and industries about a fifth. (See Table 3.)

Pollution from fossil-fuel combustion dates back well over a century to the Industrial Revolution. Coal used to heat homes and fuel factories generated a pall of smoke and haze that hung persistently over many cities in Europe and the United States. As the number of factories and homes grew, the problem worsened and many cities began to control urban smoke. But emissions of sulfur and nitrogen oxides, along with other combustion pollutants, continued to rise. Sulfur dioxide emissions began to increase rapidly in Europe after 1950 when many

* All subsequent references to tons imply metric tons.

**"Fossil-fueled power plants, industrial boilers, and nonferrous smelters lead the list of sulfur dioxide emitters."**



**Table 2: Sulfur Dioxide Emissions in Selected Countries**

| | United States | Canada | West Germany |
|---|---|---|---|
| Quantity of Emissions* | | (million metric tons/year) | |
| | 24.1 | 4.77 | 3.54 |
| Sources of Emissions | | (percent) | |
| Electric Utilities | 66 | 16 | 56 |
| Industries | 22 | 32 | 28 |
| Smelters | 6 | 45 | — |
| Homes, Businesses | 3 | 4 | 13 |
| Transportation | 3 | 3 | 3 |
| Total | 100 | 100 | 100 |

* 1980 figures for United States and Canada; 1978 for West Germany.

**Sources:** U.S. and Canada data from *United States-Canada Memorandum of Intent on Transboundary Air Pollution: Executive Summaries* (Ottawa, Canada: Environment Canada, 1983); West Germany data from Federal Minister of the Interior, "The Federal Government's Reply to the Interpellation of the Deputies: Air Pollution, Acid Rain and Death of Forests," Bonn, August 25, 1982, translation from the German by U.S. Congressional Research Service.

**Table 3: Nitrogen Oxide Emissions in Selected Countries**

| | United States | Canada | West Germany |
|---|---|---|---|
| Quantity of Emissions* | | (million metric tons/year) | |
| | 19.3 | 1.83 | 3.0 |
| Sources of Emissions | | (percent) | |
| Transportation | 44 | 61 | 45 |
| Electric Utilities | 29 | 13 | 31 |
| Industries | 22 | 20 | 19 |
| Homes, Businesses | 4 | 5 | 5 |
| Smelters, Misc. | 1 | 1 | — |
| Total | 100 | 100 | 100 |

* 1980 figures for United States and Canada; 1978 for West Germany.

**Sources:** See Table 2.

countries turned to high-sulfur oil. By 1970 annual $SO_2$ emissions had climbed to 50 million tons, two-and-a-half times mid-century levels. Similarly, $SO_2$ emissions from both the United States and Canada rose by 40 percent between the early fifties and mid-sixties.[20]

Killer pollution episodes in Donora, Pennsylvania in 1948, London in 1952, and New York City in the early sixties drove home the hazards of polluted city air. Spurred by these threats to human health, as well as by a rising tide of environmental awareness, many countries enacted pollution control laws targeting mainly sulfur dioxide and particulate concentrations in the air. Early reductions in $SO_2$ emissions came primarily by switches from high-sulfur to lower-sulfur fossil fuels. In the seventies, some countries began requiring new plants to include equipment for removing sulfur dioxide from smokestack emissions. As a result, sulfur dioxide emissions in North America peaked in the mid-sixties, and since then have fallen by 14 percent in the United States to about 24 million tons per year. In Canada, sulfur dioxide emissions have dropped back to mid-fifties levels of about 4.8 million tons per year. Emission levels have also stabilized or declined slightly in Europe, though trends vary from country to country.[21]

Government policies have paid far less attention to nitrogen oxides. This gas was not considered as great a health hazard, and since it was odorless it caused much less of a nuisance than sulfur with its rotten egg smell. Uncontrolled emissions from power plants and especially the burgeoning use of automobiles in the last three decades set $NO_x$ on a rapidly rising path. Nitrogen oxide emissions are harder to estimate than sulfur dioxide since they are determined by factors other than just the nitrogen content of the fuel. But they also are thought to have risen dramatically over the last several decades, possibly doubling in Europe between the late fifties and early seventies. In West Germany, for example, $NO_x$ emissions rose by 50 percent between 1966 and 1978. North America shows similar trends: Since the fifties nitrogen oxide emissions have roughly doubled in the United States and tripled in Canada.[22]

Trees surrounding heavy industrial polluters, such as Canada's smelters, have suffered from pollution for some time. But the extensive forest declines now unfolding are often far from major industrial and urban centers. One consequence of the drive to purify urban air over the last couple decades has been construction of tall smokestacks to better disperse pollutants into the atmosphere. These smokestacks, along with high levels of emissions, sent pollutants traveling hundreds of kilometers before returning to the earth's land and waters. The International Nickel Company's 380-meter "superstack," for example, replaced three shorter stacks in 1972. Measurements have since shown that virtually all of the sulfur and 40 percent of the heavy metals emitted travel more than 60 kilometers from the smelter.[23]

Unlike industrial emissions of carbon dioxide, which accumulate in the atmosphere, virtually all of the sulfur and nitrogen oxides that go up eventually come down in one form or another. Some return essentially unchanged as gases. Some are deposited in dry form on surfaces such as leaves and needles, where reactions with moisture can form

**"Recent publicity focused on acid rain has tended to ignore its common origins and interactive effects with other damaging pollutants."**



acids. The longer the oxides remain in the atmosphere, the more likely they are to undergo oxidation to nitric and sulfuric acid, the primary acids in acid rain. Under certain conditions, some of the nitrogen oxides will react with hydrocarbons to form ozone. Further complicating the matter, ozone can in turn speed up the transformation of sulfur and nitrogen oxides to sulfates and nitrates, the compounds in acid rain.[24]

In light of these interactions, trying to single out one pollutant as the cause for forest damage would be difficult. Trees in a given location can be affected simultaneously by several pollutants in a variety of ways. Moreover, each pollutant may affect the formation and fate of others. If ozone helps form acid rain, trees dying primarily from acid rain are dying indirectly and in part from ozone. Recent publicity focused on acid rain has tended to ignore its common origins and interactive effects with these other damaging pollutants. Divorcing acid rain from the complete pollution picture in this way may lead to ineffective strategies to control it, and more importantly, may prevent other damaging pollutants, such as ozone, from getting the attention they deserve. Nonetheless, acid deposition is of special concern because of its pervasiveness, its insidious ways of inflicting damage, and its potential long-term consequences.

Although acid rain was recognized over a century ago, only in the last three decades has the phenomenon become widespread. In broad areas of eastern North America and northern and central Europe, the annual pH of rain and snow now averages between 4 and 4.5. The pH scale, commonly used to express acidity, ranges from 0 to 14, with anything less than 7 considered acidic. The scale is logarithmic; a decrease of one unit means a 10-fold increase in acidity. (Vinegar, with a pH of about 3, for example, is 100,000 times more acidic than baking soda, with a pH of 8.) Rain falling in preindustrial times is thought to have been in the range of 5.6, slightly acidic from interactions with natural carbon dioxide in the atmosphere. Precipitation in many industrial regions is now 10-30 times more acidic than would be expected in an atmosphere free of humanity's pollution.[25]

The precise mechanism by which acid deposition may be damaging forests is not known. Sulfates and nitrates raining down as acids have drastically different effects on different forest stands, and even on different tree species in the same forest stand. Incoming acids affect interactions between the soil and living biomass of an ecosystem in complex and varying ways. Soil structure and composition, vegetation type, climate, and elevation are only some of the natural determining variables. Yet research over the last decade has uncovered some common effects of acidity that point to several pathways by which acid deposition can threaten forests.

**"Precipitation in many industrial regions is now 10-30 times more acidic than would be expected in an atmosphere free of humanity's pollution."**



Trees derive their nutrition primarily from elements such as calcium, magnesium, and potassium that are weathered from minerals in the soil. Acid deposition adds hydrogen ions to the soil, which displace these important nutrients from their sites bound to soil particles. Soils with a pH of 5 or more are seldom in danger since they have plentiful calcium carbonate (the constituents of lime) or silicates (which have abundant calcium, potassium, and/or magnesium) that effectively neutralize the acid ions. Yet soils at lower pH levels have fewer of these buffering agents. Competition from incoming acids causes the leaching of calcium and magnesium from the soil. Large areas of the southeastern United States, the Appalachian, Adirondack, and New England mountain ranges, the Canadian shield of eastern Canada, and extensive areas in Scandinavia, for example, are underlain by slightly acidic, poorly buffered soils that are especially susceptible to this effect. Although soil changes generally take place over a long period of time, soil studies in Sweden suggest that substantial leaching of nutrients from sensitive soils can occur in just a decade.[26]

Sulfates and nitrates, the other key constituents of acid deposition along with hydrogen ions, can initially have a fertilizing effect on many soils and for a time may actually boost tree growth. Forests in Scandinavia and portions of West Germany seem to have shown this effect. Yet this enhanced growth is short-lived, for eventually these "fertilizer" supplies will exceed the forests' capacity to use them. Sulfate saturation usually precedes nitrate saturation, but excess quantities of either or both simply pass through the soil, carrying vital nutrients with them. With forest productivity closely tied to nutrient availability, this leaching of soils by hydrogen, sulfate, or nitrate ions eventually reduces forest growth.[27]

Research also has shown that heavy metals—either mobilized in the soil or introduced from the atmosphere—may be involved in the forest damage now occurring. Dr. Bernhard Ulrich, a soil scientist who has studied damaged beech and spruce forests in the Solling Plateau of West Germany for nearly two decades, has hypothesized that as soils become increasingly acidic, aluminum, which is normally harmlessly bound in soil minerals, becomes soluble and toxic. The free aluminum attacks the tree's root system, making the tree less able to take up moisture and nutrients and to protect itself from insect attacks and droughts.[28]

Trace amounts of heavy metals can also enter the forest from the atmosphere. Combustion of fossil fuels, smelting, the burning of leaded gasoline, and refuse incineration are major sources of trace metals in the air. Field and laboratory research at the University of Vermont suggest that heavy metals and acid deposition act synergistically on forest systems, stunting the growth not only of trees, but of mosses, algae, nitrogen-fixing bacteria, and fungi that are essential to a forest's health. Between 1965 and 1980, metal concentra-

**"Air quality levels established to protect human health appear too lenient to protect the health of forests."**



tions have markedly increased in the soils on Camels Hump, a site of massive spruce dieback in Vermont's Green Mountains. Lead concentration doubled, while that of copper rose by 40 percent and zinc by 70 percent. These metals enter the forest with the rain and fog, which in the Vermont mountain peaks have average acidities 100 times greater than "pure" rain.[29] Researchers at the Oak Ridge National Laboratory in Tennessee have analyzed tree cores and found higher metal concentrations in recently-formed wood. Cores taken from southern Appalachian trees showed concentrations of zinc, copper, chromium, and aluminum generally high enough to be toxic.[30]

Of all the pathways by which air pollutants can affect forests, changes in the soil—whether by nutrient leaching, accumulation of heavy metals, or mobilization of toxic aluminum—are the most foreboding. In sensitive ecosystems these changes may be irreversible, thus harming not only mature trees now standing, but the seeds and seedlings that will become the forests of the next generation. The beech trees studied by Dr. Ulrich in West Germany have great difficulty regenerating, apparently because of acidity in the upper soil layers. The number of spruce and maple seedlings on Camels Hump in Vermont has declined by about half over the last two decades, and the number of spruce seedlings in the higher elevations of New York's Whiteface Mountain has dropped by 80 percent.[31]

Sulfur and nitrogen oxide gases can also enter trees directly through their leaves or needles, much as carbon dioxide is taken in for photosynthesis. These pollutants can alter the trees' metabolism and ability to produce food, and thus its productivity and growth. Forestry experts at the 1982 Stockholm Conference on Acidification of the Environment reported that tree growth can apparently decline when average yearly sulfur dioxide concentrations run as low as 25-50 micrograms per cubic meter, levels that prevail over large portions of Europe.[32] For comparison, the national annual ambient air quality standard for sulfur dioxide in the United States is 80 micrograms per cubic meter, and the European Economic Community standard is 80-120. Thus, air quality levels established to protect human health appear too lenient to protect the health of forests.

Dry sulfate and nitrate particles deposited on moist foliage can form acids that leach nutrients from leaves and needles much as they are leached from soils. West German scientists have found magnesium and calcium deficiencies in the needles of declining spruce trees in the Black Forest and Bavarian Forest of southern West Germany. They suggest that acid deposition, aided by ozone that first attacks the needle's outer surface, is weakening trees through the foliage as well as the soil.[33]

Ozone by itself has been found to damage trees when concentrations of 100-200 micrograms per cubic meter last 6-8 hours a day for several days. This is roughly 2-3 times greater than natural background levels typical for a summer day. In many rural areas of Europe, average daily concentrations are regularly in this range, and peak levels can exceed natural levels by 6-10 times. (See Table 4.) Acute stress from these episodic peaks may worsen damage caused by high average concentrations. Some scientists studying the pattern of spruce and fir dieback in the Black Forest and the state of North Rhine Westfalia contend that ozone is the leading cause there.[34]



No single hypothesis can account for the varying patterns of forest destruction observed. A reasonable explanation for the decline on one site may appear infeasible for another. While these complexities frustrate the search for a clear-cut cause and effect, they are not surprising. Apart from their unique pathways of destruction, air pollutants are most simply understood as a biological stress. Just as stress is manifest in human beings in different ways—such as ulcers or high blood pressure—air pollution stress on trees shows up in a variety of ways depending upon the tree species, soil type, and the specific pollutants involved. Pollution-induced stress weakens a biological system and makes it more susceptible to harm from natural stresses such as droughts, insect attacks, frost, and wind. In the Appalachians, strong evidence exists that the growth of trees has become more closely tied to temperature and rainfall over the past few decades, a sign of increased stress.[35] Droughts may have triggered the fir dieback in West Germany in 1976 and the spruce decline on Camels Hump in the mid-sixties. Insect infestations and root fungi have been linked to forest damage in the United States and Europe. Yet these natural factors alone seem insufficient to explain the sustained patterns of dieback and decline. Whether as a predisposing stress or a primary cause, air pollutants appear to figure prominently.

**Table 4: Summer Ozone Concentrations in Selected European Countries**

| | Upper Daily Average | Peak | Increase of Daily Average Over "Natural" Levels[1] |
|---|---|---|---|
| | (micrograms per cubic meter) | | (percent) |
| Netherlands | 80-130 | 500 | 180 |
| West Germany[2] | 100-150 | 400-500 | 200 |
| United Kingdom | 90-165 | 200-500 | 210 |
| Belgium | — | 300 | — |
| France | 70-120 | — | 160 |
| Norway | — | 200-300 | — |

[1] The midpoint of the upper daily average range is used for this calculation; "natural" ozone concentration assumed to be 60 micrograms per cubic meter.

[2] Daily average figures for West Germany are from the Black Forest; the peak values are frequently recorded in rural areas. Peaks in Black Forest typically are 110-180 micrograms per cubic meter.

**Source:** Environmental Resources Limited, *Acid Rain: A Review of the Phenomenon in the EEC and Europe* (London: Graham & Trotman Limited, 1983).

**"In the summer of 1983, the reported value of the trees that had been lost in West Germany was about $1.2 billion."**



The biosphere is not infinitely resilient. What is happening in the industrial world's fields and forests are signs that fossil-fuel combustion has ecological limits, and that exceeding them exacts a price. Unless energy and environmental strategies begin to reflect this, today's threats are bound to become tomorrow's catastrophes. Given the rapidity with which the forest destruction has unfolded, the relevant question is no longer whether proof of damage from air pollutants or acid rain is irrefutable, but whether the forests are sufficiently threatened to warrant action. Undoubtedly, German foresters would answer with an unequivocal yes. But the real test is whether nations so far spared severe losses will muster the political will to take action to avoid them. By encouraging energy efficiency, recycling, and the development of renewable energy sources, and by burning coal cleanly and only when necessary, nations will help protect their forests, crops, lakes, and people for generations. The connections are real, and so are the consequences of ignoring them.

## References


**1.** Carl O. Dunbar and Karl M. Waage, *Historical Geology* (New York: John Wiley and Sons, Inc., 1969).

**2.** Edward C. Krug and Charles R. Frink, "Acid Rain on Acid Soil: A New Perspective," *Science*, August 5, 1983.

**3.** The survey was conducted by the Allensbach Institute and was referred to in James Buchan, "Germany's Dying Forests: 'It's Just Like Being at a Graveside,' " *Financial Times*, November 19, 1983.

**4.** The Federal Minister of Food, Agriculture and Forestry, "Forest Damage Due to Air Pollution: The Situation in the Federal Republic of Germany," Bonn, November 1982; The Federal Minister of the Interior, "The Federal Government's Reply to the Interpellation of the Deputies: Air Pollution, Acid Rain and Death of Forests," Bonn, August 25, 1982, translation from the German by U.S. Library of Congress, Congressional Research Service; Der Bundesminister Für Ernährung, Landwirtschaft und Forsten, "Neuartige Waldschäden in der Bundesrepublik Deutschland," Bonn, October 1983.

**5.** Der Bundesminister Für Ernährung, Landwirtschaft and Forsten, "Neuartige Waldschäden"; Georg H. M. Krause, "Forest Effects in West Germany," presented at Symposium on Air Pollution and the Productivity of the Forest, Washington D.C., October 4-5, 1983.

**6.** George H. Tomlinson and C. Ross Silversides, *Acid Deposition and Forest Damage - The European Linkage* (Montreal: Domtar, Inc., September 1982); figure for the Erz Mountains from Bayerischen Staatsministeriums für Ernährung, Landwirtschaft und Forsten, "Waldsterben durch Luftverschmutzung," Munich, July 1983.

**7.** Eugeniusz Pudlis, "Poland's Plight: Environment Damaged from Air Pollution and Acid Rain," *Ambio*, Vol. XII, No. 2, 1983.

**8.** The Netherlands, France, and Italy referenced in Environmental Resources Limited, *Acid Rain: A Review of the Phenomenon in the EEC and Europe* (London: Graham & Trotman Limited, 1983); East Germany referenced in John J. Metzler, "Germany Battles Acid-Rain Pollution," *Journal of Commerce*, April 13, 1983; Romania referenced in "Romania Launches a Major Tree Planting & Conservation Plan," *World Environment Report*, June 2, 1980; Switzerland damage from Margaret Studer, "Swiss Tackle Air Pollution to Keep Forests Green," *The Christian Science Monitor*, October 20, 1983; reference to various accounts of damage elsewhere include G. H. Tomlinson, "Die-back of Forests - Continuing Observations," Domtar, Inc., Montreal, June 1981 (including appendices) and Studer, "Swiss Tackle Air Pollution."

**9.** Christer Agren, "Forest death in Sweden" and "Norwegian forest owners anxious about damage to woodlands," *Acid News*, January 1984.

**10.** "Volga Forests Dying from Toxic Pollution," *Washington Post*, January 6, 1984.

**11.** The Federal Minister of Food, Agriculture and Forestry, "Forest Damage due to Air Pollution"; Georg H. M. Krause, private communication, January 1984; Edgar Gaertner, "La Mort de la Foret," *Le Monde Diplomatique*, August 1983; J. M. Bradley, "What is killing the great forests of West Germany?" *World Environment Report*, August 15, 1983.

**12.** Arthur H. Johnson and Thomas G. Siccama, "Acid Deposition and Forest Decline," *Environmental Science & Technology*, Vol. 17, No. 7, 1983.

**13.** Camels Hump data documented in Thomas G. Siccama, Margaret Bliss, and H. W. Vogelmann, "Decline of Red Spruce in the Green Mountains of Vermont," *Bulletin of the Torrey Botanical Club*, April-June, 1982; broader spruce decline documented in Johnson and Siccama, "Acid Deposition and Forest Decline"; reference to commercially valuable spruce from Arthur H. Johnson, "Decline of High-Altitude Spruce-Fir Forests," presented at Symposium on Air Pollution and the Productivity of the Forest; quote is from H. W. Vogelmann, "Catastrophe on Camels Hump," *Natural History*, November 1982.

**14.** A. H. Johnson et al., "Recent Changes in Patterns of Tree Growth Rate in the New Jersey Pinelands: A Possible Effect of Acid Rain," *Journal of Environmental Quality*, October-December 1981.

**15.** Paul R. Miller et al., "Photochemical Oxidant Air Pollutant Effects on a Mixed Conifer Forest Ecosystem," U.S. Environmental Protection Agency, Environmental Research Laboratory, Corvallis, Oregon, 1977; Paul R. Miller, private communication, October 5, 1983.

**16.** Damage estimates from John Skelly, "Blue Ridge Mountains," presentation at Symposium on Air Pollution and the Productivity of the Forest"; studies of growth in pines include S. O. Phillips et al., "Eastern White Pine Exhibits Growth Retardation by Fluctuating Air Pollutant Levels: Interaction of Rainfall, Age, and Symptom Expression" and "Growth Fluctuation of Loblolly Pine Due to Periodic Air Pollution Levels: Interaction of Rainfall and Age," *Phytopathology*, June 1977.

**17.** Phillips et al., "Eastern White Pine Exhibits Growth Retardation."

**18.** Anthony C. Tennissen, *Nature of Earth Materials*, (Englewood Cliffs, New Jersey: Prentice-Hall, Inc., 1974).

**19.** Sulfur emissions from Swedish Ministry of Agriculture, *Proceedings: The 1982 Stockholm Conference on Acidification of the Environment* (Stockholm, Sweden, 1982) and Swedish Ministry of Agriculture, *Acidification Today and Tomorrow*, Stockholm, Sweden, 1982; International Nickel Company emissions from Subcommittee on Acid Rain, *Still Waters*, House of Commons, Canada, 1981; Mt. St. Helens emissions from U. S. Environmental Protection Agency, *The Acidic Deposition Phenomenon and its Effects: Atmospheric Sciences* (Vol. 1, draft), Washington D.C., May 1983; nitrogen oxide emissions from Fred Fehsenfeld, "Gas Phase and Precipitation Acidities in the Colorado Mountains," in *Acid Rain in the Rocky Mountain West*, proceedings before the Colorado Department of Health, Golden, Colorado, June 2-3, 1983.

**20.** For historical sketches of air pollution control see Erik P. Eckholm, *Down to Earth*, (New York: W. W. Norton & Co., 1982) and Edwin S. Mills, *The Economics of Environmental Quality* (New York: W. W. Norton & Co., 1978); emissions trends from Swedish Ministry of Agriculture, *Proceedings: The 1982 Stockholm Conference* and Harald Dovland and Arne Semb, "Atmospheric Transport of Pollutants," in D. Drablos and A. Tollan, eds., *Ecological Impact of Acid Precipitation*, (Oslo, Norway: SNSF Project, 1980).

**21.** *United States-Canada Memorandum of Intent on Transboundary Air Pollution: Executive Summaries* (Ottawa, Canada: Environment Canada, 1983); Swedish Ministry of Agriculture, *Proceedings: The 1982 Stockholm Conference*.

**22.** Estimates for Europe from Lars N. Overrein et al., *Acid Precipitation Effects on Forest and Fish: Final Report of the SNSF Project 1972-1980* (Oslo: The Norwegian Interdisciplinary Research Programme, 1980); estimates for West Germany from Georg H. M. Krause, private communication, January 1984; estimates for North America from *United States-Canada Memorandum of Intent*.

**23.** For a brief history of tree damage from air pollutants see George H. Tomlinson II, "Air Pollutants and Forest Decline," *Environmental Science & Technology*, Vol. 17, No. 6, 1983; INCO measurements from B. Freedman and T. C. Hutchinson, "Smelter Pollution Near Sudbury, Ontario, Canada, and Effects on Forest Litter Decomposition," in T. C. Hutchinson and M. Havas, eds., *Effects of Acid Precipitation on Terrestrial Ecosystems* (New York: Plenum Press, 1980).

**24.** Steven L. Rhodes and Paulette Middleton, "Acid Rain's Gang of Four: More than One Impact," *The Environmental Forum*, October 1983; Dovland and Semb, "Atmospheric Transport of Pollutants."

**25.** For a brief history of acid rain see Eville Gorham, "What to Do About Acid Rain," *Technology Review*, October 1982; Gene E. Likens et al., "Acid Rain," *Scientific American*, October 1979.

**26.** Bryon W. Bache, "The Acidification of Soils" and B. Ulrich, "Production and Consumption of Hydrogen Ions in the Ecosphere," in Hutchinson and Havas, eds., *Effects of Acid Precipitation on Terrestrial Ecosystems*; William W. McFee, "Sensitivity of Soil Regions to Acid Precipitation," U. S. Environmental Protection Agency, Environmental Research Laboratory, Corvallis, Ore., January 1980; reference to Swedish studies from Swedish Ministry of Agriculture, *Acidification Today and Tomorrow*.

**27.** Enhanced growth cited in Bernhard Ulrich, "Dangers for the Forest Ecosystem Due to Acid Precipitation," translated for U. S. Environmental

Protection Agency by Literature Research Company, Annandale, Va., undated, and Swedish Ministry of Agriculture, *Acidification Today and Tomorrow*; nutrient saturation and leaching discussed in G. Abrahamsen, "Effects of Acid Precipitation on Soil and Forest: Leaching of Plant Nutrients," in Drablos and Tollan, eds., *Ecological Impact of Acid Precipitation*; Dale W. Johnson, "Acid Rain and Forest Productivity," Environmental Sciences Division, Oak Ridge National Laboratory, Oak Ridge, Tennessee, undated.

**28.** Ulrich, "Dangers for the Forest Ecosystem Due to Acid Precipitation."

**29.** Richard Klein, "Testimony of Richard Klein Before the United States Senate Committee on Environment and Public Works," Hearings, June 30, 1981.

**30.** S. B. McLaughlin et al., "Interactive Effects of Acid Rain and Gaseous Air Pollutants on Natural Terrestrial Vegetation," Environmental Sciences Division, Oak Ridge National Laboratory, Oak Ridge, Tennessee, undated.

**31.** Beech regeneration from Ulrich, "Production and Consumption of Hydrogen Ions"; Camels Hump seedlings from Siccama et al., "Decline of Red Spruce in the Green Mountains of Vermont"; Whiteface Mountain referred to in Richard E. Rice, "The Effects of Acid Rain on Forest and Crop Resources in the Eastern United States," The Wilderness Society, Washington D. C., September 1983.

**32.** Swedish Ministry of Agriculture, *Proceedings: The 1982 Stockholm Conference*.

**33.** Von H. W. Zöttl and E. Mies, "Die Fichtenerkrankung in Hochlagen des Südschwarzwaldes," *Allgemeine Forst-und Jagdzeitung*, June-July 1983; Von C. Bosch et al., "Über die Erkrankung der Fichte (Picea abies Karst.) in den Hochlagen des Bayerischen Waldes," *Forstwissenschaftliches Centralblatt*, June 1983.

**34.** Damage levels for ozone cited in Swedish Ministry of Agriculture, *Proceedings: The 1982 Stockholm Conference*; references to German scientists include Von H. Mohr, "Zur Faktorenanalyse des 'Baumsterbens'—Bemerkungen eines Pflanzenphysiologen," *Allgemeine Forst-und Jagdzeitung*, June-July 1983; Von Dr. Bernhard Prinz, "Gedanken zum Stand der Diskussion über die Ursache der Waldschäden in der Bundesrepublik Deutschland," *Der Forst-und Holzwirt*, September 1983.

**35.** L. J. Puckett, "Acid Rain, Air Pollution, and Tree Growth in Southeastern New York," U. S. Geological Survey, Reston, Va, 1981; Richard L. Phipps, "Ring Width Analysis," presented at the Symposium on Air Pollution and the Productivity of the Forest.

CHAPTER 11

# Groundwater Contamination: Local Problems Become a National Issue



LAURA TANGLEY

A few years ago, more than 400,000 residents of the San Gabriel Valley in California found out the water they had been drinking for years was badly contaminated with trichloroethylene (TCE), a confirmed carcinogen in laboratory animals. Public health officials subsequently closed 39 wells supplying water to 13 cities there. More recently, TCE turned up in 36 wells on Long Island, New York, where a population of 3 million people relies on underground water supplies. And in Bedford, Massachusetts, four municipal wells that once provided 80% of the town's drinking water were closed because they, too, contained TCE as well as other toxic chemicals. Soon it was discovered that one-third of the communities in Massachusetts were drinking water from contaminated wells.

TCE, a widely used industrial degreaser, is one of the more common contaminants now known to be present in underground drinking-water supplies. At least a hundred different substances pollute groundwater somewhere in the United States today. Sources of the contamination are diverse, ranging from huge hazardous waste dumps to small gasoline stations, farms, and septic tanks. TCE, for example, reaches groundwater not only through industrial waste disposal, but also through backyard septic tanks because it is a component of many household cleaning fluids.

The groundwater contaminants themselves are diverse. They vary from simple inorganic ions, such as nitrate and chloride, to complex synthetic organic compounds, such as TCE, and include even pathogenic organisms and radionuclides. Groundwater, which supplies half the nation's population with drinking water, contains many of these contaminants at concentrations far higher than have ever been detected in the most polluted rivers, lakes, and streams. Some groundwater contaminants do not occur in surface water at all. Serious groundwater pollution problems have now been reported from all parts of the United States.

Although groundwater contamination is not a new problem, it is receiving new national attention. Currently, nearly a dozen major studies of groundwater pollution are underway, including one by the National Academy of Sciences and another by the Office of Technology Assessment. During the 98th Congress, groundwater has been a topic of debate in hearings to reauthorize several environmental statutes. This new legislative interest may have spurred, at least in part, EPA's revival of a groundwater protection strategy first drafted—but never completed—nearly four years ago. Some environmentalists say the new strategy doesn't go far enough to protect the nation's underground water supplies, but other experts point out that the fundamental nature of this resource may preclude a more centralized groundwater policy.

## Groundwater Characteristics

Invisible and long-ignored, groundwater is nevertheless a vast natural resource. Aquifers, the permeable, saturated layers of rock, sand, and gravel



Laura Tangley is Features and News Editor of *BioScience.*

that hold groundwater, underlie most of the United States (Figure 1) and supply drinking water to at least some populations in every state. Groundwater comprises more than 96% of all fresh water in the United States; estimates of its volume range from 33 quadrillion gallons to 100 quadrillion gallons. But because groundwater hydrogeology is so complex, much less is known about the resource than is known about surface water. "Still, we know a lot more than the media and government would lead us to believe," says Jay Lehr, executive director of the National Water Well Association and a hydrogeologist who has studied groundwater himself for 30 years. "They're just beginning to wake up to it now."

One well-known characteristic of groundwater, particularly relevant to its contamination, is the speed that the water moves, anywhere from a few feet to just a fraction of an inch per day. Thus there is very little mixing of groundwater, and a well, once contaminated, stays that way for hundreds or even thousands of years. This characteristic may have its advantages, too. "Whatever we've put down there isn't going anywhere very fast," says Lehr, who estimates that about two percent of the nation's underground water resources are now contaminated. When it comes to cleaning up aquifers that are already dirty, "we have the advantage of time," he says. On the other hand, because there is a long lag time between a contaminant entering the ground and showing up in a nearby well, groundwater may actually be much more polluted than anyone realizes.

Slow movement of contaminants also means that groundwater quality can differ dramatically from place to place. It is possible for an aquifer to be highly polluted in one location yet pristine just a few hundred feet away. This fact, combined with the multiplicity and diversity of both contaminants and their sources, results in the site-specificity of groundwater pollution problems. In addition, aquifer geology and natural groundwater quality vary locally, leading many researchers to believe that the resource is best managed at the state or local level.

### Multiplicity of Contaminants

Like groundwater, many groundwater pollutants tend to be invisible—as well as being odorless and tasteless. Detection of drinking-water pollution problems, then, is difficult, and often occurs only after contamination has already been suspected. Many groundwater contaminants, particularly volatile organics, do not occur in surface water. Thus, experience with these chemicals is limited, and only recently has it become possible to detect many of them at the low concentrations found in groundwater.

As of 1981, the Council on Environmental Quality (CEQ) had identified 33 toxic organic compounds commonly found in drinking-water wells (Table 1). Since then, the list of both contaminants and the states in which they are found has grown. According to Lehr, 80% of all groundwater contamination problems are caused by chlorinated compounds used as industrial solvents and degreasers, TCE and carbon tetrachloride, for example. Because monitoring groundwater is both difficult and expensive, however, researchers rarely attempt thorough analyses, and what they find is often a matter of what they are looking for. Although there is abundant anecdotal evidence indicating that groundwater contamination is serious and widespread, there is no single, national data base on the subject. Much remains unknown.

Nevertheless, groundwater contamination has fast become an urgent national concern, primarily because the resource is such an important one. Not only does underground water provide about 50% of the US population with

**Figure 1.** Groundwater resources of the United States. (Source: US Water Resources Council)

**Table 1.** Toxic organic compounds found in drinking-water wells.

| Chemical | Concentration (ppb) | State | Highest Surface Water Concentration Reported (ppb) |
|---|---|---|---|
| Trichloroethylene (TCE) | 27,300 | Pennsylvania | |
| | 14,000 | Pennsylvania | |
| | 3,800 | New York | 160 |
| | 3,200 | Pennsylvania | |
| | 1,530 | New Jersey | |
| | 900 | Massachusetts | |
| Toluene | 6,400 | New Jersey | |
| | 260 | New Jersey | 6.1 |
| | 55 | New Jersey | |
| 1,1,1-Trichloroethane | 5,440 | Maine | |
| | 5,100 | New York | 5.1 |
| | 1,600 | Connecticut | |
| | 965 | New Jersey | |
| Acetone | 3,000 | New Jersey | NI* |
| Methylene chloride | 3,000 | New Jersey | |
| | 47 | New York | 13 |
| Dioxane | 2,100 | Massachusetts | NI |
| Ethyl benzene | 2,000 | New Jersey | NI |
| Tetrachloroethylene | 1,500 | New Jersey | |
| | 740 | Connecticut | 21 |
| | 717 | New York | |
| Cyclohexane | 540 | New York | NI |
| Chloroform | 490 | New York | |
| | 420 | New Jersey | 700 |
| | 67 | New York | |
| Di-n-butyl-phthalate | 470 | New York | NI |
| Carbon tetrachloride | 400 | New Jersey | 30 |
| | 135 | New York | |
| Benzene | 330 | New Jersey | |
| | 230 | New Jersey | 4.4 |
| | 70 | Connecticut | |
| | 30 | New York | |
| 1,2-Dichloroethylene | 323 | Massachusetts | |
| | 294 | Massachusetts | 9.8 |
| | 91 | New York | |
| Ethylene dibromide (EDB) | 300 | Hawaii | |
| | 100 | Hawaii | NI |
| | 35 | California | |
| Xylene | 300 | New Jersey | |
| | 69 | New York | 24 |
| Isopropyl benzene | 290 | New York | NI |
| 1,1-Dichloroethylene | 280 | New Jersey | |
| | 118 | Massachusetts | |
| | 70 | Maine | 0.5 |
| 1,2-Dichloroethane | 250 | New Jersey | 4.8 |
| Bis (2-ethylhexyl) phthalate | 170 | New York | NI |
| DBCP | 137 | Arizona | |
| (Dibromochloropropane) | 95 | California | NI |
| | 68 | California | |
| Trifluorotrichloroethane | 135 | New York | NI |
| | 35 | New York | |
| Dibromochloromethane | 55 | New York | 317 |
| | 20 | Delaware | |
| Vinyl chloride | 50 | New York | 9.8 |
| Chloromethane | 44 | Massachusetts | 12 |
| Butyl benzyl-phthalate | 38 | New York | NI |
| gamma-BHC (Lindane) | 22 | California | NI |
| 1,1,2-Trichloroethane | 20 | New York | NI |
| Bromoform | 20 | Delaware | 280 |
| 1,1-Dichloroethane | 7 | Maine | 0.2 |
| alpha-BHC | 6 | California | NI |
| Parathion | 4.6 | California | 0.4 |
| delta-BHC | 3.8 | California | NI |

*NI = not investigated.
CEQ/1981

drinking water, it also provides 25% of the water used for all purposes, including 40% of irrigation water. Approximately 75% of US cities derive at least part of their supplies from groundwater, and virtually all rural residents rely on wells for drinking water. Groundwater use is also increasing much faster than the corresponding use of rivers, lakes, and streams. Since 1950, there has been a threefold increase in subsurface water withdrawals, and the rate continues to rise faster than that for surface waters.

Often there are no surface-water alternatives for communities that depend on groundwater. This makes the toxicity of chemicals found so far a particular cause for alarm. Most of the synthetic organics detected in drinking-water wells are poisonous at high concentrations. Although less is known about the effects of lower, long-term doses, many of the compounds are carcinogenic in humans or laboratory animals (Table 2). Unfortunately, several of them have not yet been tested for carcinogenicity.

Such testing, particularly in humans, is difficult. "Unlike bacterial or viral contamination, the health effects associated with ingestion of chemicals at very low levels over long periods of time are difficult or impossible to link to specific exposure," says Vernon Houk, director of the Center for Environmental Health at the Centers for Disease Control in Atlanta, Georgia. "These effects, which may include certain cancers and liver, neurologic, or kidney disease, have causes other than chemical exposure. Medical science at this point is unable to distinguish among the various causes of these chronic diseases."

Still, says Houk, the chemicals are dangerous, and lack of complete data should not be used as an excuse for stalling the process of getting them out of groundwater. When asked at a recent congressional hearing about the long-term health effects of TCE, Houk replied that "the point has to be made over and over that we are never going to have all of the information on these compounds

**Table 2.** Selected synthetic organic chemicals detected in drinking-water wells.

| Chemical | Evidence for carcinogenicity |
|---|---|
| Benzene | H |
| alpha-BHC | CA |
| beta-BHC | NTA |
| gamma-BHC (Lindane) | CA |
| Bis (2-ethylhexyl) phthalate | NTA |
| Bromoform | NTA |
| Butyl benzyl phthalate | NTA |
| Carbon tetrachloride | CA |
| Chloroform | CA |
| Chloromethane | NTA |
| Cyclohexane | NTA |
| Dibromochloropropane (DBCP) | CA |
| Dibromochloromethane | NTA |
| 1,1-Dichloroethane | SA |
| 1,2-Dichloroethane | CA |
| 1,1-Dichloroethylene | NTA |
| 1,2-Dichloroethylene | NTA |
| Di-n-butyl phthalate | NTA |
| Dioxane | CA |
| Ethylene dibromide (EDB) | CA |
| Isopropyl benzene | NTA |
| Methylene chloride | NTA |
| Parathion | SA |
| Tetrachloroethylene | CA |
| Toluene | NTA |
| 1,1,1-Trichloroethane | NA |
| 1,1,2-Trichloroethane | CA |
| Trichloroethylene (TCE) | CA |
| Trifluorotrichloroethane | NTA |
| Vinyl chloride | H, CA |
| Xylene | NTA |

H = Confirmed human carcinogen.
CA = Confirmed animal carcinogen.
SA = Suggested animal carcinogen.
NA = Negative evidence of carcinogenicity from animal bioassay.
NTA = Not tested in animal bioassay.
CEQ/1981

and what they do to humans. I don't believe it is fair to say that we cannot take action just because we don't have all the information we need."

Because groundwater is an important component of the earth's hydrologic cycle, its contamination threatens the environment as well as public health. Subsurface waters often join surface waters in wetlands, among the most productive and valuable of all coastal ecosystems. Groundwater also supplies about 30% of average stream flow, and during periods of drought its springs may release the only water present in an otherwise dry river or stream bed. Thus the quality of groundwater critically affects numerous terrestrial as well as aquatic ecosystems.

## Microbial Ecosystems

Recently, scientists were surprised to learn that groundwater also contains complex ecosystems of its own. Microbiologists long believed that layers of soil and rock located beneath the root zone were essentially devoid of life. But in the last few years they have discovered vast communities of deep subsurface microorganisms. The density and diversity of these communities are quite high, rivaling the biomass of microbes in many nutrient-rich lakes.

These newly discovered microbial ecosystems may well be affected by groundwater pollution. But they also may be able to alter the contaminants, either degrading or enhancing their toxicity. The tendency of microbes to degrade organic pollutants in surface water is well known. Although subsurface organisms encounter a very different environment—one that is dark, oxygen-poor, and lacking common microbial community components such as protozoa—they, too, are equipped to degrade toxic organic chemicals. Some microbiologists, including John Wilson and James McNabb of EPA's Environmental Research Laboratory in Ada, Oklahoma, have studied the possibility of toxic chemical biotransformation by common groundwater microorganisms (Table 3). They recently reported their results in *EOS*, a publication of the American Geophysical Union.

Some policymakers hope that scientists may be able to use recombinant-DNA technology to create new microbes specifically engineered to consume contaminants at a rapid rate. Microbiologists, however, urge them not to underestimate the power of an aquifer's native microbial community. C. H. Ward, who also studies microbial degradation of pollutants and codirects the National Center for Ground Water Research in Houston, Texas, argues that limited research money may be better spent on enhancing the abilities of natural microorganisms than on engineering new ones.

"I can think of very few cases where genetically engineered organisms would have any advantage over the indigenous flora," says Ward. "The trick is to change the subsurface environment so it will support those indigenous organisms at a higher rate of activity. You can do this by supplying them with limiting nutrients and oxygen," he adds. "That's one of the things we're working on here."

## Sources of Pollution

Most federal money spent so far on groundwater pollution has gone toward programs that are researching these or other methods of rehabilitating contaminated aquifers. The techniques, many of which are still experimental, tend to be expensive. "Although cleaning up contaminated groundwater is important, I'd like to see equal attention and resources going toward preventing new groundwater pollution," says James Tripp, an attorney with the Environmental Defense Fund who has extensive experience in groundwater issues.

The problem with preventing new groundwater contamination is that sources of the pollution are diverse, and they generally overlap. OTA, which is preparing perhaps the most comprehensive study of groundwater contamination to date, identifies 36 different sources of contamination. Although large hazardous-waste facilities have traditionally received more attention, the new study, scheduled to be released this spring, suggests that smaller, more numerous sources are the most serious. The highest-priority sources of groundwater contamination identified by OTA include industrial landfills, surface impoundments, subsurface percolation from septic tanks and cess pools, injection wells, agricultural fertilizers, oil and gas wells,

**Table 3.** Prospect of biotransformation of selected organic pollutants in water-table aquifers.

| Class of compounds | Aerobic water, concentration of pollutant, μg/l | | Anaerobic water |
|---|---|---|---|
| | >100 | <10 | |
| HALOGENATED ALIPHATIC HYDROCARBONS | | | |
| Trichloroethylene | none | none | possible* |
| Tetrachloroethylene | none | none | possible* |
| 1,1,1-Trichloroethane | none | none | possible* |
| Carbon Tetrachloride | none | none | possible* |
| Chloroform | none | none | possible* |
| Methylene Chloride | possible | improbable | possible |
| 1,2-Dichloroethane | possible | improbable | possible |
| BROMINATED METHANES | improbable | improbable | probable |
| CHLOROBENZENES | | | |
| Chlorobenzene | probable | possible | none |
| 1,2-Dichlorobenzene | probable | possible | none |
| 1,4-Dichlorobenzene | probable | possible | none |
| 1,3-Dichlorobenzene | improbable | improbable | none |
| ALKYLBENZENES | | | |
| Benzene | probable | possible | none |
| Toluene | probable | possible | none |
| Dimethylbenzenes | probable | possible | none |
| Styrene | probable | possible | none |
| PHENOL AND ALKYL PHENOLS | probable | probable | probable† |
| CHLOROPHENOLS | probable | possible | possible |
| ALIPHATIC HYDROCARBONS | probable | possible | none |
| POLYNUCLEAR AROMATIC HYDROCARBONS | | | |
| Two and three rings | possible | possible | none |
| Four or more rings | improbable | improbable | none |

*Possible, probably incomplete.
†Probable, at high concentration.
EOS/Vol. 64, No. 33, 16 August 1983.

and underground storage tanks (Figures 2 and 3).

Of these, two of the most serious may be the disposal pits, ponds, and lagoons, collectively called surface impoundments, and leaking underground storage tanks, says Robert Harris, codirector of Princeton University's hazardous-waste research program. According to EPA, there are nearly 200,000 active surface impoundments in the United States, and about 26,000 of them are unlined and may be leaking a wide array of toxic chemicals into underground water supplies. The impoundments may hold industrial, municipal, agricultural, mining, or oil and gas wastes. John Skinner, director of EPA's Office of Solid Waste Programs, said at a recent congressional hearing that 95% of all operating surface impoundments are located within a quarter of a mile of drinking-water supplies.

Equally or even more ubiquitous than impoundments are underground tanks that store a variety of waste as well as nonwaste substances. The storage tanks are usually made of steel, which rusts and corrodes over time. It is now known that thousands of underground storage tanks are leaking toxic chemicals into drinking-water supplies nationwide.

"Gasoline may be one of the most common causes of groundwater pollution in many parts of the country due to leakage from underground storage tanks," said EPA Assistant Administrator for Water Jack Ravan at a hearing before the toxic substances subcommittee of the Senate Environment and Public Works Committee last fall. He said that 75,000–100,000 of these tanks are now releasing 11 million gallons of gasoline into the ground each year. "The potential for contamination from this single source is not small," said Ravan. "One gallon of gasoline per day leaking into a groundwater source is enough to pollute the water of a 50,000-person community to a level of 100 parts per billion."

Petroleum companies are concerned about this problem, he added, and are trying to replace steel tanks with relatively leakproof fiberglass ones. The new tanks are expensive, however, and it will be a long time before they can be installed at every corner gas station that now has an old, corroded tank. Although gasoline, specifically, is not defined as a hazardous substance under federal regulations, two of its components, ethylene dibromide (EDB) and benzene, are toxic. Both cause cancer in laboratory animals.

### EPA Strategy

Despite the dangers of these two sources of groundwater contamination, neither underground storage tanks nor surface impoundments are controlled by any federal regulations. In addition, there is no federal legislation specifically designed to ensure the quality of groundwater. Although OTA has identified 16 federal statutes with provisions enabling agencies to protect groundwater from specific sources of pollution, no single source is comprehensively covered by any statute. Current groundwater programs are administered by dozens of offices scattered among different agencies. Often, these programs are incompletely implemented or even ignored.

In order to coordinate various federal programs and provide protection from currently unregulated pollution sources, EPA, the agency with primary responsi-

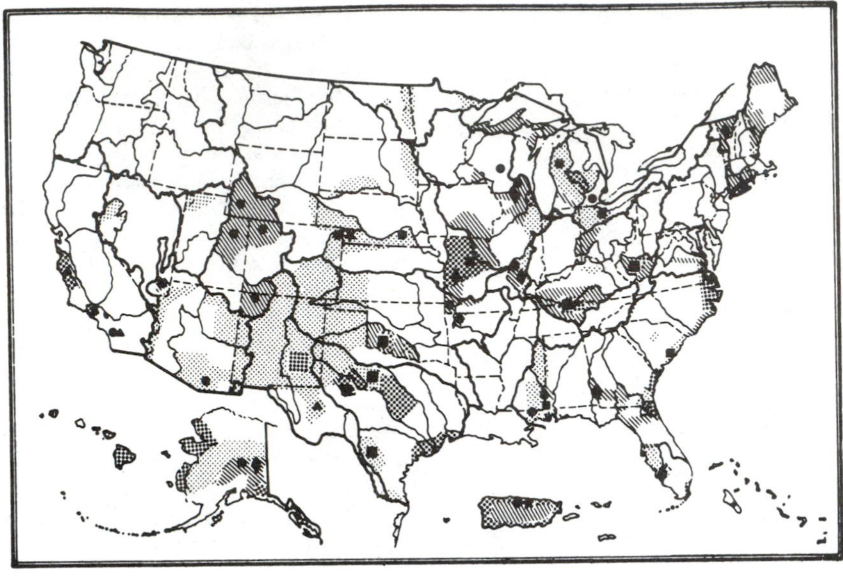


**Figure 3.** Groundwater pollution problems. (Source: US Water Resources Council)

bility for groundwater quality, has prepared a groundwater protection strategy. The agency first drafted a similar strategy in 1980. But with the change in administration the following year, that document was apparently abandoned. The new strategy stems in part from increasing pressure both from Congress and environmental organizations since then.

A draft of the strategy, due to be released in final form this spring, states that EPA will move toward regulating both underground storage tanks and surface impoundments. Its plans for underground tanks are more specific. The strategy states that EPA's Office of Toxic Substances will not only study the scope of the problem, it will also design a "regulatory program" as soon as possible. As for surface impoundments, EPA at this time plans only to study the situation to decide if new regulations are needed.

EPA acknowledges that "in some circumstances full clean-up of contamination is beyond the capability of existing technology" and that not all aquifers can be protected from all sources of contamination. Thus the agency proposes an aquifer classification system, defining as "special aquifers" those that are now either irreplaceable sources of drinking water or considered "ecologically vital." For these special aquifers, new EPA regulations may be particularly strong, including a ban on all potentially harmful activities above them. Those activities would include hazardous-waste landfills, surface impoundments, storage tanks, and even pesticide spraying. The agency would be more lenient with aquifers designated either "current and potential sources of drinking water" or, especially, "aquifers not considered potential sources of drinking water." The draft strategy states that aquifers in the third category could be exempt from provisions of existing regulations.

The classification system has received support from some environmentalists. "I'm a strong advocate of the classification approach," says EDF's James Tripp. "We have limited resources for both protecting and cleaning up groundwater. We need a framework for making decisions. Often, where a source of contamination is located turns out to be more important than what it is. Policymakers should think more in terms of watersheds and recharge areas and make sure those critical land areas are protected." Tripp adds, however, that criteria for placing an aquifer in EPA's third category should be very strict, no matter where it's located.

EPA's classification scheme is intended not only to guide the agency's own programs, but to serve as a model for states setting up groundwater protection programs as well. In fact, beyond the few specific proposals mentioned above, the strategy suggests no new federal legislation or regulations and thus leaves primary responsibility for groundwater to the states. "It is important to understand that states have traditionally stood as the first line of authority on groundwater protection and continue to today," says Ravan. Although some congressional leaders and environmentalists argue that the federal government should do more, many individuals with experience on groundwater issues agree that the site-specific nature of pollution problems means they may be better regulated at the state or local level.

"It staggers the imagination to think about how the federal government could possibly come up with programs to regulate all groundwater contaminants in all situations," says Paula Magnuson, sen-

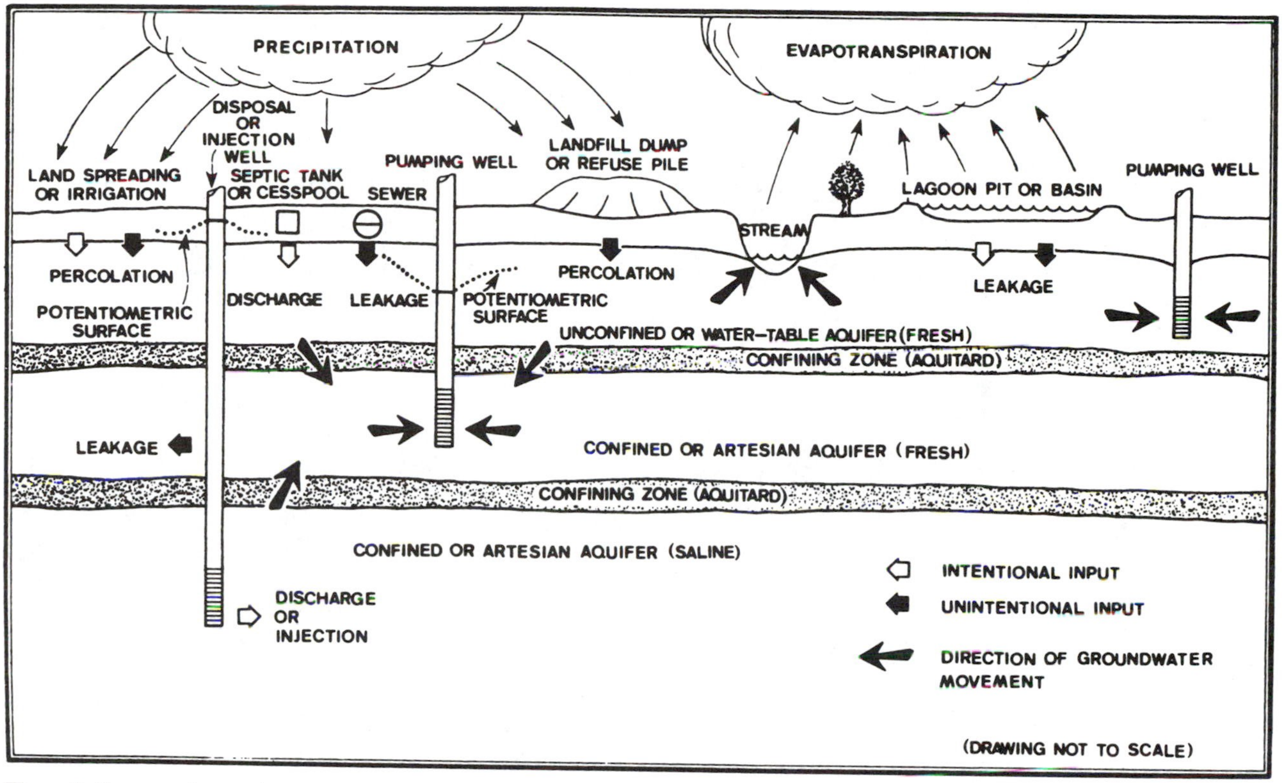

**Figure 2.** Sources of groundwater contamination. (Source: EPA)

ior scientist with Geraghty and Miller, Inc., a groundwater consulting firm in Syosset, New York. "I think it's more appropriately controlled at the state level." Says hydrogeologist Lehr, "there's no way the government could manage this problem on a national basis. It's fundamentally a local issue." He adds that the states not only have the desire to protect their own groundwater resources, but many of them are already doing a good job as well.

All the researchers *BioScience* talked to, however, agreed that the federal government still has a key role to play—"a sort of a fatherly role," according to Lehr. Specifically, they said federal agencies should provide the states with basic research, technical training, drinking-water standards, and, perhaps most important, money with which to run their groundwater programs.

For example, Florida, which has one of the best state groundwater programs, could use a lot of help from the federal government, says the director of that program, Rodney DeHan. Although he agrees that hydrogeological and contaminant conditions vary from state to state, "the nature of the contaminants and their effects on human health do not." DeHan believes that EPA should set drinking-water standards that, so far, it has issued for just a few of the dozens of contaminants found in groundwater. Training sessions on the latest monitoring and abatement technology would also be helpful, he says. But by far Florida's most critical need is money, says DeHan, who claims he cannot even hire a qualified toxicologist because the highest salary the state can afford does not compete with private industry. "My difficulty with the new strategy is that it leaves responsibility for groundwater to the states, but doesn't back that mandate with any funding," he says. "Without money, EPA's strategy is meaningless."

# CHAPTER 12

# Biological Consequences of Global Pollution of the Marine Environment

S. A. PATIN

## General characteristics of biological and ecological effects of pollution

The principal difficulty in analysing the consequences of man-made changes in the chemical composition of the marine environment is that the response of marine organisms to the presence of toxicants or other pollutants takes place simultaneously at all biological levels in the sea, from subcellular up to super-organismic levels, and even at the level of ecosystems. The result is a complex pattern of direct and indirect effects of the toxicant being investigated against the background of the natural dynamics of hydrobiological processes.

The biological and ecological consequences of the pollution of the marine environment may be systematically considered in several ways. It is expedient to distinguish between two groups of ecotoxicological situations, depending on the identity and on the intensity of the operative factor, and its association with the zone of lethal and tolerable doses.

One such situation is the direct toxicity to individual populations and communities. This is accompanied by rapid injury to the principal physiological and biochemical systems of the organism, and results in lethal toxication, elimination of individual species and populations or causes profound pathological alterations on the level of individual organisms, individual populations, and occasionally, on entire ecosystems. Impressive and occasionally tragic examples of such situations, which usually result from mechanical breakdowns or from sporadic bulk discharges of industrial and other wastes, are familiar (North *et al.*, 1965; Nelson-Smith, 1970; Mironov, 1972; Aubert, 1973). The areas of intense pollution are usually limited, and the consequences of such events do not usually affect large water basins.

The other group of effects, which are not as evident, and are much less familiar, is associated with the effect of relatively low concentrations of

*Note:* Cross-references in this paper refer to chapter numbers in the original book and not to *Global Ecology*.

Reprinted with permission from S. A. Patin (1982), *Pollution and the Biological Resources of the Oceans*, pp. 157–171. London: Butterworths.

S. A. Patin is Deputy Director, All-Union Research Institute of Marine Fisheries and Oceanography, Moscow, U.S.S.R.

pollutants on marine organisms, communities and ecosystems under conditions of chronic pollution. Owing to differences in the resistance levels of aquatic organisms in different taxonomic groups, different developmental stages and different physiological conditions in the communities, the biological reactions are extremely complex. The integral and most important responses are undoubtedly changes in the stability and homeostasis of the ecosystem (Fedorov, 1975). Such changes may be detected in the field as: decreases in the index of species diversity (Bechtel and Copeland, 1970; Stirn *et al.*, 1975); impairment of rates and mutual relationships between production processes and decomposition of organic matter (Kamshilov, 1973); anomalous dynamics of dissolved oxygen (Braginskii, 1972); changes in the dominant species in the biocoenosis. These and other ecological disturbances (Woodwell, 1970; Chuikov, 1975) are very difficult if not quite impossible to describe, much less to predict, under conditions of low-level chronic pollution.



Such consequences are 'inconspicuous' as compared with those of acute toxicity (e.g., death of a large number of fish), but are no less serious or less significant. The opposite may well be the case, since: chronic pollution is both extensive and ubiquitous; there is a correlation between the level of pollution and the distribution of biomass and productivity (as explained in Chapter 2); the concentrations of the principal toxicants remain more or less stable, etc. As a result, large-scale, man-made ecological anomalies in the World Ocean are only to be expected. The obviousness and timeliness of this view has failed so far to attract the attention of research workers to the required extent, even though the importance in dealing with the problem of pollution as a whole is evident.

Of particular importance in this connection are the production processes in the World Ocean, particularly processes of renewed formation of organic matter under conditions of chronic pollution of the euphotic zone. The problem we are faced with is not only the fundamental ecological significance of primary production as the material and energetic source of life in marine basins, but also the higher sensitivity of photosynthetic processes to the changed chemical composition of the habitat of unicellular algae (Chapters 6 and 7). The most obvious and familiar manifestations of such ecological anomalies are the numerous cases of eutrophication (hypereutrophication) of coastal and especially fresh waters as a result of abundant growth of certain species of phytoplankton in the presence of excess amounts of biogenic materials.

Less well known and less thoroughly studied are the effects of inhibition of growth, cell division and photosynthesis of marine phytoplankton as a function of the concentrations of specific toxicants or their combinations. As described earlier, the production characteristics of unicellular marine algae are highly sensitive to the effect of pollutants present in low concentrations. It should be recalled that such concentrations have in fact been repeatedly recorded in marine basins, both by ourselves and by other workers.

The probable man-made modifications of the normal course of hydrobiological processes in the sea (about which very little is as yet known) may include the disruption of the complex system of extrametabolic relationships in marine communities. Recent studies (Vinberg, 1969; Khailov,

1970) show that interorganismic metabolism, which is typically intensive and highly specific as regards the synthesis and the utilisation of metabolites, is very important as an integrating and regulating factor in marine communities.

It can be easily understood that even low concentrations of pollutants which are foreign to a system (e.g., dissolved petroleum products) may block certain metabolic interactions of the organisms across the aquatic medium and thus disturb the biocommunication systems in communities, and hence also their structural and productive characteristics. The same applies to sensory connections between aquatic organisms, chemoreception and other phenomena, which may be impaired by the presence of small amounts of industrial impurities. Such effects are complex and their significance is not yet clear, so that little is known on the subject; nonetheless, such facts as the disturbed migration of diadromous fishes (FAO, 1971) indicate that such anomalies are real and must be closely studied.

Another group of critical situations in marine communities is associated with the concentration of toxic pollutants in water–air and water–seabed interfaces, and the effect of toxicants on the most sensitive organisms and communities of neuston and benthos. While the response of marine benthos to toxic effects has been well investigated, at least in highly polluted areas (Bellan and Bellan-Santini, 1972; Stora, 1974), neuston has not so far been ecologically or toxicologically studied, despite certain indications that this community is important as the nodal point of trophic and reproductive relationships in the ecosystem, and also in view of its sensitivity to toxic factors (Polikarpov and Zaitsev, 1969). This view is supported by our own and other data on the increased sensitivity of early developmental stages of organisms. Many of these organisms pass through a neuston stage during which they are subjected to particularly strong toxic and even mechanical effects of pollutants in the neuston biotope. For example, the presence of a thin film of petroleum on the sea surface (a common enough occurrence nowadays) kills the larvae of commercially important fishes by isolating the larvae from the atmosphere and preventing the air bladder from becoming filled (Vinogradov, 1972).

Other ecological aspects of this problem include: accumulation and migration of toxicants along the food chain; inhibition of the reproduction function of organisms; additive and specific effects of the totality of the toxic factors; diversity in the resistance of organisms belonging to different taxonomic groups.

We have already noted the diverse responses of marine organisms of different taxonomic groups at various developmental stages to the presence of exogenic toxic (disturbing) factors in the aquatic medium. It must also be borne in mind that the organisms and their populations form part of a complex system of contradictory dynamic and mutually determined relationships, both with one another and with the abiotic factors present in the environment, and that any major change in any one of these factors alters the relationship of the aquatic organisms to the overall conditions of their existence. We may also recall that ecosystems are ‘random’ biological systems, in the sense that they are less deterministic than biological systems with a central direction (organisms with a central and vegetative nervous system), and for this reason their most important characteristics will be complexity, stability and similar parameters (Fedorov, 1974).

In these circumstances an evaluation, much less prediction, of the consequences of chemical pollution of communities and ecosystems may prove unrealistic, since our knowledge of the subject is clearly inadequate for solving such complex problems.

Recent theoretical treatments and modelling of the stability of complex systems, including systems exposed to man-made influences (Fedorov and Sokolova, 1972; Fedorov, 1974), may offer some promise in this respect. However, the use of these new and promising ideas and principles in an ecological study of global chemical pollution of the World Ocean must be left to future generations of workers, since at the time of writing the available information is insufficient for the determination of the stability of marine ecosystems on even a regional scale.

Despite the complexities and limitations involved (both those already mentioned and others), we must make an attempt to analyse and to evaluate, to some extent, the biological consequences of the global pollution of the World Ocean. It would be important to be able to give at least qualitative answers to a number of questions which are frequently posed in the literature and in scientific debates, but which have so far been treated only in a most general manner. A few examples of such questions are listed below:

(1) What evidence is there for the appearance of global consequences of the pollution of the marine environment?
(2) What is the nature and the intensity of both actual manifestations and those to be expected in future?
(3) What is the spatial structure and the localisation pattern of man-made interference with life in the World Ocean?
(4) What are the organisms primarily and directly affected by pollutants, and what trophic levels and marine production chains are thus affected?

The list of such questions could be extended, but the answers are more difficult, not only owing to the scarcity of experimental data, but also because a suitable methodological approach is lacking.

The amount of currently available information on marine ecosystems is insufficient for mathematical modelling to give a quantitative description and evaluation of large-scale, man-made ecological and biological abnormalities. It is more realistic to approach the problem by comparative analysis and systematisation of all known biogeochemical and ecotoxicological information which is available on the subject. This should include a comparison between the experimentally established threshold and toxic concentrations of the main pollutants with the actually recorded levels of their concentrations in marine basins. In this way it is possible to obtain rough answers to a few of the above questions as to the nature of the biological consequences of chemical pollution of the marine environment.

The validity of the conclusions arrived at as a result of the comparison of toxicological (experimental) and biogeochemical (natural) data will depend, in the first place, on the validity of extrapolation of the experimental results to conditions actually occurring in nature. It should be recalled in this connection that in organising our experiments we attempted to work under conditions which resemble, as far as possible, the natural conditions of organisms in the marine environment (as regards the ratio between the biomass or the number of individuals and the environment, temperature, salinity, etc.). In addition, experiments with certain forms of plankton were set up *in situ*. As a result, the extrapolation of experimental data becomes more realistic, but a complete similarity between the experimental and the natural conditions is practically impossible to attain. This limitation is perfectly natural, and is operative in all experimental ecological studies of the reactions and responses of biological systems to the action of some environmental factor.

*Tables 10.1–10.3* summarise data on the operative pollutant concentrations acting on the principal groups of marine organisms, based on all the results which were discussed above. They also give the concentration levels of chemical components in sea water for pelagial waters, neritic zones and under conditions of local pollution.

The most important general conclusion which may be drawn from the data presented here is that the active concentration ranges of the principal toxicants are of the same order of magnitude as those actually found in the euphotic layer of sea water in neritic zones and in inland seas. The minimum toxic concentration levels of a number of toxicants (petroleum and petroleum products, DDT, PCB, mercury, copper) may even be equal to their concentrations actually found in pelagial waters of the World Ocean, while the threshold concentrations are actually attained in almost every case.

These conclusions, which are based on all the available data concerning the major groups of marine organisms and the most frequently occurring pollutants, indicate that the present-day pollution level has a *real toxic effect* on marine flora and fauna. We shall discuss this conclusion—which is obviously of major importance to the problem of pollution as a whole—in more detail, with special emphasis on the diversity of the responses of components of marine biocoenoses to man-made alterations in the composition of the environment.

## 'Ecological targets' of disturbing factors in marine communities

The data in *Tables 10.1–10.3* indicate that the amplitudes of threshold concentrations and even more so of toxic concentrations for most groups of marine organisms are rather extensive. In isolated cases the toxicity levels comprise 5 or 6 orders of magnitude. As has been repeatedly stressed above in the interpretation of the experimental data, this is undoubtedly a manifestation of group features of the response of aquatic organisms to toxicants of various identities. At the same time, the diversity of the results is also undoubtedly due to the different experimental conditions under which they were obtained. The differentiation of groups of organisms by their specific resistance to any given pollutant is therefore difficult. An ecologically more effective approach is a comparative analysis of minimum

**Table 10.1** LEVELS OF TOXIC* AND THRESHOLD† CONCENTRATIONS (IN mg/ℓ) OF ADDITIONS OF MERCURY, CADMIUM AND LEAD IN SEA WATER FOR THE MAIN GROUPS OF MARINE ORGANISMS

| *Organismic group and habitat zone* | *Mercury concentration* | | *Cadmium concentration* | | *Lead concentration* | |
|---|---|---|---|---|---|---|
| | *Toxic* | *Threshold* | *Toxic* | *Threshold* | *Toxic* | *Threshold* |
| Unicellular algae | $10^{-4}$–1 | $10^{-4}$–$10^{-2}$ | $10^{-2}$–1 | $10^{-3}$–$10^{-1}$ | $10^{-2}$–$10^{-1}$ | $10^{-2}$–$10^{-1}$ |
| Macrophytes | $10^{-2}$–10 | $\leqslant 10^{-2}$ | – | – | – | – |
| Protozoa | – | $10^{-3}$–$10^{-2}$ | – | – | $10^{-1}$–$10^{3}$ | – |
| Crustaceans | $10^{-4}$–10 | $10^{-5}$–$10^{-3}$ | $10^{-2}$–10 | $<10^{-2}$–$10^{-1}$ | $10^{-2}$–$10^{2}$ | $10^{-2}$ |
| Molluscs | $10^{-3}$–10 | $10^{-3}$–10 | $10^{-1}$–10 | $<10^{-2}$–$10^{-1}$ | ~1 | – |
| Annelids | $10^{-3}$–10 | – | 1–$10^{2}$ | <1 | $10^{-1}$–$10^{2}$ | $10^{-1}$–10 |
| Fishes | $10^{-3}$–$10^{2}$ | $10^{-4}$–$10^{-2}$ | $10^{-1}$–10 | $10^{-2}$–$10^{-1}$ | $10^{-2}$–$10^{2}$ | $10^{-2}$–10 |
| All groups (minimum levels) | $10^{-4}$–$10^{-2}$ | $10^{-5}$–$10^{-2}$ | $10^{-2}$–1 | $10^{-3}$–1 | $10^{-2}$–1 | $10^{-2}$–$10^{-1}$ |
| Levels in euphotic layer | | | | | | |
| Pelagic zone of World Ocean | $10^{-5}$–$10^{-4}$ | | $10^{-5}$–$10^{-4}$ | | $10^{-5}$–$10^{-3}$ | |
| Neritic zone and inland seas | $10^{-4}$–$10^{-3}$ | | $10^{-3}$–$10^{-2}$ | | $10^{-2}$–$10^{-1}$ | |
| Zones of local pollution | $10^{-3}$–$10^{-2}$ | | $>10^{-2}$ | | $10^{-1}$–1 | |

*Toxic concentrations: limits of content in medium at which the measured indicators, mainly biological, reliably decreased by more than 50 per cent of control in experiments lasting for not less than 2–4 days.
†Threshold concentrations: minimum levels in medium at which the measured indicators (biological and physiological–biochemical) decreased by 50 per cent of control in experiments comparable in length with ontogenesis of the given organism.

**Table 10.2** LEVELS OF TOXIC AND THRESHOLD CONCENTRATIONS* (IN mg/ℓ) OF ADDITIONS OF COPPER, ZINC AND ARSENIC IN SEA WATER FOR THE MAIN GROUPS OF MARINE ORGANISMS

| *Organismic group and habitat zone* | *Copper concentration* | | *Zinc concentration* | | *Arsenic concentration* | |
|---|---|---|---|---|---|---|
| | *Toxic* | *Threshold* | *Toxic* | *Threshold* | *Toxic* | *Threshold* |
| Unicellular algae | $10^{-2}$–1 | $10^{-3}$–$10^{-2}$ | $<10^{-1}$–10 | $<10^{-1}$–1 | >1 | $10^{-1}$–1 |
| Macrophytes | $10^{-2}$–$10^{2}$ | $<10^{-2}$ | $10^{-1}$–$10^{-2}$ | – | – | – |
| Protozoa | $10^{-2}$–$10^{2}$ | – | $<10^{-1}$–$10^{3}$ | – | – | – |
| Crustaceans | $10^{-3}$–$10^{2}$ | $<10^{-2}$ | $10^{-2}$–$10^{2}$ | $<10^{-2}$–$10^{-1}$ | – | – |
| Molluscs | $10^{-2}$–10 | $10^{-2}$–$10^{-1}$ | $10^{-1}$–$10^{2}$ | $10^{-1}$ | – | – |
| Annelids | $10^{-1}$–10 | $10^{-3}$–$10^{-1}$ | $10^{-2}$–10 | $<10^{-2}$–$10^{-1}$ | – | – |
| Fishes | $10^{-2}$–1 | $10^{-2}$–$10^{-1}$ | $10^{-2}$–$10^{2}$ | $10^{-2}$–1 | 1–10 | – |
| All groups (minimum levels) | $10^{-3}$–$10^{-1}$ | $10^{-3}$–$10^{-2}$ | $<10^{-2}$–$10^{-1}$ | $<10^{-2}$–$10^{-1}$ | – | – |
| Levels in euphotic layer | | | | | | |
| Pelagic zone of World Ocean | $10^{-3}$–$10^{-2}$ | | $10^{-3}$–$10^{-1}$ | | $10^{-4}$–$10^{-2}$ | |
| Neritic zone and inland seas | $10^{-3}$–$10^{-1}$ | | $10^{-2}$–$10^{-1}$ | | – | |
| Zones of local pollution | $10^{-1}$–1 | | $10^{-1}$–1 | | – | |

*See footnote to *Table 10.1*

**Table 10.3** LEVELS OF TOXIC AND THRESHOLD CONCENTRATIONS* (IN mg/ℓ) OF ORGANIC MATTER IN SEA WATER FOR THE MAIN GROUPS OF MARINE ORGANISMS

| *Organismic group and habitat zone* | *Concentrations of dissolved oil products* | | *Concentrations of organochlorine toxicants* (DDT, PCB, etc.) | | *Concentrations of detergents* | |
|---|---|---|---|---|---|---|
| | *Toxic* | *Threshold* | *Toxic* | *Threshold* | *Toxic* | *Threshold* |
| Bacteria | $\sim 10^3$ | $10$–$10^2$ | – | – | – | – |
| Unicellular algae | $10^{-1}$–$10^3$ | $10^{-2}$–$10^2$ | $10^{-5}$–$10^{-1}$ | $10^{-5}$–$10^{-3}$ | $1$–$10^3$ | $10^{-1}$–$10^2$ |
| Macrophytes | $10^2$–$10^4$ | $10^2$–$10^3$ | – | – | $1$–$10^2$ | $<1$–$10^2$ |
| Crustaceans | $10^{-1}$–$10^5$ | $10^{-2}$–$10^2$ | $10^{-5}$–$10^{-1}$ | $10^{-5}$–$10^{-3}$ | $1$–$10^3$ | $<1$–$10$ |
| Molluscs | $10^{-1}$–$10^5$ | $<1$–$10^4$ | $10^{-2}$–$10$ | $10^{-3}$–$1$ | $10^{-1}$–$10^3$ | – |
| Annelids | – | – | – | – | $10^{-1}$–$10$ | – |
| Fishes | $10^{-2}$–$10^5$ | $10^{-2}$–$10$ | $10^{-4}$–$1$ | $<10^{-4}$–$10^{-3}$ | $1$–$10^3$ | – |
| All groups (minimum levels) | $10^{-1}$–$10^2$ | $10^{-2}$–$10$ | $10^{-5}$–$10^{-2}$ | $10^{-3}$–$10^{-5}$ | $10^{-1}$–$1$ | – |
| Levels in euphotic layer | | | | | | |
| Pelagic zone of World Ocean | $10^{-3}$–$10^{-2}$ | | $10^{-6}$–$10^{-4}$ | | $10^{-2}$ | |
| Neritic zone and inland seas | $10^{-2}$–$10$ | | $>10^{-4}$ | | $10^{-3}$–$1$ | |
| Zones of local pollution | $10^{-1}$–$10^2$ | | – | | $1$–$10$ | |

*See footnote to *Table 10.1*.

toxic and/or threshold concentrations. Such data were obtained under more uniform experimental conditions (mostly in long-term experiments); moreover, the concentration levels of the toxicants used in such experiments are closest to those recorded under real conditions of chemical pollution.

We have seen (Chapter 9) that the ranges of minimum concentrations corresponding to the incipient appearance of toxic effects in organisms of various taxonomic and ecological groups are relatively narrow and show only minor differences. This does not mean that there is no selective resistance to toxic effects among different groups of marine organisms or even within one particular group. Significant differences in the responses to different toxic effects were noted even for closely related species. *A fortiori*, such differences will also be displayed by groups of organisms which differ in their organisational and ecological levels.

Even if these limitations and diversification of experimental data are allowed for, it is evident that groups such as macrophytes and segmented worms are particularly resistant to most kinds of toxicants. These are probably the most resistant components of marine communities, as confirmed by observations of evolution with time of the populations and specific structures of biocoenoses in the zones of local pollution of marine basins (Nelson-Smith, 1970).

Since the biological reactions and responses of different groups of aquatic organisms to toxic factors are different, there is reason to accept the existence of what might be called ‘ecological targets’ in the sea—i.e., the existence of species, populations or even higher-level groupings, as well as of links in the production chain which are particularly sensitive to the effects of certain pollutants. It is these ecosystem components and production stages which ultimately determine the effects and consequences of man-made changes in the material composition of marine biotopes.

It is well known that the concept of targets of harmful effects has recently proved very fruitful in radiobiology in investigating the action of ionising radiation on cellular and sub-cellular structures of living organisms (Timofeev-Resovskii *et al.*, 1968). In our view, there is no reason why an identical approach should not be adopted in ecotoxicological studies of mechanisms and effects of toxicity at various organisational levels of marine life, with allowance for the specificity of hydrobiological processes.



As more data become available, and as future investigations are carried out, the concept of 'ecological targets' for injurious factors may be expected to become better defined and more useful in ecotoxicological research. Even now, the available data are sufficient to identify certain groups of organisms, growth stages and links in the marine food chain, which may be related to the concept of 'ecological target' for a given toxic factor.

An interesting feature is the high sensitivity of the productive characteristics of marine phytoplankton, particularly, photosynthetic activity, to even low pollution levels. Unlike other groups of organisms (other than macrophytes), unicellular algae are not typically eliminated in a toxic zone, but their production is suppressed. Production is ecologically the most important process, and one which determines all transformation, material transfer and energy transfer effects in marine ecosystems.

There is reason to believe that all persistent abnormalities of photosynthetic generation of new organic material in the photic layer ultimately affect structural, functional (including trophic) and other important vital and evolutionary parameters of marine communities and ecosystems. Thus the sensitivity of phytoplankton to the action of pollutants does not merely result in decreased production in the presence of low toxicant concentrations, but also in an abundant growth and development of certain species of monocellular algae. These eutrophication effects are in response to the excessive enrichment of the environment in biogenic substances, including organic pollutants of different composition and origin.

Numerous examples have been reported of ecological abnormalities caused by suppression or by the enhancement of primary production in the presence of pollutants in marine basins (Bartsch, 1970; Moore, 1970; Stirn, 1970; Woodwell, 1970; O'Sullivan, 1971; Warren, 1971; Aubert and Aubert, 1972; James and Head, 1972; Topachevskii and Polikarpov, 1973). Such effects, caused by man-made eutrophication, are regional in size; examples are the Baltic Sea (Bykova, 1971; Fonselius, 1972; Nikolayev, 1974), the North Sea (James and Head, 1972; Folkard and Jones, 1974) and the Adriatic Sea (Stirn, 1970). A significant, persistent decrease in primary production, with accompanying disruption of trophic interrelationships and structures of biocoenoses of abundant species of phytoplankton and zooplankton, due to pollution, was also observed in coastal and estuarine regions of India (Gopalakrishnan, 1972), Hawaiian Islands (Clutter, 1972) and elsewhere. Such anomalies may affect not merely isolated seas, but also extensive oceanic regions such as the Northern Atlantic, where a marked decrease in the number of species of phytoplankton and a decrease in zooplankton biomass has been observed over the past 22 years (Glover *et al.*, 1972). A negative correlation was observed in the same region between the level of chemical pollution of surface waters and chlorophyll *a* content (Oradovskii *et al.*, 1975).

The above examples of primary production impairment, as well as literature data and our own results discussed above, indicate that primary producers are the target for polluting factors in the marine environment. It should be emphasised that the principal danger is not so much the algicidal effect, i.e., the elimination of isolated species of monocellular algae, as the disruption of the rhythm and nature of the process of photosynthesis and production of new organic matter in the euphotic layer of seas and oceans. In other words, a kind of pathology of the productive process in marine ecosystems results under conditions of chronic pollution.

As far as the marine fauna is concerned, there is probably no single ecological or taxonomic group of aquatic organisms which would be most sensitive to one or more environmental toxic factors. The non-specific biological effects of these factors (at least as far as the toxicants studied here are concerned) also prevents any differentiation between the degree of resistance of the various physiobiochemical and other systems of vital activities of marine animals. Planktonic crustaceans tend to be more sensitive to the effects of toxicants, probably owing to the accumulation of large amounts of pollutants due to their filter-feeding mechanism.

On the strength of the above data, and of those given in *Appendixes 1–9*, it is possible to assign to each individual toxicant a number of organisms which are least resistant to it. Thus, for instance, *Acartia tonsa*, *Eurytemora affinis* and certain other crustaceans are highly sensitive to mercury and copper (*Appendixes 5* and *6*); *Acartia clausi*, *Paracalanus parvus*, *Oithona nana* and *Penilia avestria* are highly sensitive to petroleum products (*Appendix 1*), etc. However, these and other organisms, although being highly sensitive to toxic effects, must not be considered as ‘ecological targets’ of any given toxic factor, since we are unable to identify any specific reason for their low resistance to toxicants (such as some specific toxication mechanism or selective accumulation and deposition of the toxicant in organs and tissues).

It is quite probable, moreover, that other species of aquatic organisms, which have not yet been studied (and which are far more numerous than those which have), display a similar or even higher sensitivity to toxic factors. We have seen that such sensitivity varies greatly with the experimental conditions. In other words, the available comparative data on the resistance of various species of marine animals to toxicants are too few to establish any objective criteria for the identification of any one or more groups of organisms as specifically reacting to any particular toxic factor.

In spite of the above, such ‘ecological targets’ for toxic factors may well exist among the fauna of marine basins. We may recall in this context the general tendency of small marine organisms to display a lower resistance to toxicants (Chapter 9).

A study of *Figures 9.7–9.11* fails to reveal the existence of any sharp zone of dimensional characteristics of organisms which are the most sensitive to pollutants. Nevertheless, the actual concentrations of toxicants present in marine regions (*Tables 10.1–10.3*) indicate that it is small species and forms of aquatic organisms (average size of about 1 mm or less) that are most affected. This dimensional population includes many abundant species of planktonic filter-feeders and embryonic and post-embryonic growth stages of most species of marine fauna. It is these biocoenotic components, which act as concentrators of pollutants, that are highly sensitive to toxic factors. Accordingly, we are entitled to consider such species and forms of animals as 'ecological targets' of toxic factors in the sea.

In contrast, other workers are of the opinion that small organisms, with a large production-to-biomass ratio, are the most stable components of ecosystems (Budyko, 1975), whereas sudden changes in the environment primarily affect higher organisms with a considerable degree of trophic specialisation (Kamshilov, 1974). Obviously, the high reproduction rate of the small forms favours their adaptation to changing environments in which they are more capable of genotypic fixation of the acquired features of phenotypic adaptation of organisms and cells to the new conditions of existence.

However, this view does not seem to contradict our own conclusions or the data indicating a lower resistance to toxicants of marine organisms of small body size. This is because the physiological and biochemical compensation of a given change in the environment by the organism may take place only if the change remains within the range of tolerable doses. However, we have here extreme, opposite situations, in which the high reproduction rate of small organisms does not ensure a genotypic fixation of adaptive characteristics or compensation mechanisms of toxic effects.

It should also be borne in mind that Budyko's (1975) conclusions regarding the high resistance of small organisms apply to the changes of natural, non-specific factors of the environment, such as temperature, salinity and oxygenation. Strictly speaking, there is no *a priori* reason why these conclusions should be extrapolated to apply to the effects of man-made factors, especially of toxicants such as DDT and PCB, which have no natural analogues in the sea. It should also be clear that even if such changes occur in the natural background, e.g., in the background concentration of heavy metals, the compensation mechanisms of organisms for such effects may well be altogether different from the natural transformations which accompany a temperature or a salinity adaptation to non-specific factors.

Finally, an important conclusion in this context refers to the high sensitivity of aquatic organisms in their early developmental stages to harmful effects, and to toxicants in particular. This has been repeatedly demonstrated and can be regarded as conclusively established. It is supported both by the experimental data discussed above and by the published results of field observations (Sprague and Duffy, 1971; Kostyuchenko, 1973; Gopalakrishnan, 1974). Thus, for instance, the proportion of pelagic dead eggs of sea-water fish in oil pollution zones may be as high as 80 per cent, which is much higher than in relatively clean water basins (Kostyuchenko, 1973); the correlation coefficient between the proportion of dead eggs and the concentration of petroleum products in the Black Sea is 0.78–0.94 for three species of commercially important fishes (Kostyuchenko, 1975).

Similar data have been reported for fresh waters. This is new evidence that the early developmental stages and small aquatic organisms of all species and forms in fact serve as 'ecological targets' for individual pollutants.

The elimination of higher organisms as a result of relatively rapid, unfavourable changes occurring in the environment, and the disappearance of commercially valuable fishes from polluted aquatic basins may possibly be the result not so much of the direct effect of the pollutants on adult individuals as the consequence of the harm inflicted on populations in their early, more sensitive stages. Large aquatic organisms, such as fish, necessarily respond to the presence of toxic factors in the environment by appropriate modifications of their physiological and biochemical mechanisms, which result in the establishment of a new functional level. Such reactions and transformations in the homeostatic regulation are fast enough and radical enough to permit many species of invertebrates and certain fish species to survive for a long time in unfavourable (toxic) environments. We may recall in this connection that the accumulated concentrations of mercury in fish and invertebrates inhabiting marine regions with a high concentration of this metal have on occasions proved to be lethal to man (Ui, 1969; Jernelov and Lann, 1971). Eggs and larvae of marine animals would be unable to survive or to develop normally in the presence of such high concentrations (several hundred μg/ℓ).

This brief analysis of ecotoxicological data indicates that the components of marine communities which are the most sensitive to chemical pollution and whose response to it is the most rapid (ecological targets for toxic factors) are small species and forms of aquatic organisms. The reasons for this situation were explained in Chapter 9, but yet another reason is connected with the existence of two kinds of phenotypic reactions of organisms to changes in the environment—homeostatic regulation and anastatic adaptation (Khlebovich and Berger, 1975). It may be assumed that the evolutionary earlier anastatic regulation, which takes place at the level of cells and tissues, is typical of lower organisms which are usually small. Our knowledge of the mechanisms of anastatic regulation is still incomplete, but they are probably incapable of optimising the vital functions to the same extent as the more sophisticated system of homeostatic regulation in higher animals in the presence of equal doses of harmful factors.

## Character and scale of biological consequences of chemical pollution of the marine environment

Direct observations confirm that man-made alterations of the marine environment are on a global scale. However, such changes are not evidence of large-scale biological and ecological abnormalities in the World Ocean, but are merely one of their features. Direct proofs of such abnormalities can at present be given only for regions of severe chronic pollution, for which the results of several years' observations of the abundance, biomass and productivity of the principal groups of marine organisms are available. Examples of local and regional man-made impairment of hydrobiological phenomena in the seas were given above. Since each ocean may be considered as a single, large-scale biogeocoenosis (Bogorov, 1970), we may conclude that any permanent changes in communities inhabiting certain regions will ultimately affect the populations in other regions and the World Ocean as a whole. However, this consideration, which is of a general nature, is only one of many which must be borne in mind when dealing with the possibility of global biological consequences of pollution of the marine environment.



More realistic evidence in favour of the large-scale harmful effect of chemical pollution on marine flora and fauna is the coincidence between the minimum toxic and threshold concentrations of individual pollutants to the principal groups of marine organisms, and the actually recorded concentrations of these pollutants in seas and oceans (*Tables 10.1–10.3*). This conclusion, which is based on observations actually carried out in sea basins, seems to be particularly convincing, since it is supported by numerous ecotoxicological and biogeochemical data.

We may also recall the general pattern of pollution of seas and oceans (Chapter 2) and, in particular, the large coincidence of the zones of maximum pollution and the zones of the most abundant biomasses and highest productivity in the World Ocean. This situation, which is illustrated in *Figure 10.1*, is particularly marked in neritic regions at moderate latitudes of the northern hemisphere (shaded areas in *Figure 10.1*). In these latitudes the specific diversity of marine animals has undergone a considerable decrease (Zenkevich, 1948), accompanied by a decreased stability of biocoenoses to stress and toxic effects.

One of the factors which aggravates the consequences of pollution is the localisation of toxic pollutants at the water–atmosphere interface and water–seabed interface, i.e., in the hyponeuston and benthos biotopes, respectively. This widespread effect is always dangerous, since it affects the most sensitive early growth stages of numerous species of marine fauna. The particularly high pollution levels of the so-called contact zones (Patin *et al.*, 1976) include not only hyponeuston and benthos biotopes but also sea-shores and estuarine regions which are highly important in the life and productivity of marine communities.

Experimental data and a number of field observations clearly confirm the low resistance to toxicants of marine animals at early developmental stages. These and similar data obtained for marine plants (Polishchuk, 1971) indicate that the individual development of almost all aquatic organisms (except for bacteria and protozoa, about which very little is known) includes *critical periods*, during which the organisms are particularly sensitive to harmful or lethal effects of certain environmental factors. Such periods undoubtedly coincide with early developmental stages, but may also occur during other stages—for example, the moulting period of crustaceans or the spawning of fish.

The length of the critical periods and the relative proportion of the individuals actually in the critical stage are clearly different for different species and populations. Thus, for instance, in the case of single-spawning fish, the adult forms after spawning constitute a minor proportion of the

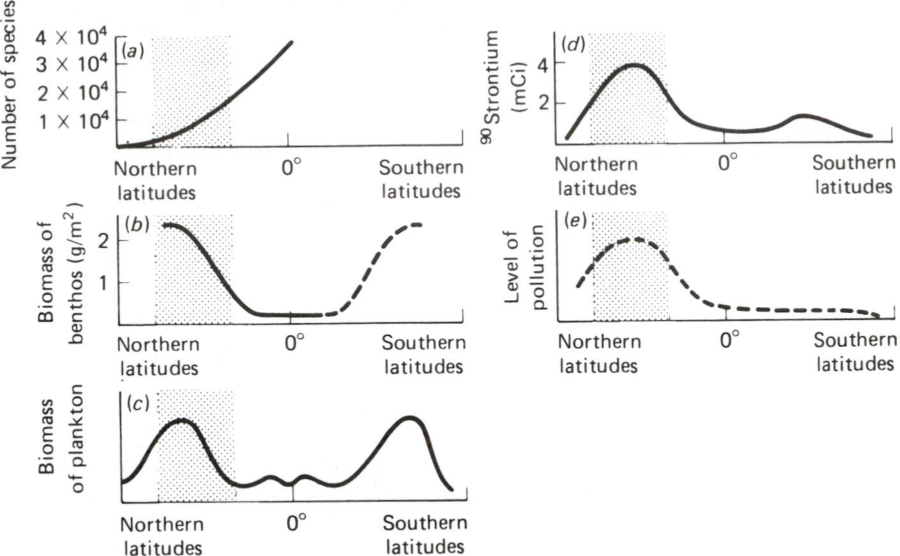


**Figure 10.1** Zonality of the distribution of some biological characteristics of the World Ocean and levels of pollution of the surface (euphotic) layer of sea water. (a) Number of species of marine animals (Zenkevich, 1948); (b) biomass of benthos (Vinogradov, 1963); (c) biomass of plankton (Zenkevich, 1960); (d) cumulative global stores of $^{90}$strontium (Patin, 1965a); (e) hypothetical curve of change in levels of chemical pollution (crosshatch zones indicate latitudinal belt of increased pollution)

population represented by embryonic and larval forms (Bocharov, 1975). During such periods, even if these are relatively short, the response and the abundance of the population under conditions of pollution will depend on the resistance of the most sensitive, early stages of ontogenesis. It is important to note in this context that the high reproduction rate of small animals, which involves a certain adaptation potential, is responsible for the less favourable situation of such organisms as compared with larger organisms; the relative duration of breeding and growth of such organisms in embryonic and larval stages in their entire life cycle is much longer than in the case of larger organisms. This is additional evidence for the low resistance to toxicants of small organisms, and for the idea according to which they represent 'ecological targets' for the pollutants present in the sea. The abundance of larval forms (e.g., of Copepoda) is illustrated by the fact that the biomass of the larvae may be up to 10 times larger than that of adults (Bocharov, 1975).

It is known that zooplankton comprises not only holoplanktonic forms, pelagic eggs and larvae of fish and certain other animals, but also very large quantities of planktotrophic larvae of benthic crustaceans, worms, molluscs and other invertebrates. As distinct from deep-sea waters, in shallow near-shore waters indirect growth due to the presence of pelagic larvae becomes a mass phenomenon (Mileikovskii, 1976; Konstantinov, 1972).

This parallels the increasing gradient of pollution levels (Chapter 2). Thus, benthos populations are subjected to repeated harmful pollution effects during their most sensitive embryonic and larval stages and in the most highly polluted plankton and hyponeuston biotope (the euphotic layer of the neritic zone).

The above data make it possible to draw certain very approximate but nonetheless revealing conclusions as to the extent and nature of 'non-obvious' effects of the global pollution of the World Ocean.

One such conclusion is based on the reduced growth and reproduction rate of unicellular marine algae and their communities at levels of pollution equal or nearly equal to those in the surface layer of the neritic zone (Chapter 7). In view of the fact that some of the results discussed above have been obtained *in situ* during long-term experiments, and since direct evidence for the impairment of production and of structural characteristics of marine phytoplankton in environments altered by man is available (Stirn, 1970; James and Head, 1972; Kasymov, 1975; Kiryushina *et al.*, 1975; Oradovskii, 1975; Patin and Ibragim, 1975), we are entitled to speak of a global inhibition of photosynthesis, at least as concerns the most polluted neritic zone of the World Ocean. Thus, for instance, a global 10 per cent decrease of primary production (which is not an exaggerated estimate under the present-day conditions of pollution, and in view of the experimental data given above and their interpretation) will result in a correspondingly lower production rate at other trophic levels, up to the nekton, for which—as compared with the 'control' situation of pollution-free water—it may amount to 20 million tons of raw substance per year, including commercially important fishes. This evaluation, which is based on the known values of the biomass and World Ocean production, and on the assumption of balanced trophic levels (Bogorov, 1970), clearly shows the large-scale dimensions of the 'latent', and so far little studied consequences of global pollution of the marine environment. For the sake of comparison, it may be noted that the global commercial production of all aquatic organisms is about 70 million tons catch per year (Moiseev, 1969).

Obviously, the trophic cycle is never fully balanced, and its deviation from equilibrium shows large variations under different conditions (Karpevich, 1976). However, for the World Ocean as a whole, the biomass distribution of consumers, including fish, ultimately depends on the amount of primary production (Parin, 1968). It should also be noted that phytoplankton in eutrophic waters constitutes not only the staple food of microzooplankton, but is also the initial substrate of the so-called detrital food chain and dissolved organic matter. It has been shown by Sorokin (1975) that the latter two components, together with the associated bacterioplankton, play a trophically important part in marine ecosystems by supplying food to benthic and deep-sea communities, and providing a kind of buffering stock of nutritive substances in the World Ocean.

It is clear from the above that a constant, large-scale inhibition of primary production may not affect the food supply of herbivorous zooplankton, but will ultimately impair the overall trophic structure of marine ecosystems which ensure the realisation of photosynthetic processes.

Other ecological aspects of man-made alteration of the rate of photosynthesis in the World Ocean and the oxygen and $CO_2$ balance in the biosphere (Barinov, 1972; Davitaya, 1975) are outside the scope of this book, but we may point out the possible biospheric consequences of the man-made deficiency of oxygen, which is probably responsible for the decrease in the intensity of the magnetic field, noted during the past few years, with its unfavourable influence on life on Earth (Barinov, 1972). Man-made abnormalities in the photosynthesis of marine phytoplankton may (and probably do) also impair the processes of photosynthetic aeration of water in certain zones and the intensity of oxidative mineralisation and self-purification of water basins.

The above rough estimate of the global inhibition of production in the World Ocean should be considered as highly conservative, since toxic effects and consequences of pollution at all trophic levels are not considered. According to Cousteau (1971), the life intensity in seas and oceans has decreased as a result of pollution by more than 30 per cent during the past 30 years, as evidenced by direct observations of marine flora and fauna. Such an estimate is necessarily both approximate and subjective, but the important point is that direct observations in fact confirm the views according to which the biological consequences of pollution of the marine environment in fact occur on a global scale.

## Conclusions

(1) The minimum threshold values and toxic concentrations of the most frequently occurring pollutants for the main groups of marine organisms, except for certain macrophytic algae, worms and bacteria, coincide with the maximum concentrations of these substances under conditions of regional and global pollution.

(2) Aquatic organisms which are most sensitive to the effect of pollutants, and which may be referred to as 'targets of toxic factors', include monocellular algae, the photosynthesis of which becomes inhibited in the presence of low concentrations of many toxicants in the environment, as well as certain species of microzooplankton filter-feeders and early developmental stages of benthic and nektonic organisms (particularly in hyponeuston and benthos biotopes, in which all pollutants are locally concentrated).

(3) Analysis of the results of ecotoxicological and biogeochemical investigations indicates that the concentrations of the most frequently occurring pollutants actually present in sea water in fact have a significant, large-scale (global) deleterious effect on the biotic population of the World Ocean. The probable decrease in the annual production of nekton, including commercially important fish, produced by large-scale marine pollution is at least 20 million tons of biomass, as compared with the optimum production in a 'pure' World Ocean.

## References


Aubert, M. and Aubert, J. (1973). Pollutions marines et amenagement des rivages. C.E.R.B.O.M., p. 308. Nice, France.

Barinov, G.V. (1972). Biosphere rhythms and the problem of preserving oxygen balance. *Zh. obshch. Biol.*, 33(5), 771–778.

Bartsch, A.F. (1970). Water pollution—an ecological perspective. *J. Wat. Pollut. Control Fed.*, 42(5), Part 1, 819–823.

Bocharov, Yu.S. (1975). Some problems of toxicology in early stages of ontogenesis of animals. *Zh. obshch. Biol.*, 36(6), 847–858.

Bogorov, V.G. (1970). Quantitative evaluation of animal and plant populations of the ocean. DAN SSR. 162, 1181–1184.

Budyko, M.I. (1975). Ecological factors of evolution. *Zh. obshch. Biol.*, 36(1), 36–47.

Bykova, A.N. (1971). *Pollution of the Baltic Sea*, p. 20. Moscow: TSNHTEIRKH.

Clutter, R.J. (1972). Subtle effects of pollution on inshore tropical plankton. In *Marine Pollution and Sea Life*, pp. 435–439. London: Fishing News (Books) Ltd.

Cousteau, J.Y. (1971). Statement on global marine degradation. *Biol. Conserv.*, 4(1), 61–66.

Davitaya, F.F. (1975). *Atmosphere and Biosphere—Past, Present and Future*, p. 36. Leningrad: Gidrometeoizdat.

Fedorov, V.D. (1974). Stability of ecosystems and its measurement. *Izv. Akad. Nauk SSSR, Ser. Biol.*, 3, 402–415.

Fedorov, V.D. and Sokolova, S.A. (1972). Stability of planktonic community and some environmental characteristics. *Okeanologiya*, 12(6).

Folkard, A.R. and Jones, P.G.W. (1974). Distribution of nutrient salts in the southern North Sea during early 1974. *Mar. Pollut. Bull.*, 5(12), 181–185.

Fonselius, S.H. (1972). On biogenic elements and organic matter in the Baltic. *Ambio Spec. Rep.*, 1.

Glover, R.S., Robinson, G.A. and Colebrook, J.M. (1972). Plankton in the North Atlantic: An example of the problems and analyzing variability in the environment. In *Marine Pollution and Sea Life*, pp. 439–443. London: Fishing News (Books) Ltd.

Gopalakrishnan, V. (1974). Biological processes involved in pollution of coastal aquaculture waters. In *Proc. Indo-Pacific Fisheries Council, 15th Session, Wellington, New Zealand*, 18–27 Oct. 1972. pp. 1–7. FAO: Bangkok.

James, A. and Head, P.C. (1972). The discharge of nutrients from estuaries and their effect on primary productivity. In *Marine Pollution and Sea Life*, pp. 163–166. London: Fishing News (Books) Ltd.

Jernelov, A. and Lann, H. (1971). Mercury accumulation in food chains. *Oikos*, 22(3), 403–406.

Kamshilov, M.M. (1974). *Evolution of the Biosphere*, p. 253. Moscow: Nauka.

Karpevich, A.F. (1976). *Theory and Practice of Acclimatization of Aquatic Organisms*, p. 432. Moscow: Pishchevaya Promyshlennost.

Kasimov, A.G. (1975). Change in biological productivity of the Caspian Sea due to industrial pollution. In *2nd All-Union Scientific Conference on Problems of Water Toxicology, Baku*, pp. 38–39.

Khlebovich, V.V. and Berger, V.Ya. (1975). Some aspect of the study of phenotypic adaptations. *Zh. obshch. Biol.*, 36(1), 11–25.

Kiryushina, L.P., Trufanova Z.A. and Kopylova, Z.N. (1975). Group effect of toxicants on phyto- and zooplankton. In *Problems of Water Toxicology*, Part 1, pp. 107–109. Petrozavodsk.

Konstantinov, A.S. (1972). *General Hydrobiology*, p. 472. Moscow: Vysshaya Shkola.

Kostyuchenko, L.G. (1973). Development of ichthyoplankton in oil polluted regions. In *All-Union Symposium on Investigation of the Black and Mediterranean Sea, Utilization and Protection of their Resources*, Vol. 4, pp. 57–58. Kiev.

Mileikovsky, S.A. (1976). Effect of anthropogenic pollution of marine coastal and estuarine waters on the development, distribution, and settlement of pelagic larvae of bottom invertebrates. *Trudy Inst. Okeanol.*, 105. 249–270.

Mironov, O.G. (1972). *Biological Resources of the Sea and Oil Pollution*, p. 105. Moscow: Pishchevaya Promyshlennost.

Moiseev, P.A. (1969). *Biological Resources of the World Ocean*, p. 339. Moscow: Pishchevaya Promyshlennost.

Moore, N.W. (1970). The ecological impact of pollution. *IUCN Publs New. Ser.*, 17, 76–81.

Nelson-Smith, A. (1975). The problem of oil pollution of the sea. In *Advances in Marine Biology*, pp. 215–306. London: Academic Press.

Nelson-Smith, A. (1975). Oil pollution and marine ecology. *Proc. Challenger Soc.*, 4(6), 267.

Nikolayev, N.I. (1974). Basic tendencies in biology of the present-day Baltic. *Okeanologiya*, 14(6), 52–56.

Orakovskii, S.G., Simonov, A.I. and Yushchak, A.A. (1975). Investigation of the character of distribution of chemical pollutants in the Gulf Stream zone and their effect on primary production of oceanic waters. *Met. Gidrol.*, 2, 48–58.

O'Sullivan, A.J. (1971). Ecological effects of sewage discharge in the marine environment. *Proc. R. Soc.*, B117(1048), 331–351.

Parin, N.V. (1968). *Ichthyofauna of the Ocean Epipelagic Zone*, p. 186. Moscow: Nauka.

Patin, S.A. (1976). Ecological aspects of global pollution of the marine environment. *Okeanologiya*, 16(4), 621–626.

Patin, S.A. (1977). Chemical pollution and its effect on marine organisms and communities. In *Oceanology. Biology of the Ocean, Vol. 2. Biological Productivity of the Ocean*, pp. 322–333. Moscow.

Patin, S.A., Aivazova, L.E. and Tkachenko, V.N. (1976). Effect of oil, DDT and heavy metals on some species of phytoplankton of brackish seas. In *Symposium of Countries of the Council of Economic Cooperation on the Problem of Pollution of Brackish Seas, Gdnya*, pp. 319–324.

Patin, S.A. and Ibragim, A.M. (1975). Effect of trace elements of a group of metals on primary production and phytoplankton of the Mediterranean and Red Seas. *Okeanologiya*, 15(4), 886–890.

Polishchuk, R.A. (1971). Effect of inorganic salts of Zn, Cu, and Hg on processes of photosynthesis and respiration of algae of the Mediterranean Sea. In *Research Expeditions in the Mediterranean Sea*, pp. 88–98. Kiev.

Sorokin, Yu.I. (1975). Heterotrophic microplankton as components of marine ecosystems. *Zh. obshch. Biol.*, 36(5), 716–730.

Sprague, J.B. and Duffy, J.R. (1971). DDT residues in Canadian Atlantic fishes and shellfishes. *J. Fish. Res. Bd. Can.*, 28(1), 59–63.

Stirn, J. (1970). Further contributions to the study of the bioproductivity in polluted marine ecosystems. *Revue int. Oceanogr. Med.*, 18–19, 21–27.

Timofeev-Resovskii, N.V., Ivanov, V.I. and Korogodin, V.I. (1968). *Application of Random Principle in Biology*, p. 225. Moscow: Atomizdat.

Ui, J. (1971). Mercury pollution of sea and fresh water, its accumulation into water biomass. *Rev. int. Oceanogr.*, 22–23, 79–128.

Warren, C.E. (1971). *Biology and Water Pollution Control*, p. 434. Philadelphia: W.B. Saunders Company.

Woodwell, G.M. (1970). Effects of pollution on the structure and physiology of ecosystems. *Science*, 168(3930), 429–431.

Zenkevich, L.A. (1948). Biologicheskaya struktura okeana. *Zool. Zh.*, 27(2), 80–95.

Zenkevich, L.A. (1960). Spetsialnaya kolichestvennaya kharakteristika glubokovodnoi zhizni v okeane. *Izv. Adad. Nauk SSSR, seriya geograficheskaya*, 2, 52–65.

CHAPTER 13

# Soil Erosion: Quiet Crisis in the World Economy



LESTER R. BROWN AND EDWARD C. WOLF

## Introduction

Over the past generation world food output has more than doubled. Coming at a time when little new land was brought under the plow, this was an impressive achievement. But we can now see that this remarkable feat has a high price: Some of the agricultural practices that boosted food production have also led to excessive soil erosion.

Spurred by both population growth and rising affluence, world demand for food climbs higher each year. In the face of this continuously expanding demand and the associated relentless increase in pressures on land, soil erosion is accelerating. Anson R. Bertrand, a senior official of the U.S. Department of Agriculture, has described the situation in the United States: "The economic pressure—to generate export earnings, to strengthen the balance of payments and thus the dollar—has been transmitted more or less directly to our natural resource base. As a result soil erosion today can be described as epidemic in proportion."[1]

Bertrand's perceptive, though sobering, linkage between economic forces and soil deterioration applies to other countries as well. Most countries feel pressure to feed their own people, rather than to expand food exports. But demands placed on soils are increasing worldwide. Each year the world's farmers must now attempt to feed 81 million more people, good weather or bad.[2]

Soil erosion is a natural process, one that is as old as the earth itself. But today soil erosion has increased to the point where it far exceeds the natural formation of new soil. As the demand for food climbs, the world is beginning to mine its soils, converting a renewable resource

This paper is an updated version of the chapter "Conserving Soils" that appeared in *State of the World 1984* (W. W. Norton & Co., 1984).



Lester R. Brown is President and Senior Researcher, Worldwatch Institute.

Edward C. Wolf is a Researcher, Worldwatch Institute.

into a nonrenewable one. Even in an agriculturally sophisticated country like the United States, the loss of soil through erosion exceeds tolerable levels on some 44 percent of the cropland. Indeed, the U.S. crop surpluses of the early eighties, which are sometimes cited as the sign of a healthy agriculture, are partly the product of mining soils.

The incessant growth in demand for agricultural products contributes to soil erosion in many ways. Throughout the Third World farmers are pushed onto steeply sloping, erosive land that is rapidly losing its topsoil. Elsewhere, such as the American Midwest, many farmers have abandoned ecologically stable, long-term rotations, including hay and grass, as well as row crops, in favor of the continuous row cropping of corn or other crops. In other areas farming has extended into semiarid regions where land is vulnerable to wind erosion when plowed.

The loss of topsoil affects the ability to grow food in two ways. It reduces the inherent productivity of land, both through the loss of nutrients and degradation of the physical structure. It also increases the costs of food production. When farmers lose topsoil they may increase land productivity by substituting energy in the form of fertilizer, or through irrigation to offset the soil's declining water absorptive capacity. Farmers losing topsoil may experience either a loss in land productivity or a rise in costs. But if productivity drops too low or costs rise too high, farmers are forced to abandon their land.

Grave though the loss of topsoil may be, it is a quiet crisis, one that is not widely perceived. And unlike earthquakes, volcanic eruptions or other natural disasters, this humanmade disaster is unfolding gradually. Often the very practices that cause excessive erosion in the long run, such as the intensification of cropping patterns and the plowing of marginal land, lead to short-term production gains, creating an illusion of progress and a false sense of security.

Over most of the earth's surface, the thin mantle of topsoil on which agriculture depends is six to eight inches thick. Although the depletion of this thin layer of soil may compromise economic progress and political stability even more than dwindling oil reserves, nowhere has the depletion of topsoil gained the attention paid to the depletion of oil reserves. Fifteen years ago, the public was largely unaware of the rate of oil depletion, but that changed with the oil price hikes of 1973 and 1979.

Governments everywhere have responded to the growing scarcity of oil, but such is not the case for soil. With only occasional exceptions, national agricultural and population policies have failed to take soil depletion into account. In part, the contrasting awareness of oil and soil depletion is the understandable product of differing levels of information. Estimates are regularly made for oil reserves, adjusting annually both for depletion through production and for new discoveries. Such a procedure does not exist for world soil reserves. Indeed, not until topsoil has largely disappeared and food shortages have developed or famine threatens does this loss become apparent.

**"Nowhere has the depletion of topsoil gained the attention paid to the depletion of oil reserves."**



Projections of the world supply and demand of food made in the late sixties and early seventies did not anticipate the slowdown in growth of world food output over the last decade. Nor did they anticipate the sustained decline since 1970 of per capita food production in Africa and the Andean countries of Latin America. One reason for this shortcoming may have been the failure to incorporate the effect of soil erosion on food production. Projections of world food production always incorporate estimates of future cropland area, but what has been lacking thus far has been any effort to project changes in inherent productivity of the projected cropland area.

To help remedy this shortcoming in world food supply projections we have undertaken an estimate of the worldwide loss of topsoil from cropland. Making such an estimate is not a simple matter of tabulating data from individual countries; unfortunately, few have attempted to measure their topsoil loss. We do know that if the current rate of topsoil loss through erosion continues there will be a sharp reduction in the amount of topsoil available by the end of the century. Largely because of population growth, we project a 19 percent decline in cropland per person between 1984 and the end of the century. But our projections of the amount of topsoil per person, assuming current rates of soil erosion continue, shows a decline of 32 percent from 1984 to 2000. The difference between the decline in cropland area per person and topsoil per person is significant, and this difference will affect food production trends.[3]

The United States is the only major food producer that has taken a systematic inventory of its topsoil. The Soil Conservation Service of the Department of Agriculture assesses soil resources and rates of erosion every five years. The first assessment was completed in 1977, the second in 1982. Data from the two studies indicates little change in the rate of excessive erosion of some 1.7 billion tons per year over this five-year span.[4]

Any estimate of world soil losses depends on fragments of data that exist for various parts of the world, as well as assumptions to fill in the gaps in the global data fabric. Our estimates draw on many sources. Direct evidence includes national soil surveys compiled for some countries; local studies by soil scientists within various countries; data from United Nations agencies such as the Food and Agriculture Organization and the United Nations Environment Program; and studies, both published and unpublished, from international aid agencies including the World Bank, the Agency for International Development and the Swedish International Development Authority. More indirect sources of evidence include hydrological studies of river silt loads; rates of reservoir sedimentation; meteorological studies of atmospheric dust flows; private communication with soil scientists and government officials throughout the world; and lastly, personal experiences and observations when traveling.



Our estimate of world topsoil loss from cropland is not highly refined and by no means final. Though inexact, it is presented here to draw attention to a process that will eventually undermine the world economy if not arrested. Without some sense of how fast soils are being lost, it will be difficult to mobilize the resources to save them.

## The Causes of Soil Erosion

The apparent increase in soil erosion over the past generation is not the result of a decline in the skills of farmers but rather of the pressures on farmers to produce more. In an integrated world food economy, the pressures on land resources are not confined to particular countries; they permeate the entire world. Many traditional agricultural systems that were ecologically stable as recently as mid-century, when there were only 2.5 billion people in the world, are breaking down as world population moves toward 5 billion.

Over the millennia, as the demand for food pressed against available supplies farmers devised ingenious techniques for extending agriculture onto land that was otherwise unproductive while still keeping erosion in check and maintaining land productivity. These techniques include terracing, crop rotations, and fallowing. Today, land farmed through these specialized techniques still feeds much of humanity. Although these practices have withstood the test of time, they are breaking down in some situations under the pressure of continuously rising demand.

In mountainous regions such as those in Japan, China, Nepal, Indonesia, and the Andean countries, construction of terraces historically permitted farmers to cultivate steeply sloping land that would otherwise quickly lose its topsoil. Centuries of laborious effort are embodied in the elaborate systems of terraces in older settled countries. Now the growing competition for cropland in many of these regions is forcing farmers up the slopes at a pace that does not permit the disciplined construction of terraces of the sort their ancestors built, when population growth was negligible by comparison. Hastily constructed terraces on the upper slopes often begin to give way. These in turn contribute to landslides that sometimes destroy entire villages, exacting a heavy human toll. For many residents of mountainous areas in the Himalayas and the Andes, fear of these landslides has become an integral part of daily life.

Research in Nigeria has shown how much more serious erosion can be on sloping land that is unprotected by terraces. Cassava planted on land of a 1 percent slope lost an average of 3 metric tons per hectare each year, comfortably below the rate of soil loss tolerance.

**"Throughout the Third World increasing population pressure and the accelerating loss of topsoil seem to go hand in hand."**



On a 5 percent slope, however, land planted to cassava eroded at a rate of 87 tons per hectare annually—a rate at which a topsoil layer of six inches would disappear entirely within a generation. Cassava planted on a 15 percent slope led to an annual erosion rate of 221 tons per hectare, which would remove all topsoil within a decade. Intercropping cassava and corn reduced soil losses somewhat, but the relationship of soil loss and slope remained the same.[5]

Throughout the Third World increasing population pressure and the accelerating loss of topsoil seem to go hand in hand. Soil scientists S. A. El-Swaify and E. W. Dangler have observed that it is in precisely those regions with high population density that "farming of marginal hilly lands is a hazardous necessity. Ironically, it is also in those very regions where the greatest need exists to protect the rapidly diminishing or degrading soil resources." It is this vicious cycle, set in motion by the growing human demands for food, feed, fiber, and energy, that makes mounting an effective response particularly difficult.[6]

In other parts of the world farmers have been able to cultivate rolling land without losing excessive amounts of topsoil by using crop rotations. Typical of these regions is the midwestern United States, where farmers traditionally used long-term rotations of hay, pasture, and corn. Fields planted in row crops, such as corn, are most susceptible to erosion. By alternating row crops with cover crops like hay, the average annual rate of soil erosion was kept to a tolerable level. Not only do crop rotations provide more soil cover, but the amount of organic matter that binds soil particles together remains much higher than it would under continuous row cropping.

As world demand for U.S. feedstuffs soared after World War II and as cheap nitrogen fertilizer reduced the need for legumes, American farmers throughout the Midwest, the lower Mississippi Valley, and the Southeast abandoned crop rotations to grow corn or soybeans continuously. The risks associated with this shift in cropping patterns have long been known. Research undertaken in Missouri during the thirties showed an increase in soil erosion from 2.7 tons per acre (1 acre equals 0.4 hectares) annually when land was in a corn-wheat-clover rotation to 19.7 tons per acre when the same land was planted continuously to corn. (See Table 1.) The lower rate is within established erosion tolerance levels, whereas the higher rate would lead to the loss of one inch of topsoil in less than a decade. Much of the decline in inherent soil fertility that occurs under row crops is being masked by advances in technology, particularly by the increasing use of chemical fertilizer.

**Table 1: Cropping Systems and Soil Erosion**

| Cropping System | Average Annual Loss of Soil |
|---|---|
| | (tons/acre) |
| Corn, wheat, and clover rotation | 2.7 |
| Continuous wheat | 10.1 |
| Continuous corn | 19.7 |

**Source:** M. F. Miller, "Cropping Systems in Erosion Control," Missouri Experiment Station, Bulletin 366 (1936), reprinted in National Agricultural Lands Study, *Soil Degradation: Effects on Agricultural Productivity* (Washington, D.C.: 1980).

Fallowing has permitted farmers to work the land both in semiarid regions and in the tropics, where nutrients are scarce. In vast semi-arid areas—such as Australia, the western Great Plains of North America, the Anatolian plateau of Turkey, and the drylands of the Soviet Union—where there is not enough moisture to support continuous cultivation, alternate-year cropping has evolved. Under this system land is left fallow without a cover crop every other year to accumulate moisture. The crop produced in the next season draws on two years of collected moisture.

In some situations this practice would lead to serious wind erosion if strip-cropping were not practiced simultaneously. Alternate strips planted to crops each year serve as windbreaks for the fallow strips. This combination of fallowing and strip-cropping permitted wheat production to continue in the western U.S. Great Plains after the Dust Bowl years.[7]

Rising demand for food has reduced the area fallowed in key dryland farming regions. As world wheat prices climbed sharply during the mid-seventies, U.S. summer fallow land dropped from 17 million hectares in 1969 to 13 million hectares in 1974.[8] This decline led Kenneth Grant, head of the U.S. Soil Conservation Service, to warn farmers that severe wind erosion and dust bowl conditions could result. He cautioned farmers against the lure of record wheat prices and short-term gains that would sacrifice the long-term productivity of their land. By 1977, the National Resources Inventory showed that wind erosion in wheat-growing states such as Texas and Colorado far exceeded the tolerance levels.[9]

At the same time, the amount of fallowed land in the Soviet Union was also being reduced. During the late sixties and early seventies, the Soviets were consistently fallowing 17-18 million hectares each year in the dryland regions. But after the massive crop shortfall and heavy imports of 1972, the fallowed area was reduced by one-third. By the early eighties Soviet officials were returning more land to fallow in an effort to restore land productivity.[10]

In the tropics—such as parts of Africa south of the Sahara, Venezuela, the Amazon Basin, and the outer islands of Indonesia—fallowing is used to restore the fertility of the soil. In these areas more nutrients are stored in vegetation than in the soil. When cultivated and stripped of their dense vegetative cover, soils of the humid tropics quickly lose their fertility. In response to these conditions, farmers have evolved a system of shifting cultivation: They clear and crop land for two, three, or possibly four years and then systematically abandon it as crop yields decline. Natural vegetation soon takes over the abandoned field. Moving on to fresh terrain, farmers repeat the process. When these cultivators return to their starting point after 20-25 years, the soil has regained enough fertility to support crop production for a few years.

Mounting population pressures in the tropics are forcing shifting cultivators to shorten these rotation cycles. As this happens, land productivity falls. A 1974 World Bank study reported that in Nigeria "fallow periods under shifting cultivation have become too short to restore fertility in some areas."[11] In some locales the original cropping cycle of 10-15 years has already been reduced to 5. Since 1950, the cropped area in Nigeria has multiplied 2.5 times as new land, largely marginal, has been added and fallow cycles have been shortened. Together, these two trends have offset the gains from the increased use of chemical fertilizer, the adoption of improved varieties, and the expansion of irrigation. Cereal yields are no higher than they were in the early sixties.

As population pressure has intensified in the river floodplains of northern Thailand over the last few decades, rice farmers have migrated to the nearby uplands where they practice rain-fed cropping. Early migrants adopted a "slash and burn" system with a cycle of eight to ten years, which seemed quite stable. As migration has continued, however, the forest fallow cycle has been shortened to between two to four years in many areas. In analyzing this situation, John M. Schiller concluded that "soil erosion problems are becoming clearly manifest in some areas and the effect of increased runoff is causing increased flooding in the lowlands and the siltation of dams. A potentially very unstable physical, economic, and social situation is developing in the affected areas."[12]

Similar pressures on land are evident in tropical Latin America. According to U.N. Food and Agriculture Organization researchers: "There is abundant evidence in certain regions of Venezuela that, with growing population pressure, the fallow period is becoming increasingly shorter so that soil fertility is not restored before recropping. This leads to a fall in the organic content and the water holding capacity of the soil. Soil structure deteriorates and compaction becomes more common . . . in other words, with the population of modern times, formerly stable, shifting cultivation systems are now in a state of breakdown."[13]

Another source of accelerated soil erosion in recent years has been the shift to larger farm equipment, particularly in the Soviet Union and United States. In the United States, for example, the shift to large-scale equipment has often led to the abandonment of field terraces constructed to reduce runoff on sloping lands. In dryland farming regions, tree shelter belts that interfere with the use of large-scale equipment have also been removed. The enlargement of fields to accommodate huge tractors and grain combines also reduces border areas that have traditionally served as checks on erosion.

This transformation of agricultural practices has been fueled by the growing worldwide demand for U.S. feed crops, particularly corn and soybeans, and by the availability of cheap chemical fertilizer. Demand growth, in turn, has been amplified by population growth that has hastened the deterioration of traditional agriculture in many countries. As a result, agricultural systems throughout the world are now experiencing unsustainable levels of soil loss.

## Dimensions of the Problem

One of the first scientists to assess the dimensions of world soil erosion was geologist Sheldon Judson, who estimated in 1968 that the amount of river-borne soil sediment carried into the oceans had increased from 9 billion tons per year before the introduction of agriculture, grazing, and other activities to 24 billion tons per year. Judson observed: "There is no question that man's occupancy of the land increases the rate of erosion. Where that occupation is intense and is directed to the use of land for cultivated crops, the difference is one or more magnitudes greater than when the land is under a complete natural vegetative cover, such as grass or forest." His estimates indicate that humans have become an important geologic agent, accelerating the flow of soil to the oceans.[14]

Although detailed information on soil erosion at the local level is available for only a few countries, data on the sediment load of the world's major rivers and on the wind-borne movement of soil over the oceans do provide a broad-brush view of soil erosion at the continental level. The most recent figures on river sediment flow show the world's major rivers carrying heavy loads of soil to the oceans. Data compiled in 1980 by three Chinese scientists working for the Yellow River Conservancy Commission in Beijing indicated that river was carrying 1.6 billion tons of soil to the ocean each year. (See Table 2.) Hydrologists estimate that on average one-fourth of the soil lost through erosion in a river's watershed actually makes it to the ocean as sediment. The other three-fourths is deposited on footslopes, in reservoirs, on river floodplains or other low-lying areas, or in the riverbed itself, which often causes channel shifts.[15]

Close behind the Yellow River, in terms of silt load, is the Ganges of India, which deposits 1.5 billion tons of soil into the Bay of Bengal each year. The Mississippi, the largest U.S. river, carries 300 million tons of soil into the Gulf of Mexico each year, far less than the Yellow

**"The Ganges of India deposits 1.5 billion tons of soil into the Bay of Bengal every year."**



**Table 2: Sediment Load of Selected Major Rivers**

| River | Countries | Annual Sediment Load |
|---|---|---|
| | | (million metric tons) |
| Yellow | China | 1,600 |
| Ganges | India | 1,455 |
| Amazon | several | 363 |
| Mississippi | United States | 300 |
| Irrawaddy | Burma | 299 |
| Kosi | India | 172 |
| Mekong | several | 170 |
| Nile | several | 111 |

**Source:** S. A. El-Swaify and E. W. Dangler, "Rainfall Erosion in the Tropics: A State of the Art," in American Society of Agronomy, *Soil Erosion and Conservation in the Tropics* (Madison, Wisc.: 1982).

or the Ganges. Yet it represents topsoil from the agricultural heartland and is thus a source of major concern for U.S. agronomists.[16]

Scientists have recently documented that vast amounts of wind-borne soil are also being deposited in the oceans as sediment. Island-based air sampling stations in the Atlantic, along with recent satellite photographs, indicate clearly that large quantities of soil dust are being carried out of North Africa over the Atlantic. Visible from satellites, these huge plumes of fine soil particles from the arid and desert expanses of North Africa at times create a dense haze over the eastern Atlantic. Estimates of the amount of African soil being carried west in this way, reported in four studies between 1972 and 1981, range from 100-400 million tons annually, with the latest report being at the upper end of the range.[17]

A 1983 *Science* article reported a similar loss of soil from Asia, soil that is carried eastward over the Pacific. Air samples taken at the Mauna Loa Observatory in Hawaii from 1974 through 1982 indicate a continuous movement of soil particles from the Asian mainland, with a peak annual flow consistently occurring in March, April, and May, a time that coincides with a period of strong winds, low rainfall, and plowing in the semiarid regions of North Asia. Scientists at Mauna Loa can now tell when spring plowing starts in North China.[18]

Although soil erosion data are not available for most countries, a rough estimate of the excessive worldwide loss of topsoil from croplands is needed. Without such an estimate, assessments of the world food prospect are unrealistic. The estimate developed in the following pages is the best that we can construct from the information now available. If other governments were to follow the U.S. lead and take careful inventories of their soil resources to determine the rate of excessive erosion, then they would have the information needed to formulate realistic agricultural and population policies.



The United States is one of the few countries that has analyzed soil losses in detail. As directed by the Rural Development Act of 1972, the Soil Conservation Service undertook an exhaustive inventory of land use and soil loss in 1977. Based on some 200,000 data samplings, it yielded remarkably detailed information on local soil loss throughout the United States. The first inventory formed the basis for the comprehensive appraisal of the nation's soil and water resources mandated by the Soil and Water Resources Conservation Act of 1977. A second inventory, conducted in 1982, expanded the survey to nearly one million sample points nationwide, the most comprehensive look at soil resources ever completed by any country.[19]

Rates of soil loss revealed by the National Resources Inventory can be related roughly to the tolerable level, a rate that would not impair long-term productivity. Calculated at from one to five tons per acre annually, depending on soil and climatic conditions, this figure represents the maximum level of soil erosion that will permit a high level of crop productivity to be sustained economically and indefinitely. The 1982 inventory showed that 44 percent of U.S. cropland was losing topsoil in excess of its soil loss tolerance level. The loss of soil at this excessive rate from the U.S. cropland base of 421 million acres totaled 1.7 billion tons, with over 90 percent coming from less than one-quarter of the cropland.[20]

India is one of the few other countries to compile a national estimate of soil loss. In 1975, Indian agricultural scientists collected data on local soil erosion from each of the research stations in the national network maintained by the Indian Council for Agricultural Research. Using these figures, they estimated that 6 billion tons of soil are eroded from India's croplands each year.[21] From this and from an estimate that 60 percent of the cropland is eroding excessively, the excessive topsoil loss can be calculated by subtracting from the total a tolerance level of five tons per acre. This yields an excessive topsoil loss from Indian cropland of 4.7 billion tons per year, more than twice the U.S. level. This estimate rests on far less data than does the figure for the United States but it is based on information from agricultural scientists familiar with local soils and it is corroborated by data on siltation of hydroelectric reservoirs, river sediment loads, and other indirect indicators.

The Soviet Union, which has the world's largest cropland area, may be losing more topsoil than any other country. Although detailed information on the extent of the loss is not available, numerous sources—including Soviet research reports, public statements by scientists and government officials, and the observations of visitors from abroad—indicate the severity of the problem. Papers published

**"Scientists at Mauna Loa (Hawaii) can now tell when spring plowing starts in North China."**



by the Soil Erosion Laboratory at the University of Moscow, for example, indicate a severe and worsening erosion situation.

During the early eighties the official Soviet press carried statements by soil scientists pleading with the agricultural bureaucracy to address the loss of topsoil. And in early 1981, Dr. Vladimir Borovsky, a prominent soil scientist and director of the Kazakh Institute of Soil Science, publicly charged the Academy of Agricultural Sciences with neglect of soil problems. In a broadcast on Moscow radio, Borovsky argued that Soviet agriculture will be retarded without effective soil management. His warnings have received some support at the highest levels of Soviet government, with Politboro member Mikhail S. Gorbachev urging planners to heed the advice of soil scientists.[22] But in the face of pressures to expand production and reduce the food import deficit—now the world's largest—soil scientists are often ignored and responsible soil management practices are cast aside.

As in the United States, erosion has been spurred by the shift to large, heavy equipment and the enlargement of fields, which eliminated many natural boundary constraints on erosion of soil by both wind and water. Each year an estimated half-million hectares of cropland are abandoned because they are so severely eroded by wind that they are no longer worth farming. One scholar of Soviet environmental policies and trends, Thane Gustafson, observes, "Fifty years of neglect have left a legacy of badly damaged soils."[23]

Although there are no official figures on soil erosion, an estimate of Soviet soil losses based on the local data that are available can be compared with the situation in the United States, where detailed erosion information has been collected. Two Soviet scientists, P. Poletayev and S. Yashukova, writing on environmental protection and agriculture in a Soviet economics journal in 1978, reported that "two-thirds of the plowed land in the Soviet Union has been subjected to the influence of various forms of erosion." Knowing the area affected by erosion, only the rate of erosion need be determined to estimate the total topsoil loss.[24]

Like the United States, the Soviet Union has an extensive dryland farming area and a substantial irrigated area. The European Soviet Union, which accounts for a large share of total farm output, has moisture levels similar to the U.S. Midwest. In terms of rainfall intensity, topography, and erodibility of prevailing soil types, nothing indicates that soil erosion in the Soviet Union would be markedly less than in the United States. Where cropping patterns are concerned, the Soviet Union relies much more heavily on small grains, whereas the United States relies relatively more on row crops, such as corn and soybeans.

Much of the Soviet grain land, however, remains bare during the winter and early spring, when rainfall is heaviest in many regions of the country. In a paper presented in the United States in 1983, P.S. Tregubov of the Dokuchaev Soil Institute in Moscow reported that land left in bare fallow to be sown to winter crops sustained losses that far exceeded the rate of new soil formation. He observed that "spring was found to be the most dangerous period because soils are characterized by fluidity after snow thawing." To document this, Tregubov cited long-term experiments showing a mean annual soil loss on bare fallow of 59 tons per hectare annually in the Baltic Sea shore regions, 46 tons per hectare in the Rostov region, and 32 tons per hectare in the Transcaucasian region. By comparison, in the American states with the most severe erosion rates in 1982, Texas lost nearly 40 tons per hectare, Colorado lost 32 tons, and Iowa, just over 30 tons.[25]

These data and observations suggest it is not unreasonable to assume that Soviet soils are eroding at least as rapidly as those in the United States. If 44 percent of the land is affected by erosion at the same rate as in the United States, which may be a conservative assumption, the excessive loss of topsoil from Soviet croplands is over 2.5 billion tons per year.

In China, the fourth major food-producing country, river siltation is now a nationally recognized threat—one that has reached dimensions unmatched elsewhere. Dust storms in the north and the siltation of major rivers indicate the heavy soil loss. Observations by outsiders who have been called in to help assess soil conditions indicate that the erosion rate in China is at least as great as that in India, where more detailed data are available.

A comparison of the sediment load of the Yellow River in China with the Ganges in India indicates the relative magnitude of soil loss through erosion faced by these two population giants. The Ganges, with a drainage basin of 1.1 million square kilometers, carries an annual sediment load of 1.46 billion tons of soil, while the Yellow River, which has a drainage basin of 668,000 square kilometers, carries 1.6 billion tons of soil to the ocean each year. These numbers suggest that the rate of soil loss in China is substantially greater than in India. For the purposes of constructing a rough global estimate, it can be assumed that the erosion rate on China's cropland exceeds India's by 30 percent. Given China's smaller cropland area, this means that China's excessive loss of topsoil from its croplands totals 4.3 billion tons per year.

For most Third World countries information on soil erosion is largely indirect, such as data on sedimentation of reservoirs and river silt loads. Other indirect sources include information on cropland abandonment as a result of severe erosion and crop reports showing long-term declines in yields. Among the most graphic sources are reports by agricultural scientists, development technicians, and other observers. (See Table 3.)

**Table 3: Observations of Soil Erosion in the Third World**

| Country | Observation | Source |
|---|---|---|
| Nepal (Katmandu) | "Local inhabitants . . . all concur that the problem is more severe now than a generation ago." | *Mountain Research and Development,* (Boulder, CO) 1982 |
| Peru | "Erosion is estimated to affect between 50 and 60 percent of the surface of the whole country." | *Mountain Research and Development,* (Boulder, CO) 1982 |
| Indonesia (Java) | "Soil erosion is creating an ecological emergency in Java, a result of overpopulation, which has led to deforestation and misuse of hillside areas by land-hungry farmers. Erosion is laying waste to land at an alarming rate, much faster than present reclamation programs can restore it." | U.S. Embassy, Jakarta, 1976 |
| Ethiopia | "There is an environmental nightmare unfolding before our eyes . . . over 1 billion tons of topsoil flow from Ethiopia's highlands each year." | U.S. AID Mission, Addis Ababa, 1978 |
| South Africa | "The province of Natal, incorporating Kwazulu, is losing 200 million tons of topsoil annually." | John Hanks, Institute of Natural Resources, Natal, 1980 |
| Bolivia | "Recent aerial photographs have shown the rapid extension of desert-like conditions caused by wind erosion." | Hélène Rivière d'Arc, Institut des Hautes Études d'Amérique Latine, Paris, 1980 |
| Iran | "The area of abandoned cultivated land has doubled in recent years." | Harold Dregne, Texas Tech University, 1971 |

Altogether, the excessive loss of topsoil from cropland in the four major food-producing countries, which have 52 percent of the world's cropland and account for over half of its food production, is estimated at 13.2 billion tons per year. To obtain a rough idea of excessive soil erosion for the world as a whole, an assumption must be made about other countries. If the rates of soil erosion for the rest of the world are similar to those of the "big four"—which is a conservative assumption given the pressures on land in the Third World—then the world is now losing an estimated 25.4 billion tons of soil from croplands in excess of new soil formation. (See Table 4.)

**Table 4: Estimated Excessive Erosion of Topsoil From World Cropland**

| Country | Total Cropland | Excessive Soil Loss |
|---|---|---|
| | (million acres) | (million tons) |
| United States | 421 | 1,700 |
| Soviet Union | 620 | 2,500 |
| India | 346 | 4,700 |
| China | 245 | 4,300 |
| Total | 1,632 | 13,200 |
| Rest of World | 1,506 | 12,200 |
| Total | 3,138 | 25,400 |

**Source:** Worldwatch Institute estimates.

Because of the shortsighted way one-third to one-half of the world's croplands are being managed, the soils on these lands have been converted from a renewable to a nonrenewable resource. Assuming an average depth of remaining topsoil of seven inches, or 1,120 tons per acre, and a total of 3.1 billion acres of cropland, there are 3.5 trillion tons of topsoil with which to produce food, feed, and fiber. At the current rate of excessive erosion, this resource is being depleted at 0.7 percent per year—7 percent each decade. In effect, the world is mining much of its cropland, treating it as a depletable resource, not unlike oil.

When most of the topsoil is lost on land where the underlying formation consists of rock or where the productivity of the subsoil is too low to make cultivation economical, it is abandoned. More commonly, however, land continues to be plowed even though most of the topsoil has been lost and even though the plow layer contains a mixture of topsoil and subsoil, with the latter dominating. Other things being equal, the real cost of food production on such land is far higher than on land where the topsoil layer remains intact.

## The Erosion of Productivity

Whenever erosion begins to exceed new soil formation, the layer of topsoil becomes thinner, eventually disappearing entirely. As the topsoil layer is lost, subsoil becomes part of the tillage layer, reducing the soil's organic matter, tilth, and aeration, and adversely affecting other structural characteristics that make it ideal for plant growth.

This overall deterioration in soil structure is usually accompanied by a reduced nutrient retention capacity, which lowers productivity further. Additional chemical fertilizer can often compensate for the loss of nutrients, but the deterioration of soil structure is difficult to remedy.

The effects of erosion on productivity are not easily measured since they are usually gradual and cumulative. In an effort to understand the erosion/productivity relationship better, the U.S. Secretary of Agriculture in 1980 appointed a National Soil Erosion-Soil Productivity Research Planning Committee. Among other things, committee members began gathering data from past experiments to establish an empirical foundation for predicting the effect of continuing soil loss on crop yields and production costs. They reported that when corn was grown continuously on a plot in Iowa from which the topsoil had been removed, yields were only 20 percent of those on a control plot. In an experiment in Missouri, corn yields on a desurfaced plot were 47 percent of those on the control plot. In this case, the subsoil was a clay loam—a higher-quality subsoil than is commonly the case.[26]

In an experiment in East Texas, cotton yields on land with the topsoil removed averaged only 32 percent of the control plot's. And in Minnesota, yields on severely eroded soils were roughly two-thirds those on slightly eroded soils.[27] A 1979 experiment on piedmont soils in Georgia designed to measure the effects of erosion on corn yields showed that severely eroded, moderately eroded, and uneroded soils averaged 36, 75, and 92 bushels respectively. On these soils, researchers estimated that each centimeter of topsoil lost through water erosion reduced the average corn yield by 2.34 bushels per acre.[28]

Leon Lyles, an agricultural engineer with the U.S. Department of Agriculture, has provided probably the most comprehensive collection of research results on the effect of soil erosion on land productivity. Drawing on the work of U.S. soil scientists both within and outside government, Lyles compared 14 independent studies, mostly undertaken in the corn belt states, to summarize the effects of a loss of one inch of topsoil on corn yields. His survey found that such a loss reduced yields by as little as 3 bushels per acre to as much as 6.1 bushels per acre. (See Table 5.) In percentage terms, the loss of an inch of topsoil reduced corn yields at these 14 sites by an average of 6 percent. Results for wheat, drawing on 12 studies, showed a similar relationship between soil erosion and land productivity. The loss of an inch of topsoil reduced wheat yields 0.5-2.5 bushels per acre. In percentage terms, the loss of an inch reduced wheat yields an average of 6 percent, exactly the same as for corn. (See Table 6.)

All the studies on soil erosion and land productivity that Lyles cited showed that the excessive loss of topsoil lowered yields measurably, although the extent of yield reduction varied. And, as noted, his compilation of studies showed a remarkable similarity in the effect of soil erosion on the yield of wheat, a crop usually grown under lower rainfall conditions, and that of corn, usually grown in areas of higher

**Table 5: Effect of Topsoil Loss on Corn Yields**

| Location | Yield Reduction Per Inch of Topsoil Lost | | Soil Description |
|---|---|---|---|
| | (bushels/acre) | (percent) | |
| East Central, Illinois | 3.7 | 6.5 | Swygert silt loam |
| Fowler, Indiana | 4.0 | 4.3 | Fowler, Brookston, and Parr silt loams |
| Clarinda, Iowa | 4.0 | 5.1 | Marshall silt loam |
| Greenfield, Iowa | 3.1 | 6.3 | Shelby silt loam |
| Shenandoah, Iowa | 6.1 | 5.1 | Marshall silt loam |
| Bethany, Missouri | 4.0 | 6.0 | Shelby and Grundy silt loams |
| Columbus, Ohio | 3.0 | 6.0 | Celina silt loam |
| Wooster, Ohio | 4.8 | 8.0 | Canfield silt loam |

**Sources:** Various reports cited in Leon Lyles, "Possible Effects of Wind Erosion on Soil Productivity," *Journal of Soil and Water Conservation,* November/December 1975.

**Table 6: Effect of Topsoil Loss on Wheat Yields**

| Location | Yield Reduction Per Inch of Topsoil Lost | | Soil Description |
|---|---|---|---|
| | (bushels/acre) | (percent) | |
| Akron, Colorado | 0.5 | 2.0 | Weld silt loam |
| Geary County, Kansas | 1.3 | 6.2 | (not available) |
| Manhattan, Kansas | 1.1 | 4.3 | Smolan silty clay loam |
| Columbus, Ohio | 1.3 | 5.3 | Cropped soil |
| Oregon | 1.0 | 2.2 | Deep soil |
| Oregon | 2.5 | 5.8 | Thin soil |
| Palouse area, Washington | 1.6 | 6.9 | (not available) |

**Sources:** Various reports cited in Leon Lyles, "Possible Effects of Wind Erosion on Soil Productivity," *Journal of Soil and Water Conservation,* November/December 1975.

rainfall. Recent, more detailed research on three soil types in Minnesota shows that the effect of erosion varies with soil type and depth. It specifically notes that on some deeper soils, such as the Kenyon soils that are 76 inches deep, the near-term effects of erosion are negligible.[29] For the world as a whole soils of this depth constitute a small share of the total, an exception to the more typical 6-8 inches of topsoil found on most cropland.

Perhaps the most detailed analysis to date of the long-term effects of soil erosion on land productivity and food production costs is one undertaken for the Southern Iowa Conservancy District. Conducted by an interdisciplinary team of three scientists, this analysis assumed that soil erosion would continue at recent rates. The researchers classified the degree of erosion into three phases: soils that are slightly eroded, with no appreciable mixing of subsoil and topsoil in the plow layer; those that are moderately eroded, with some mixing of subsoil into the plow layer; and severely eroded soils, where the topsoil is largely gone and the plow layer is predominantly subsoil.

In 1974, the base year, 2.1 million of the district's 3.5 million acres of cropland fell into one of the three erosion phases, with the largest acreage being in the moderately eroded category. Assuming a continuation of the same rate of erosion, this would also be true for the year 2000. But by 2020, the researchers predicted, the largest share would be in the severely eroded category. As soils progress from the moderately to severely eroded category, the amount of nitrogen, phosphorus, and potash needed to grow corn increases by 38 pounds per acre. (See Table 7.) Closely paralleling this would be an increase in fuel requirements for tillage. As erosion proceeds, soils become more compact and difficult to till. The actual fuel increase varied widely by soil type, but on the average the severely eroded soils would require 38 percent more fuel for tillage than the slightly eroded soils.

Soil erosion would not only raise the costs of production by increasing the amount of fertilizer and fuel used, it would also reduce yields. For corn, a shift from slight to moderate erosion would reduce the average corn yield by 16 bushels per acre, while going from the moderate to severe category would lower yields another 7 bushels per acre. (See Table 8.) Although the soybean yield decline was much smaller, it was proportionate, since soybean yields are roughly one-third those of corn.

Although there are few reliable data on the effect of soil erosion on land productivity for most countries, some insights into the relationship can be derived from these U.S. studies. Given the consistency of the decline in productivity across a wide range of soil types and crops, it would not be unreasonable to assume that a similar relationship between soil erosion and land productivity exists in other coun-

**Table 7: Increase in Fertilizer Needs for Corn as Soil Erodes, Southern Iowa**

| Change in Erosion Phase | Nitrogen | Phosphate | Potash |
|---|---|---|---|
| | | (pounds per acre) | |
| Slight to Moderate | 10 | 2 | 6 |
| Moderate to Severe | 30 | 1 | 7 |

**Source:** Paul Rosenberry, Russell Knutson, and Lacy Harmon, "Predicting Effects of Soil Depletion From Erosion," *Journal of Soil and Water Conservation*, May/June 1980.

**Table 8: Reduction in Yields of Key Crops as Soil Erodes, Southern Iowa**

| Change in Erosion Phase | Reduction in Yield Per Acre | | |
|---|---|---|---|
| | Corn | Soybeans | Oats |
| | | (bushels) | |
| Slight to Moderate | 16 | 5 | 9 |
| Moderate to Severe | 7 | 3 | 4 |



**Source:** Paul Rosenberry, Russell Knutson, and Lacy Harmon, "Predicting Effects of Soil Depletion From Erosion," *Journal of Soil and Water Conservation*, May/June 1980.

tries, for the basic agronomic relationships are the same. Indeed, research on West African soils shows land productivity there to be even more sensitive to topsoil loss than in North America. The loss of 3.9 inches of topsoil in West Africa cut corn yields by 52 percent. Yields of cow peas, a leguminous crop, were reduced by 38 percent. This marked decline may attest to the fragility of tropical soils.[30]

## Erosion's Indirect Costs

When farmers lose topsoil they pay for it in reduced soil fertility, but unfortunately the costs of erosion are not confined to the farm alone. As soil is carried from the farm by runoff, it may end up in local streams, rivers, canals, or irrigation and hydroelectric reservoirs. The loss of topsoil that reduces land productivity may also reduce irrigation, electrical generation, and the navigability of waterways.

The increase in the amount of irrigated land in the world went hand in hand with efforts to raise food supplies during the third quarter of this century. Often the centerpiece of national development strategies throughout the Third World, multipurpose dams represented enormous investments and an important part of the capital stock of new nations. Typical of these was the Mangla Reservoir in Pakistan. The designers of the reservoir projected a life expectancy for the dam of at least a century. What they did not reckon on was the effect of mounting population pressure on the watershed feeding the reservoir. A combination of the axe and the plow, as land-hungry peasants push up the hillsides, is leading to a rate of siltation that will probably fill the reservoir with silt at least 25 years earlier than projected. (See Table 9.) One recent estimate predicts it will be filled within half a century.[31]

In the Philippines, scores of hydroelectric and irrigation reservoirs have been constructed, many of them with assistance from international development agencies. Here, as in Pakistan, the combination of watershed deforestation and steep slopes being cleared for cultivation is yielding record siltation rates. A report of the Agency for International Development on the prospects for the Ambuklao Dam

**"The loss of topsoil may also reduce irrigation, electrical generation, and the navigability of waterways."**



**Table 9: Siltation Rates in Selected Reservoirs**

| Country | Reservoir | Annual Siltation Rate | Time To Fill With Silt |
|---|---|---|---|
| | | (metric tons) | (years) |
| Egypt | Aswan High Dam | 139,000,000 | 100 |
| Pakistan | Mangla | 3,700,000 | 75 |
| Philippines | Ambuklao | 5,800 | 32 |
| Tanzania | Matumbulu | 19,800 | 30 |
| Tanzania | Kisongo | 3,400 | 15 |

**Source:** S. A. El-Swaify and E. W. Dangler, "Rainfall Erosion in the Tropics: A State of the Art," in American Society of Agronomy, *Soil Erosion and Conservation in the Tropics* (Madison, Wisc.: 1982).

notes that "the cutting of timber and the subsequent loss of water retention capacity of land surrounding the reservoir has resulted in massive silting of the reservoir, reducing its useful life from 60 to 32 years."[32]

One reason for the excessively rapid siltation rates is that multipurpose dams are designed by engineers who sometimes fail to recognize the impoundments they build as part of a watershed, which often drains an area of several thousand square miles. The Anchicaya Dam in Colombia is a classic example. Engineers expressed little concern with the siltation problem, even though when the project began farmers were already invading the upper reaches of the watershed that feeds the dam. Within two years of its completion, the dam had already lost a quarter of its storage capacity because of siltation.[33]

In India, the indirect costs of water-eroded soil are summed up well by B. B. Vohra, Chairman of the National Committee on Environmental Planning. He observes that the "premature siltation of our 500,000 odd ponds and of the 487 reservoirs of our major and medium irrigation and multipurpuse projects on which the community has invested over 100 billion rupees during the last three decades is a particularly serious matter." He notes that siltation rates are now commonly several times as high as the rate that was assumed when the projects were designed. (See Table 10.) Vohra observes that not only is the life expectancy of these projects being severely reduced, but "in most cases there will be no alternative sites for dams once the

**Table 10: India: Siltation Rates in Selected Reservoirs**

| Reservoir | Assumed Rate | Observed Rate | Ratio of Observed to Assumed Erosion |
|---|---|---|---|
| | (in acre-feet) | | |
| Bhakar | 23,000 | 33,475 | 1.46 |
| Maithon | 684 | 5,980 | 8.74 |
| Mavurakshi | 538 | 2,000 | 3.72 |
| Nizam Sugar | 530 | 8,725 | 16.46 |
| Panchet | 1,982 | 9,533 | 4.81 |
| Ramganga | 1,089 | 4,366 | 4.01 |
| Tungabhadra | 9,796 | 41,058 | 4.19 |
| Ukai | 7,448 | 21,758 | 2.92 |

**Sources:** Adapted from S. A. El-Swaify and E. W. Dangler, "Rainfall Erosion in the Tropics: A State of the Art," in American Society of Agronomy, *Soil Erosion and Conservation in the Tropics* (Madison, Wisc.: 1982), and Center for Science and Environment, *The State of India's Environment 1982* (New Delhi: 1982).

existing ones are rendered useless." A dam site is often unique. Once lost, it cannot be replaced. For India, what is at stake, according to Vohra, "is the loss of the irreplaceable potential—for irrigation, for electricity and for flood control—that these storages represent."[34]

The list of countries with soil-silting disasters goes on and on. The names change but the conditions are common. Whether in Nigeria, Indonesia, Pakistan, or Mexico, the same basic principles of soil physics are at work. When soil on sloping land is farmed improperly, it begins to move under the impact of rain and ends up in places where it usually does more harm than good.

The third major indirect cost of soil erosion is the loss of navigability. Perhaps the most dramatic case occurs in the Panama Canal. The combination of deforestation and the plowing of steeply sloping land in the watershed area by landless *campesinos* is leading to an unprecedented siltation of the lakes that make up part of the Canal. If the trends of the late seventies and early eighties continue, the capacity of the Panama Canal to handle shipping will be greatly reduced by the end of the century, forcing many ocean-going freighters that have relied on its 10,000 mile shortcut to make the trip via Cape Horn.[35]

Within the United States, soil once used in the Midwest to grow corn now clogs the Mississippi waterways. One of the largest items in the budget of the Army Corps of Engineers is the dredging of inland waterways, particularly in the lower Mississippi River. Vast quantities of soil reach the Gulf of Mexico to become ocean sediment, but substantial amounts are deposited on the way, making large-scale dredging imperative if this major artery connecting U.S. farms with world markets is to continue to function.

## The Global Balance Sheet

As world food demand has begun its second doubling since mid-century, pressures on land have become so intense that close to half the world's cropland is losing topsoil at a rate undermining its long-term productivity. Since agriculture is the foundation of the global economy, this loss of topsoil, if unarrested, will undermine the economy itself. Nonetheless, few countries, industrial or developing, are responding effectively to this emerging threat to economic sustainability.

Newspaper headlines that describe widening food deficits and chronic hunger in many Third World countries also describe a world finding it difficult to live within its means. Eager to maximize food output today, we are borrowing from tomorrow. The loss of over 25 billion tons of topsoil from our cropland each year is the price we pay for shortsighted agricultural policies designed to boost food output at the expense of soils, and of failed or nonexistent population policies.

In addition to the unprecedented growth in world food demand since mid-century, new demands on the world's land resources have recently emerged. When growth in the world fish catch slowed abruptly after 1970, it forced the world's consumers to turn more to land-based protein sources. With the depletion of oil reserves and the associated rise in oil prices during the seventies, the world is now turning to agriculture to produce more of the world's energy, as well as food.

Several countries have turned to agriculture as a source of liquid fuels, most importantly Brazil, which in 1984 devoted some 3.2 million acres of land to the production of sugarcane for distillation into alcohol fuel. The United States, meanwhile, used some 1.5 million acres of corn for fuel alcohol. Although this combined area is only a fraction of 1 percent of the world's cropland, it is growing steadily and should continue to do so as the transition from oil to renewable energy sources continues. Agronomist R. Neil Sampson observes that, "Seldom has such a totally new set of competitive forces been unleashed on the land as those that appear on the horizon in the declining decades of the petroleum era."[47]

The long-term social threat posed by uncontrolled soil erosion raises profound questions of intergenerational equity. If our generation persists in mining the soils so that we may eat, many of our children and their children may go hungry as a result. Agricultural economist Lloyd K. Fischer of the University of Nebraska observes that the quality of our diet in the future will be "substantially lower and the costs dramatically higher if the management of our land and water resources is not improved." He notes further that "we must cease to behave as if there were no tomorrow, or tomorrow will be bleak indeed for those who must spend their lives there."[48]

Soil erosion is a physical process, but its consequences are economic. As soils are depleted through erosion, the productivity of laborers working the eroding land becomes more difficult to raise. In agrarian societies, deterioration of this resource base makes it more difficult to raise income per person. Further, as growth in food output slows, so does overall economic output. In largely rural, low-income societies with rapid population growth, this can translate into declining per capita income, as it already has for a dozen countries in Africa.

Over the long term, world agricultural trade patterns and the international debt structure will be altered. As soils are depleted, countries are forced to import food to satisfy even minimal food needs. Scores of countries in the Third World and Eastern Europe find their international indebtedness further aggravated by their chronic dependence on imported food. And the loss of topsoil will force an energy-for-topsoil substitution as it increases the need for fertilizer and fuel for tillage. Other things being equal, land with less topsoil requires more energy to produce our food.

Soil erosion will eventually lead to higher food prices, hunger, and quite possibly, persistent pockets of famine. Although the world economy has weathered a severalfold increase in the price of oil over the past decade, it is not well equipped to cope with even modest rises in the price of food. Although the immediate effects of soil erosion are economic, the ultimate effects are social. When soils are depleted and crops are poorly nourished, people are often undernourished as well.

In efforts to conserve soil, the world is faring poorly. There are few national successes, few models that other countries can emulate. The United States has the technology, the detailed information on its soil losses and the resources to launch an exemplary soil conservation program, but it lacks leadership. Within the Third World Kenya is the only country to launch a successful national program to conserve its soils. In this respect, soil conservation contrasts sharply with oil conservation, where scores of countries have compiled impressive records in recent years. Almost everywhere dependence on petroleum is declining as it is used more efficiently. But there is no parallel with soil conservation, even though soil is a far more essential resource.

The effect of price on the depletion of soil and oil resources also varies. Higher prices for oil raise the amount that can be ultimately recovered, but higher prices for food may simply lead to more intensive land use and faster topsoil loss. And the depletion of oil reserves will make the substitution of energy for cropland more difficult, rendering the remaining soil even more valuable.

The United States unilaterally attempts to balance the world's supply and demand of agricultural commodities by withholding land from production during times of surplus. But little or no effort has been made to coordinate the farm supply management programs that divert land from production and the conservation programs designed to reduce soil erosion. U.S. farm programs have demonstrated that land can be withheld from production for supply management rea-

**"Within the Third World Kenya is the only country to launch a successful national program to conserve its soils."**



sons. Unfortunately, no effort has been made to ensure that the most erosive land was set aside. As policy analyst Kenneth Cook observes, the United States has "no policy to use the good land in preference to the worst. Indeed, with respect to matching export demand to the needs of U.S. farmers and to the needs of people and resources in the developing world, we do not have responsible policy at all. We have a simple-minded sales quota."[49]

The United States now has an opportunity to integrate soil conservation and supply management programs. With farm program costs out of control and public support for traditional farm price support programs diminishing, Congress may be unable to legislate a new farm program in 1985 unless it directly incorporates soil and water conservation with supply management and price supports. In effect, the broad base of public support for soil conservation could be used to divert highly erosive cropland to other uses, such as fuelwood production or grazing. This would bring the production of key farm commodities down to a level that would support prices needed to make agriculture profitable. This unique opportunity for the United States ought not to be wasted. Merging the two policies, however, requires a degree of agricultural leadership that does not now exist in the United States. A recent study by the American Farmland Trust emphasizes that progress toward conserving soil awaits leaders who accept "the nondegradation of agricultural resources as a central goal of national policy." The AFT report recommends that a national strategy include a cropland reserve program for highly erodible land, an effort to cut programs that subsidize cultivation of especially fragile lands, and a reorientation of USDA technical assistance to promote cost-effective conservation measures.[50]

Although no country's soil is adequately secured, there are occasional signs of hope. One is the trend toward reduced tillage in the United States, triggered by farmers' desire to reduce fuel consumption and operating costs. So far the farmers turning to reduced tillage are not usually the ones with the most rapidly eroding soils. But reduced tillage may become an economically attractive first line of defense against erosion, particularly given the high cost of constructing terraces and adopting long-term rotations and other traditional approaches to soil conservation.

Another hopeful example is Kenya's ambitious national soil conservation program. Less than a decade old, Kenya's program shows not only that conservation is compatible with small-scale farming and a large rural population, but also that conservation improvements can boost farmers' incomes. A similar approach could work in many Third World countries.

Yet another encouraging development is the response to the erosion threat by the international scientific community, as evidenced by several recent conferences and specially commissioned studies. The International Congress of Soil Science, which met in New Delhi in 1982, focused on the need for a world soils policy. In early 1983, the Soil Conservation Society of America convened an international conference on soil erosion and conservation; some 145 scientists from around the world presented papers. And the American Society of Agronomy recently published proceedings of an international symposium on soil erosion in the tropics.[51]

In the absence of successful efforts to stem the loss of topsoil, the social effects of erosion will probably first be seen in Africa, in the form of acute food shortages and higher mortality rates, particularly for infants. Africa's record population growth and rampant soil erosion, and the absence of an effective response to either, combine to ensure that the continent will be at the forefront of this unfolding global crisis. What is at stake is not merely the degradation of soil, but the degradation of life itself.

Historically, soil erosion was a local problem. Individual civilizations whose food systems were undermined in earlier times declined in isolation. But in the integrated global economy of the late twentieth century, food—like oil—is a global commodity. The excessive loss of topsoil anywhere ultimately affects food prices everywhere.

## Notes

1. Anson R. Bertrand, "Overdrawing the Nation's Research Accounts," *Journal of Soil and Water Conservation*, May/June 1980.

2. Population growth calculated from Population Reference Bureau, *1984 World Population Data Sheet* (Washington, D.C.: 1984).

3. Authors' estimates based on Table 10-9 "World Soil Resources and Excessive Soil Loss, 1980, with Projections to 2000," from Lester R. Brown et al., *State of the World 1984* (New York: W.W. Norton & Co., 1984).

4. U.S. Department of Agriculture (USDA), Soil Conservation Service, and Iowa State University Statistical Laboratory, *Basic Statistics 1977 National Resources Inventory*, Statistical Bulletin No. 686 (Washington, D.C.: 1982); USDA, Soil Conservation Service, "Preliminary 1982 National Resources Inventory," (Washington, D.C.: unpublished printout, April 1984).

5. P. O. Aina et al., quoted in R. Lal, "Effective Conservation Farming Systems for the Humid Tropics," in American Society of Agronomy, *Soil Erosion and Conservation in the Tropics*, ASA Special Publication No. 43 (Madison, Wisc.: 1982).

6. S. A. El-Swaify and E. W. Dangler, "Rainfall Erosion in the Tropics: A State-of-the-Art," in American Society of Agronomy, *Soil Erosion*.

**7.** These and other conservation practices introduced in the thirties are discussed in Donald Worster, *Dust Bowl* (New York: Oxford University Press, 1979), especially in Chapter 14, "Making Two Blades of Grass Grow."

**8.** Figures on summer fallow from U.S. Department of Agriculture (USDA), Economic Research Service, *Economic Indicators of the Farm Sector: Production and Efficiency Statistics, 1981* (Washington, D.C.: U.S. Government Printing Office, 1983).

**9.** Kenneth Grant, "Erosion in 1973-74: The Record and the Challenge," *Journal of Soil and Water Conservation*, January/February 1975; USDA, Soil Conservation Service, *Basic Statistics*.

**10.** USDA, Economic Research Service, "World Agriculture Outlook and Situation," Washington, D.C., June 1983.

**11.** Wouter Tims, *Nigeria: Options for Long-Term Development* (Baltimore, Md.: Johns Hopkins University Press, for the World Bank, 1974).

**12.** John M. Schiller et al., "Development of Areas of Shifting Cultivation in North Thailand 'Thai-Australia Land Development Project,'" in American Society of Agronomy, *Soil Erosion*.

**13.** R. F. Watters, *Shifting Cultivation in Latin America* (Rome: United Nations Food and Agriculture Organization, 1971).

**14.** Sheldon Judson, "Erosion of the Land, or What's Happening to Our Continents," *American Scientist*, July/August 1968.

**15.** Jiang Degi et al., "Soil Erosion and Conservation in the Wuding River Valley" (Beijing, China: Yellow River Conservancy Commission, April 1980), cited in El-Swaify and Dangler, "Rainfall Erosion in the Tropics"; U.S. Water Resources Council, *The Nation's Water Resources 1975-2000, Volume 2: Water Quantity, Quality and Related Land Considerations* (Washington, D.C.: U.S. Government Printing Office, 1978).

**16.** Ganges data from American Society of Agronomy, *Soil Erosion*; Mississippi data from U.S. Water Resources Council, *The Nation's Water Resources 1975-2000*.

**17.** J. M. Prospero et al., "Atmospheric Transport of Soil Dust from Africa to South America," *Nature*, February 12, 1981.

**18.** Josef R. Parrington et al., "Asian Dust: Seasonal Transport to the Hawaiian Islands," *Science*, April 8, 1983.

**19.** USDA, Soil Conservation Service, *Basic Statistics*, and "Preliminary 1982 National Resources Inventory."

**20.** *Ibid.*

**21.** K. G. Tejwani, Land Use Consultants International, New Delhi, private communication, July 3, 1983; Centre for Science and Environment, *The State Of India's Environment 1982* (New Delhi: 1982).

**22.** Quoted in Vera Rich, "Soil First," *Nature*, February 12, 1982.

**23.** Abandoned cropland figure cited is from U.S. Central Intelligence Agency, *U.S.S.R. Agricultural Atlas* (Washington, D.C.: 1974); Thane Gustafson, "Transforming Soviet Agriculture: Brezhnev's Gamble on Land Improvement," *Public Policy*, Summer 1977.

**24.** P. Poletayev and S. Yashukova, "Environmental Protection and Agricultural Production," *Ekonomika Sel'skogo Khozyaystva* (Moscow), November 1978.

**25.** P. S. Tregubov, "Soil Rainstorm Erosion and Its Control in the U.S.S.R.," presented to the International Conference on Soil Erosion and Conservation, Honolulu, Hawaii, January 16-22, 1983; U.S. erosion rates are from USDA, "Preliminary 1982 National Resources Inventory."

**26.** National Soil Erosion-Soil Productivity Research Planning Committee, USDA, Science and Education Administration, "Soil Erosion Effects on Soil Productivity: A Research Perspective," *Journal of Soil and Water Conservation*, March/April 1981.

**27.** *Ibid*.

**28.** G. W. Langdale et al., "Corn Yield Reduction on Eroded Southern Piedmont Soils," *Journal of Soil and Water Conservation*, September/October 1979.

**29.** W. E. Larson, F. J. Pierce, and R. H. Dowdy, "The Threat of Soil Erosion to Long-Term Crop Production," *Science*, February 4, 1983.

**30.** Research by R. Lal of the International Institute of Tropical Agriculture, Ibadan, Nigeria, cited by El-Swaify and Dangler, "Rainfall Erosion in the Tropics." Even more pronounced yield reductions are investigated in J. S. C. Mbagwu, R. Lal, and T. W. Scott, "Effects of Desurfacing on Alfisols and Ultisols in Southern Nigeria I: Crop Performance," unpublished.

**31.** El-Swaify and Dangler, "Rainfall Erosion in the Tropics." The effects of environmental degradation on the lifespan of the Mangla Reservoir are discussed in Erik Eckholm, *Losing Ground: Environmental Stress and World Food Prospects* (New York: W.W. Norton & Co., 1976).

**32.** U.S. Agency for International Development, *Environmental and Natural Resource Management in Developing Countries: A Report to Congress* (Washington, D.C.: 1979).

**33.** Eckholm, *Losing Ground*.

**34.** B. B. Vohra, *Land and Water Management Problems in India*, Training Volume 8 (New Delhi: Ministry of Home Affairs, 1982).

**35.** Dr. Frank Wadsworth, "Deforestation—Death to the Panama Canal," in U.S. Department of State, Office of Environmental Affairs, *Proceedings of the U.S. Strategy Conference on Tropical Deforestation*, June 12-14, 1978 (Washington, D.C.: 1978).

**47.** Authors' estimate of sugarcane acreage for alcohol production in Brazil based on James Bruce, "Brazil's Alcohol Export Prospects Shrink But Domestic Mart Continues Growth," *Journal of Commerce*, April 5, 1983. Authors' estimate of corn acreage for alcohol production in the U.S. based on data from Janet Livezey, Economic Research Service, USDA, private communication, August 15, 1983. R. Neil Sampson, "Energy: New Kinds of Competition for Land," presented to the 35th Annual Meeting of the Soil Conservation Society of America, Dearborn, Michigan, August 6, 1980.

**48.** Lloyd K. Fischer, discussion comments, in Harold G. Halcrow, Earl O. Heady, and Melvin L. Cotner, eds., *Soil Conservation Policies, Institutions and Incentives* (Ankeny, Iowa: Soil Conservation Society of America, 1982).

**49.** Ken Cook, "Surplus Madness," *Journal of Soil and Water Conservation*, January/February 1983.

**50.** American Farmland Trust, *Soil Conservation in America: What Do We Have to Lose?* (Washington, D.C.: 1984).

**51.** Symposia papers of the 12th International Congress of Soil Science, February 8-16, 1982, were published as *Desertification and Soils Policy* (New Delhi: Indian Society of Soil Science, 1982). Proceedings of the International Conference on Soil Erosion and Conservation, January 16-22, 1983, should be available in 1984; abstracts of papers presented at the conference can be obtained from the Department of Agronomy and Soil Science, University of Hawaii at Manoa, Honolulu, Hawaii. Proceedings of the ASA symposium were published in American Society of Agronomy, *Soil Erosion.*

CHAPTER 14

# Putting Food on the World's Table: A Crisis of Many Dimensions

LESTER R. BROWN

Measured just in terms of output, the past generation has been one of unprecedented progress in world agriculture. In 1950 the world's farmers produced 623 million tons of grain. In 1983 they produced nearly 1.5 billion tons. This increase of nearly 900 million tons was all the more remarkable because it occurred when there was little new cropland to bring under the plow.[1]

On closer examination, this 33-year span breaks into two distinct eras—before and after the 1973 oil price increase. Modern agriculture thrives on cheap energy, and the age of cheap energy came to an end in 1973. For 23 years, world food output expanded at over 3 percent per year, and, although there was concern about rapid population growth, there was a comfortable margin in the growth of food production over that of population. Since 1973, however, annual growth has been less than 2 percent, and the world's farmers have been struggling to keep pace with population.

The global increase in world food output also obscures wide variations in individual geographic regions. In North America, production has steadily outstripped demand, generating ever-larger surpluses. In the Soviet Union, output has fallen behind demand over the past decade, making the country the largest grain importer in history. And in Africa, which has a population of 512 million and which has to feed 14 million additional people each year, food production per person has fallen steadily since 1970. Despite a tripling of grain imports since then, hunger has become chronic, an enduring part of the African landscape.

The 1983 drought in North America and Africa must be considered against this backdrop. The principal effect of the precipitous decline in the North American harvest was a reduction in stocks and a rise in food and feedstuff prices. In Africa, where national food reserves are virtually nonexistent, the drought translated into widespread hunger and, in a score of countries, the threat of famine.[2]

## The Global Loss of Momentum

As the world recovered from World War II, hopes for improvement in world agriculture were high. An accumulating backlog of agricultural technologies such as hybrid corn and chemical fertilizers was waiting to be applied on a massive scale. Between 1950 and 1973, world grain production more than doubled, to nearly 1.3 billion tons. Although output expanded more rapidly in some regions than in others, all regions shared in the growth. This rising tide of food production improved nutrition throughout the world, helping to boost life expectancy in the Third World from less than 43 years in the early 1950s to over 53 years in the early 1970s.[3]

---

Reprinted with permission from *Environment 26(4)*: 15–20, 38–43 (1984). 

Lester R. Brown is President and Senior Researcher, Worldwatch Institute.

This period of broad-based gains in nutritional improvement came to an end in 1973. After the oil price hike that year, the growth in world grain output slowed. Since 1973 world grain production has expanded at less than 2 percent yearly, barely keeping pace with population (see Table 1). Although the period since the 1979 oil price hike is too short to establish a trend, $30-a-barrel oil may well slow growth further.

Since 1973, attention has focused on the impact of petroleum prices on food supply, but demand has also been affected. On the supply side, rising oil prices have increased the costs of basic agricultural inputs—fertilizer, pesticides, and fuel for tillage and irrigation—thus acting as a drag on output. On the demand side of the equation, escalating oil prices combined with ill-conceived national economic policies have contributed to a global economic slowdown so severe since 1979 that it has brought world growth in per capita income to a virtual halt.

Had incomes continued to rise at the same rate after 1973 as they did before, prices of food commodities would have been stronger, thus supporting a more vigorous growth in farm investment and output. Agricultural underinvestment in Third World countries has also contributed to the loss of momentum, but the central point is that the rise in oil prices, affecting both food supply and demand, has brought the era of robust growth in world food output to an end.

Oil is not the only resource whose questionable supply is checking the growth in food output. The loss of topsoil through erosion is now acting as a drag on efforts to produce more food. And the scarcity of water is also beginning to affect food production prospects. Since World War II, the world's irrigated area has more than doubled, but the flurry of dam building of the past generation has now subsided. With occasional exceptions, most of the remaining potential projects are more difficult, costly, and capital-intensive.[4]

In some situations, irrigated agriculture is threatened by falling water tables. The southern Great Plains, where much of the U.S. growth in irrigated area over the last two decades has occurred, provides a disturbing example. Irrigation there depends almost entirely on water from the Ogallala Aquifer, an essentially nonreplenishable fossil water reserve. As the water table in this vast agricultural area begins to fall with the depletion of the aquifer, the cost of irrigation rises. Already some farmers in eastern Colorado and northern Texas are converting to dryland farming. For the 32 counties in the Texas Panhandle, the U.S. Department of Agriculture projects that irrigation will be largely phased out by 1995.[5]

A somewhat analogous situation exists in the Soviet southwest, where the excessive diversion of river water for irrigation is reducing the water level of the Aral and Caspian seas. This has many long-term negative consequences, including a diminished fish catch and the gradual retreat of the water line from coastal cities that depend on it for transportation.[6] Given the strong internal pressures within the Soviet Union to produce more food, however, the diversion is continuing.

A second major threat to irrigated agriculture is the often-intense competition for water between farming, industry, and cities. In the U.S. Southwest, the irrigated area is actually declining in states such as Arizona, where Sunbelt migration is swelling cities that are bidding water away from farmers. Nationally the net area under irrigation is projected to continue growing over the rest of

**Table 1**
**WORLD OIL PRICE AND GRAIN PRODUCTION TRENDS, TOTAL AND PER CAPITA, 1950–83**

| | | | ANNUAL GROWTH | |
|---|---|---|---|---|
| PERIOD | OIL PRICE PER BARREL | GRAIN PRODUCTION | POPULATION | GRAIN PRODUCTION PER PERSON |
| | *(dollars)* | | *(percent)* | |
| 1950–73 | 2 | 3.1 | 1.9 | 1.2 |
| 1973–79 | 12 | 1.9 | 1.8 | 0.1 |
| 1979–83 | 31 | 1.0[1] | 1.7 | 0.7 |

[1]Severe drought in the United States and Africa and record idling of cropland under U.S. farm programs reduced the 1983 world harvest well below trend. Thus, the slowdown in grain production is overstated.
**SOURCES**: International Monetary Fund, *Monthly Financial Statistics*, various issues; U.S. Department of Agriculture, *World Indices of Agricultural and Food Production, 1950–82* (unpublished printout) (Washington, D.C.: 1983); United Nations, *Monthly Bulletin of Statistics*, various issues.

the century, but at a more modest rate.

New research indicates water scarcities are also emerging in Africa. South Africa, adding 720,000 people each year, is fast running out of new irrigation sites. A 1983 report of the President's Council in South Africa identified the scarcity of fresh water as a constraint on that country's demographic carrying capacity.[7]

The worldwide loss of momentum outlined above will not be easily restored. Although agricultural mismanagement abounds, particularly in the Third World and Eastern Europe, it has not worsened appreciably over the years. Nor can the situation be explained by any farmers' loss of skills. The explanation lies in the more difficult circumstances facing farmers everywhere. In the mid-1980s it is far more difficult to raise world food output at a consistent 3 percent per year than it was during the 1950s or 60s. The cheap energy that permitted farmers to override easily the constraints imposed by the scarcity of land, soil nutrients, or water is simply no longer available.

## Population, Land, and Fertilizer

The changing relationship between world population size, cropland area, and energy supplies bears heavily on the human prospect over the remainder of this century and beyond. Increasingly, the energy used in agriculture will be in the form of chemical fertilizer. As population grows, cropland per person shrinks and fertilizer requirements climb. And erosion that has robbed soils of nutrients is forcing farmers to use more fertilizers.

Even urbanization is raising demand, since, as people move to cities, it is harder to recycle the nutrients in human and household waste. Yet the combination of rising energy costs and diminishing returns on the use of additional fertilizer raises doubts that adequate food supplies can be produced in the future at prices the world's poor can afford.

The central importance of the population/land/fertilizer relationship is a recent phenomenon. Before 1950 increases in food output came largely from expanding the cultivated area, but with the scarcity of fertile new land and the advent of cheap chemical fertilizer this changed. Between 1950 and 1983 world fertilizer use climbed from 15 million to 114 million tons, nearly an eightfold increase within a generation.[8] In effect,

Bagging fertilizer at the Don Hercules Fertilizer Plant in Lahore, Pakistan. Fertilizer manufacturing is one of the world's major industries today, for as population grows, cropland per person shrinks and fertilizer requirements climb (World Bank photo by Tomas Sennett).

Men plowing a rice paddy in Sri Lanka. While a doubling of irrigated areas has figured prominently in Third World agricultural advances over the last two decades, such increases will ultimately be of little consequence if the nutrients to support the higher yields are lacking (World Bank photo by Tomas Sennett).



as fertile land became harder to find, farmers learned to substitute energy in the form of chemical fertilizer for land. Fertilizer factories replaced new land as the principal source of growth in food production.

The hybridization of corn and the dwarfing of the wheat and rice varieties that have been at the heart of Third World agricultural advances over the last two decades figured prominently, of course, in the growth in world food output. So, too, did the doubling of irrigated area. But the effectiveness of all these practices depends heavily on the use of chemical fertilizer. Without an adequate supply of plant nutrients, high-yielding cereal varieties hold little advantage over traditional ones. Likewise, an increase in irrigation is of little consequence if the nutrients to support the higher yields are lacking.

The response of crops to the use of additional fertlizer is now diminishing, particularly in agriculturally advanced countries. Some countries, such as Argentina and India, still apply relatively little fertilizer, and so have quite high response ratios. But worldwide the return on the use of additional fertilizer is on the way down. Although the biological constraints on fertilizer responsiveness can be pushed back with continued plant breeding, further declines seem inevitable.

Fertilizer manufacturing is one of the world's major industries. In an advanced agricultural country such as the United States, expenditures on fertilizer total some $10 billion per year.[9] Three basic nutrients—nitrogen, which is obtained from the air, and phosphate and potash, both mined from underground deposits—account for the great bulk of world chemical fertilizer production. The industrial fixing of atmospheric nitrogren in the form of ammonium nitrate, ammonium sulphate, urea, or other forms of nitrogen fertilizer is an energy-intensive process. Although natural gas is the preferred fuel and feedstock in the nitrogen fertilizer industry, oil figures prominently in the mining, processing, and transportation of phosphate and potash.

High energy prices have begun to shift nitrogen fertilizer production from the traditional industrial country producers, such as the United States and some in Western Europe, to countries with energy surpluses. Investment in this industry has been particularly attractive to oil-exporting countries that are flaring excess gas produced in conjunction with oil. To the extent that fertilizers are manufactured in countries such as Saudi Arabia, Iran, or Kuwait with gas that would otherwise be wasted, future price increases may be curbed. The Soviet Union, in a situation similar to the gas-surplus countries in the Middle East, is also investing heavily in nitrogen fertilizer-production capacity.[10]

The distribution of phosphate rock, the principal source of phosphate fertilizer, poses a particular problem since reserves are concentrated in Florida and Morocco. With production concentrated around the Atlantic but with the world's population and future needs for phosphate mainly in Asia, high transportation costs—and thus high fertilizer prices in Asian villages—are inevitable.

With population growth projected to continue, the cropland available per person will continue to decline and the fertilizer needed to maintain consumption will continue to rise. At some point, biological constraints on crop yields will make the substitution of fertilizer for cropland increasingly difficult and costly. When this is combined with the projected long-term rise in real cost of the oil and natural gas used to manufacture, distribute, and apply chemical fertilizer, the difficulty in restoring the steady upward trend in per capita grain production of 1950–73, when it climbed from 248 kilograms in 1950 to 326 kilograms in 1973, becomes clear.

## Real Production Trends

When measuring growth, economists adjust current prices for the rate of inflation in order to distill out the real gains in production. Something similar is needed in agriculture, where growth in output is inflated by agricultural practices that are not sustainable. Such an adjustment would shed light on the longer-term outlook by distinguishing between gains that are real and those that are made at the expense of future output.[11]

Similarly, adjustments should be made for the output from sloping land that was once in ecologically stable, long-term rotations of row crops with grass and hay, but that is now in row crops continuously, for topsoil loss in these situations has become excessive. If American farmers were to take the steps needed to protect their topsoil, U.S. farm output and exports would be substantially less in the short run, but they would be sustainable over the long term. Elimination of this agronomic deficit through a national soil conservation program that reintroduced the traditional practices cited above might also eliminate the troublesome short-term commodity surpluses that depress farm prices and income.

In addition to agronomic deficits, many of the world's farmers are also incurring economic deficits. Nowhere is this more evident than in the United States, where net farm income has narrowed almost to the vanishing point. Between 1973, when the world oil price began its astronomical climb, and 1982, farmers were caught in a squeeze between depressed commodity prices and the soaring costs for fuel, fertilizer, and equipment combined with high interest rates.

In 1982, many American farmers sold their products for less than they cost to produce. Between 1950 and 1982, U.S. farm output more than doubled, but net farm income in real terms (1967 dollars) fell from $19 billion in 1950 to scarcely $6 billion in 1982.[12] This precipitous decline occurred while the incomes of other Americans were rising steadily.

Farmers were able to sustain the heavy losses of the late 1970s and early 1980s only by going deeply into debt, borrowing against soaring land values. But the boom in land speculation came to an end in 1981, and land prices fell the following two years. As a result, many farmers suddenly lost their equity and faced bankruptcy.

Economic conditions fostering speculation in land have driven land values to a lofty level that bears little relationship to the land's productive capacity. Given the economics of the early 1980s, buying U.S. farmland now with the hope of paying for it from the produce would be wishful thinking.[13] Nevertheless, it was these spiraling land values that, until 1981, enabled many farmers to borrow and to stay in business.[14]

As farmers have borrowed against the soaring prices of their land and other assets, not only have they supported themselves and their families, they have also subsidized food consumers everywhere. Borrowing against the inflated paper value of farmland has led to artificially low food prices in recent years. And just as productivity increases cannot go on forever when topsoil is being eroded, borrowing that is unrelated to the real value of the land cannot continue indefinitely—a lesson many rural banks and farmers are unfortunately learning.

If farmers are to continue to produce, prices of farm products will need to rise. Without such an increase, the more vulnerable farmers and those who have attractive employment options or who are approaching retirement will stop producing, eventually reducing output and moving prices upward to a more realistic level.

Although U.S. data might allow the conversion of current farm output to real output by adjusting for soil erosion, similar information does not exist for most countries. And it is difficult to measure the extent to which farm output has been inflated in recent years by the growing indebtedness of farmers. If these adjustments could be made, however, it seems clear that the real world food output would be far below current consumption.

## Dependence on North America

With grain, as with oil or any other basic resource, excessive world dependence on one geographic region for supplies is risky. As the North American share of world grain exports has increased, it has surpassed the Middle Eastern share of oil exports and made the world more dependent on one region for its food than ever before.

This extraordinary dependence on one geographic region for grain supplies is a historically recent phenomenon and gives North America a politically and economically strategic role in the world food economy. Many of the world's cities, particularly those in the Third World, are fed largely with U.S. and Canadian wheat. Much of the world's milk, meat, and eggs are produced with U.S. feedgrains and soybeans.

As recently as the late 1930s, Western Europe was the only grain-deficient region and Latin America was the world's leading grain supplier, exporting some nine million tons per year. North America and Eastern Europe (including the Soviet Union) each exported five million tons of grain annually. Even Asia and Africa had modest exportable surpluses.[15]

By 1950 the shift from regional grain surpluses to deficits was well under way, and the outlines of a new world grain trade pattern were beginning to emerge. Today, with North America's unchallenged dominance as a grain supplier, international grain trade bears little resemblance to that of the 1930s (see Table 2).

As North American agricultural growth gained momentum after World War II, U.S. and Canadian exports of grain climbed from 23 million tons in 1950 to 138 million tons in 1982, though they dropped back to 122 million tons in 1983 as a strong dollar and lethargic world economy weakened the buying powers of other countries. Feedgrains—principally corn, sorghum, and barley—have made up an ever larger share of the total. Today, North America is not only the world's breadbasket, but its feed bag as well.

**Table 2**
**THE CHANGING PATTERN OF WORLD GRAIN TRADE, 1950–83[1]**

| REGION | 1950[2] | 1960 | 1970 | 1980 | 1983[3] |
|---|---|---|---|---|---|
| | *(million metric tons)* | | | | |
| North America | +23 | +39 | +56 | +131 | +122 |
| Latin America | + 1 | 0 | + 4 | − 10 | − 3 |
| Western Europe | −22 | −25 | −30 | − 16 | + 2 |
| E. Europe and Soviet Union | 0 | 0 | 0 | − 46 | − 39 |
| Africa | 0 | − 2 | − 5 | − 15 | − 20 |
| Asia | − 6 | −17 | −37 | − 63 | − 71 |
| Australia and New Zealand | + 3 | + 6 | +12 | + 19 | + 9 |

[1] Plus sign indicates net exports; minus sign, net imports.
[2] Average for 1948–52.
[3] Preliminary.

**SOURCES:** United Nations Food and Agriculture Organization. *Production Yearbook* (Rome: various years); U.S. Department of Agriculture, *Foreign Agriculture Circular*, August 1983; author's estimates.

While the United States was expanding its feedgrain exports, the shipments of soybeans grew even more rapidly. Although soybeans originated in China, they have thrived in the United States, doing far better than in their country of origin. They have also found an economic niche in the world livestock economy, with soybean meal becoming the principal protein supplement in livestock and poultry feed. Today the United States produces over 60 percent of the world's soybean crop and accounts for two-thirds of soybean exports.

The reasons for North America's emergence as the world's dominant supplier of feedgrains and feedstuffs are many. On the supply side, the United States inherited a prime piece of agricultural real estate. In contrast to Latin America, where agricultural lands are concentrated in the hands of large hacienda owners, or Eastern Europe, where state farms and collectives dominate, U.S. and Canadian agriculture are centered on the family farm. Although large by international standards, they are nonetheless family farms and have all the attendant advantages of a strong link between effort expended by those working the land and the rewards of doing so.

The restructuring of world grain trade over the last generation has resulted in part from the soil erosion problems discussed earlier and in part from differential population growth rates, as a comparison of North America and Latin America shows.[16]

Today the countries with significant exportable surpluses of grain can be counted on the fingers of one hand—the United States, Canada, Australia, Argentina, and France. Of these, the United States accounts for over half and, with Canada, covers close to 70 percent of the total.

The rest of the world's dependence on these supplies varies widely. A few countries, both industrial and developing, import more food than they produce; among these are Algeria, Belgium, Costa Rica, Japan,

U.S. and Canadian agriculture are centered on the family farm. Although large by international standards, they have all the attendant advantages of a strong link between effort expended by those working the land and the rewards of doing so. Today, North America is not only the world's breadbasket, but its feed bag as well (photo by Vernon Sigl).

Lebanon, Libya, Portugal, Saudi Arabia, Switzerland, and Venezuela. Others that may shortly move into this category include Egypt, Senegal, and South Korea.[17]

This overwhelming dependence on one region, and on one country in particular, brings with it an assortment of risks. To begin with, both the United States and Canada are affected by the same climatic cycles. A poor harvest in one is often associated with a poor harvest in the other. When reserves are low, even a modest fluctuation in the region's exportable grain surplus can send price tremors through the world food economy.

An inadvertent agricultural policy miscalculation can also be costly. This was amply demonstrated in 1983 when miscalculations in the U.S. Department of Agriculture led to the idling of more cropland than had been projected, which was followed by a severe drought that further reduced harvests. Within a matter of weeks, concerned countries watched the world grain surplus change to a potential grain deficit. The U.S. corn crop was cut in half, effectively eliminating the world feedgrain surplus.[18]

When food supplies are tight, a North American grain export embargo, whether economically or politically inspired, can drive food prices upward everywhere outside the region. In 1973, for example, President Nixon embargoed soybean exports because of shortages at home. Although this helped curb food price rises within the United States, it worsened inflationary pressures elsewhere.

During the same period, American millers and bakers were pressing for restrictions on grain exports, holding out the prospect of soaring bread prices if wheat exports were not restricted. Unfortunately, the world market conditions that would lead a principal exporter to restrict outgoing supplies are precisely the conditions that are most damaging to importing countries.

In mid-July of 1975, the Canadian Wheat Board banned further exports of wheat until the size of the harvest could be ascertained. Similarly, the United States, yielding to political pressures generated by rising domestic food prices, limited grain exports to the Soviet Union and Poland in the late summer and early fall of 1975. Levied in 1972, in 1974, and again in 1975, such restrictions on exports became common when global grain supplies were tight. Perhaps more unsettling, these export controls were adopted despite the return to production of the previously idled U.S. cropland.

As with oil, exports of grain have been restricted for political purposes. In 1973, the Department of State compiled a "hit list" of Third World countries whose U.N. voting records were not compatible with U.S. interests so that they could be denied food assistance. More recently, President Carter imposed a partial embargo on exports of grain to the Soviet Union following its invasion of Afghanistan, and President Reagan delayed negotiating a new five-year grain agreement with the Soviet Union after the imposition of martial law in Poland.[19]

Countries that rely on North American food should take heed of the philosophical debate emerging within the United States about the wisdom of mining the nation's soils to meet the ever-growing world demand. Both agricultural analysts and environmentalists argue that the country should make whatever adjustments in its agricultural practices are needed to protect the resource base, even though this would reduce that exportable surplus.

Some argue that it makes little sense to sacrifice a resource that has been a source of economic strength since colonial days merely to buy a few billion barrels of oil. And some contend that the current generation of farmers has no right to engage in the agronomic equivalent of deficit financing, mortgaging the future generations to come.

The current trend is fraught with risks, both for those whose livelihoods depend on sustained land productivity and for those in countries dependent on food imports that eventually will dry up if the mining of soil continues. Even for the importers, reduced supplies in the short term and less pressure on North American soils would be better than losing the region's export capacity over the long term.

## Food Security Indicators

One of the most useful indicators of the world food situation is the food security index, which incorporates both grain carry-over stocks and the grain equivalent of idled cropland. This combines the world's two basic reserves of food and expresses them as days of consumption, a concept readily understood by policy makers everywhere.

The two components of the index differ in important ways. Carry-over stocks, the grain in storage when the new crop begins to come in, are readily accessible and require only time for shipping arrangements to be made and for transport. Idled cropland, on the other hand, can take a year or more to be converted into food by farmers.

Carry-over stocks are held for the most part by exporting countries—the United States, Canada, Australia, Argentina, and France—largely as a service to importers. Other countries, particularly large ones such as India,

Vegetable farmer in Brazil. The scenario long predicted by ecologists—that mounting population pressures in northeastern Brazil leading to deforestation and unsustainable agricultural practices would lead to a major food crises in the area—is now unfolding (World Bank photo by Yosef Hadar).

maintain grain stocks as well, but these are usually designed specifically for their own use.[20]

Maintaining adequate grain stocks is expensive not only because of the cost of grain elevators but also because stored grain represents an investment. So when interest rates are high, the cost of carrying grain is also high. And, even with the best of storage facilities, there is always some loss involved, thus adding to the cost of maintaining the reserves.

Over time, for a reserve of grain to be adequate it should expand in tandem with world consumption. In 1960, for example, world reserves of 200 million tons were more than ample, representing nearly one-fourth of world consumption. In 1983, however, the same stocks would represent only one-eighth of world grain consumption. A level of grain reserves that was adequate in 1960 would be grossly inadequate in 1983.

The idled cropland component of the food security index consists of cropland set aside under farm programs just in the United States. Only occasionally have other countries intentionally idled cropland, and the acreages have usually been negligible. During the 1960s and early 1970s, idled U.S. cropland averaged close to 50 million acres, enough to produce an estimated 60 million tons of grain (see Table 3).

As growth in world food outputs slowed after 1973, the United States returned cropland to production from 1974 through 1977 in an attempt to rebuild stocks. Then, when grain reserves began to recover in the late 1970s, some land was taken out of production in both 1978 and 1979. With reserves beginning to drop again in 1979, all land was released for production in 1980.

The Reagan administration, wishing to reduce government intervention in the marketplace, declined to idle any cropland in 1981 and idled only a modest acreage in 1982, even though world reserves had been rebuilt following two consecutive bumper harvests in the United States and a worldwide economic recession that dampened the growth in demand. In 1983, faced with the most severe rural depression since the 1930s, the administration overreacted by devising two programs to encourage farmers to divert land to nonproductive uses.

The result was the largest diversion of acreage in U.S. history—over 70 million acres. Combined with a severe drought in the principal U.S. feedgrain- and soybean-producing areas, this led to a precipitous decline in the feedgrain harvest of over 40 percent. This in turn reduced the prospective carry-over stocks to one of the lowest levels in some years.[21]

Whenever grain stocks and the grain equivalent of idled U.S. cropland drop below 50 days of world consumption, grain prices customarily rise and become highly unstable. In 1973 and 1974, when the index dropped to 50 and 41 days, grain prices were nearly double traditional levels. The index again fell below 50 days of consumption in 1980, partly

**Table 3**
**INDEX OF WORLD FOOD SECURITY, 1960–83**

| YEAR | WORLD CARRY-OVER STOCKS OF GRAIN | *RESERVES* GRAIN EQUIV. OF IDLED U.S. CROPLAND | TOTAL | WORLD CONSUMPTION |
|---|---|---|---|---|
| | | *(million metric tons)* | | *(days)* |
| 1960 | 200 | 36 | 236 | 104 |
| 1965 | 142 | 70 | 212 | 81 |
| 1970 | 164 | 71 | 235 | 75 |
| 1971 | 183 | 46 | 229 | 71 |
| 1972 | 143 | 78 | 221 | 67 |
| 1973 | 148 | 25 | 173 | 50 |
| 1974 | 133 | 4 | 137 | 41 |
| 1975 | 141 | 3 | 144 | 43 |
| 1976 | 196 | 3 | 199 | 56 |
| 1977 | 194 | 1 | 195 | 53 |
| 1978 | 221 | 22 | 243 | 62 |
| 1979 | 197 | 16 | 213 | 54 |
| 1980 | 183 | 0 | 183 | 46 |
| 1981 | 221 | 0 | 221 | 56 |
| 1982[1] | 260 | 13 | 273 | 66 |
| 1983[2] | 191 | 92 | 283 | 68 |

[1]Preliminary.
[2]Projection.

**SOURCES:** Reserve stocks from U.S. Department of Agriculture (USDA), *Foreign Agriculture Circular*, October 1983; cropland idled in the United States data from Randy Weber, USDA, private communication, August 1983.

as a result of drought in the United States.

This time, however, prices were not nearly as volatile as before, perhaps in part because interest rates were at record highs, making investors less inclined to speculate in commodities. By late 1983, prices, particularly of feedgrains, began to rise, largely because of the unprecedented reduction in the U.S. grain harvest—the product of government miscalculation and drought.

The food security index measures the adequacy of food supplies at the global level and thus the broad potential for responding to national shortages, but it says nothing about conditions within individual countries. Here the best indicator, of course, is the nutritional state of a country's population. At issue is whether a growing child gets enough food to develop his or her full physical and mental potential or whether a worker gets enough food to be fully productive.

Per capita food availability for a country indicates what the national average is, but not whether an individual is adequately nourished. Assessing nutritional adequacy requires some knowledge of how the national food supply is distributed. But a lack of data on distribution makes it very difficult to estimate the extent of malnutrition, thus leaving the subject open to continuing debate.

The only time a decline in nutrition shows up officially is when it is severe enough to affect mortality. When this happens, a country is facing famine, the most obvious and severe manifestation of food insecurity. Using this criterion, inadequate though it is, developments over the past decade have not been encouraging.

From the postwar recovery years until the early 1970s, famine virtually disappeared from the world. Except in China, which now admits to a massive famine in 1960–61, when it was largely isolated, the world enjoyed a remarkable respite from famine for a quarter of a century. Whenever famine did threaten, the United States intervened with food aid, even when it required nearly one-fifth of the U.S. wheat crop two years in a row, as it did following monsoon failures in India in 1964 and 1965.[22]

By the early 1970s, however, food deficits were widening and famine was unfolding in several African countries and in the Indian subcontinent (see Table 4). Several famines claimed hundreds of thousands of lives, providing a grim reminder of the fragility of food security even in an age of advanced technology. Most were the product of drought and a failure of international food relief mechanisms.

During the late 1970s, world reserves were rebuilt and, except for strife-torn Kampuchea, famines subsided—only to return in 1983, a year of widespread climatic anomalies (see "El Niño and World Climate: Piecing Together the Puzzle," *Environment*, April 1984). The capacity of poor countries with falling per capita food production and deteriorating soils to withstand drought and floods has lessened. As a result, more countries than ever before face the possibility of famine in early 1984. Among the threatened countries are Bolivia and Peru in Latin America and over a score of countries in Africa. An FAO team of agronomists assessing the food situation in Africa in late 1983 identified 22 countries where crisis seemed imminent.[23]

**Table 4**
**COUNTRIES EXPERIENCING FAMINE SINCE 1950**

| YEAR | LOCATION | ESTIMATED DEATHS |
|---|---|---|
| 1960–61 | China | 8,980,000 |
| 1968–69 | Nigeria (Biafra) | 1,000,000 |
| 1971–72 | Bangladesh | 430,000 |
| 1972 | India | 830,000 |
| 1973 | Sahelian Countries | 100,000 |
| 1972–74 | Ethiopia | 200,000 |
| 1974 | Bangladesh | 330,000 |
| 1979 | Kampuchea | 450,000 |
| 1983 | Ethiopia | 30,000 |

**SOURCE**: Worldwatch Institute estimates derived from various official and unofficial sources.

Since all governments gather mortality data, though not equally well, it is possible to document these severe food shortages. But equally troubling is the number of people suffering from chronic malnutrition—the vast middle ground between those who are well nourished and those who are starving. Their numbers are difficult to measure and therefore easy to ignore. Indian economist Amartya Sen observes that his government "has been able to ignore this endemic hunger because that hunger has neither led to a run on the market, and chaos, nor grown into an acute famine with people dying of starvation. Persistent, orderly hunger does not upset the system."[24]

## A Crisis of Many Dimensions

There is no simple explanation of why efforts to eradicate hunger have lost momentum or why food supplies for some segments of humanity are less secure than they were, say, 15 years ago. Declines in food security involve the continuous interaction of environmental, economic, demographic, and political variables. Some analysts see the food problem almost exclusively as a population issue, noting that wherever population growth rates are low, food supplies are generally adequate.

Others view it as a problem of resources—soil, water, and energy. Many economists see it almost exclusively as a result of underinvestment, while agronomists see it more as a failure to bring forth new technologies on the needed scale. Still others see it as a distribution problem. To some degree, it is all of these.

In an important sense, the food problem of the mid-1980s is the result of resource depletion. The depletion of oil reserves, the loss of topsoil through erosion, and the growing competition for fresh water are central to understanding trends in the world food economy. As world population expands, the shrinking cropland area per person and the reduction in average soil depth by erosion combine to steadily reduce per capita availability of topsoil for food production. Because the energy that farmers substitute for soil lost to erosion is becoming increasingly costly, the production costs everywhere are on the rise.

In a basic arithmetic sense, the food problem is a population problem. If world population were now growing at 1 percent instead of nearly 2 percent, there would still be an ample margin for improving diets, as there was from 1950 to 1973. As noted, however, the annual growth in food production has unfortunately fallen from the rather comfortable 3 percent of that period to a rate that barely matches that of population.

Those who see the food problem primarily as a distribution problem argue that if all the world's food were equitably distributed among its people there would be no hunger and malnutrition. This argument is technically sound, but it represents a degree of global abstraction that is not very helpful in formulating policies. It would mean, for example, that much of the world would be even more dependent on U.S. farmers than they are today.

The only long-term solution to hunger in most Third World countries is more internal production. And improving distribution is not just a matter of a better transport system or subsidized food distribution programs. It requires dealing with fundamental sources of political conflict such as land reform and with economic policies that encourage employment.

Achieving a more satisfactory balance between the world demand and supply of food requires attention to both sides of the equation. On the demand side, the success of efforts to upgrade diets may depend on an emergency program to slow world population growth. Currently, farmers must produce enough additional food each year to feed an annual increment of 79 million people—people who must be provided for in years of good weather and bad.

Meaningful improvements in diet over the rest of the century will depend, too, on gains in per capita income, particularly in the Third World. Unfortunately those gains are

Construction of an anti-erosion dike in Upper Volta. This agricultural project is directed primarily to increasing the production of cotton and cereals in the area (World Bank photo by Yosef Hadar).

narrowing or disappearing. During the 1970s, 18 countries in Africa experienced a decline in per capita income. In most instances these national declines appear to be continuing in this decade; the list of countries where incomes have fallen thus far during the present decade is far longer than it was in the last one.

Also on the demand side is the question of how available food supplies are distributed. The most vulnerable segments of society in times of food scarcity are the rural landless in Third World countries. Reducing the size of this extremely vulnerable, rapidly growing landless population will require far more vigorous family planning and land reform programs than most countries have so far been able to mount.

On the supply side, the scarcity of new cropland, the continuing loss of topsoil, the scarcity of fresh water, the end of cheap energy, and diminishing returns on chemical fertilizer combine to make expanding food production progressively more difficult. In addition, food-deficit Third World countries are now struggling with a heavy external debt load that is continuing to mount. Foreign exchange scarcities are reducing imports both of agricultural inputs and of food.

At a time when concessional food aid is needed more than ever, U.S. programs are at the lowest level of nearly a generation. From their launch in 1954 through the late 1960s, U.S. food programs expanded, reaching a high of 15.3 million tons of grain in 1966. Enough to feed 90 million people in its peak years, this aid was an important defense against hunger in many countries. Preliminary estimates for 1983, however, indicate food aid shipments had dwindled to some 4 million tons of grain.

Other assistance programs, such as the World Food Programme, have been developed and expanded, but they are small by comparison and cannot begin to offset the U.S. decline. In a modest effort to enhance food security in low-income countries, the International Monetary Fund launched a new facility in 1981 that would provide short-term supplementary financing to food-deficit Third World countries suffering from temporary rises in food import bills.[25]

In addition to the traditional problems, political conflicts are creating an instability in the Middle East, Africa, and Latin America that makes agricultural progess difficult. This new hindrance to progress is reflected in the World Food Programme's emergency relief efforts. The share of emergency food relief going to victims of human-caused disasters (refugees and displaced persons) has come to dominate the program in recent years, after starting at a relatively modest level. By 1982, less than one-third of the program's resources were devoted to helping the groups who were the principal recipients in the early years, those affected by drought or caught in sudden natural disasters such as floods or earthquakes.

Other disruptions borne of desperation are cropping up in food-deficit areas. In late 1983, reports from northeastern Brazil described hordes of hungry rural people pouring into the towns demanding food, water, and jobs. After the town of Crato was invaded on three different occasions, merchants began voluntarily distributing rations of beans, sugar, and flour to forestall sacking and looting. Merchants in another town reportedly lost some 60 tons of food to looters.[26]

Although drought and recession-induced unemployment are leading to hunger, the long-term forces converging to create these conditions in the Brazilian northeast are essentially the same as those that have led to the food crises in so many African countries: record population growth, underinvestment in agriculture, and physical deterioration in the countryside. Ecologists have for years warned that mounting population pressures in northeastern Brazil leading to deforestation and unsustainable agricultural practices would turn the area into a desert. That grim scenario is now unfolding.

Trends in Africa since 1970 are a harbinger of things to come elsewhere in the absence of some major changes in population policies and economic priorities. After rising from 1950 to 1970, per capita grain production in Africa has declined rather steadily.[27] The forces that have led to this decline in Africa are also gaining strength in the Andean countries of Latin America, in Central America, and in the Indian subcontinent. Whether the declining food production now so painfully evident in Africa can be avoided elsewhere will be determined in the next few years.

The issue is not whether the world can produce more food. Indeed, it would be difficult to put any foreseeable limits on the amount the world's farmers can produce. The question is at what price they will be able to produce it and how this relates to the purchasing power of the poorer segments of humanity.

The environmental, demographic, and economic trends of the 1970s and early 1980s indicate that widespread improvements in human nutrition will require major course corrections. Nothing less than a wholesale reexamination and reordering of social and economic priorities—giving agriculture and family planning the emphasis they deserve—will get the world back on an economic and demographic path that will reduce hunger rather than increase it.



## NOTES

1. Food production data in this chapter are drawn primarily from U.S. Department of Agriculture (USDA), Economic Research Service (ERS), *World Indices of Agricultural and Food Production, 1950-82*, unpublished printout (Washington, D.C., 1983), and from USDA, Foreign Agricultural Service (FAS), *Foreign Agriculture Circulars*, various commodities, Washington, D.C., published monthly.

2. See, for example, Jay Ross, "Africa: The Politics of Hunger" (series), *Washington Post*, June 25-30, 1983.

3. Davidson R. Gwatkin, *Signs of Change in Developing-Country Mortality Trends: The End of an Era?* Development Paper No. 30 (Washington, D.C.: Overseas Development Council, February 1981).

4. United Nations Food and Agriculture Organization (FAO), *Production Yearbook* (Rome: annual, various years).

5. Kenneth B. Young and Jerry M. Coomer, *Effects of Natural Gas Price Increases on Texas High Plains Irrigation, 1976-2025*, Agricultural Report No. 448 (Washington, D.C.: USDA, Economics, Statistics, and Cooperatives Service, 1980).

6. Marshall I. Goldman, *The Spoils of Progress* (Cambridge, Mass.: M.I.T. Press, 1972); Dr. John Gribben, "Climatic Impact of Soviet River Diversions," *New Scientist*, December 6, 1979; Grigorii Voropaev and Aleksei Kosarev, "The Fall and Rise of the Caspian Sea," *New Scientist*, April 8, 1982.

7. Republic of South Africa, *Report of the Science Committee of the President's Council on Demographic Trends in South Africa* (Capetown: The Government Printer, 1983).

8. FAO, *FAO 1977 Annual Fertilizer Review* (Rome: 1978); Paul Andrilenas, USDA, ERS, private communication, December 9, 1982.

9. USDA, *Agricultural Statistics 1982* (Washington, D.C.: U.S. Government Printing Office, 1982).

10. USDA, ERS, "Fertilizer Outlook and Situation," Washington, D.C., December 1982.

11. For example, as farm commodity prices climbed in the mid-1970s, U.S. farmers brought land under the plow that was not suited to cultivation. By 1977 the Soil Conservation Service had identified 17 million acres of land in crops that were losing topsoil so rapidly they would eventually be stripped of all productive value. The agency recommended that farmers convert this land to grass or forests to preserve its production capacities. USDA, *Soil and Water Resources Conservation Act (RCA), Summary of Appraisal, Parts I and II, and Program Report Review Draft 1980* (Washington, D.C., 1980). To reach a figure of real, not just current, U.S. agricultural output, the yield from these 17 million acres, roughly 4 percent of the U.S. cropland total, should be subtracted from overall output.

12. *Economic Report of the President* (Washington, D.C.: U.S. Government Printing Office, 1983).

13. Say, for example, someone had invested in prime midwestern farmland at $2,000 an acre in 1981 and planted it in corn that yielded 110 bushels per acre. With a mortgage at 15 percent, the annual interest payment would be $300 an acre. Yet at the 1981 price of $2.40 a bushel, the total income from each acre would have been $264, not enough to pay the interest, much less the principal or any of the production costs.

14. While net farm income has declined markedly, farm debt has soared. As recently as 1973, net farm income exceeded $30 billion compared with a farm debt of $65 billion, a ratio of roughly one to two. By 1983, net farm income totaled $22 billion, while the farm debt had climbed to $215 billion—close to 10 times income. *Economic Report,* note 12 above.

15. Lester R. Brown, *Man, Land and Food: Looking Ahead at World Food Needs*, Foreign Agriculture Economic Report No. 11 (Washington, D.C.: USDA, ERS, 1963).

16. During the late 1930s, Latin America had a larger grain export surplus than North America, but the region's more rapid rate of population growth soon changed this. Indeed, if the regions had grown at the same rate since 1950, North America's population in 1983 would be so large that it would consume the entire grain harvest, leaving little or none for export. And North America, too, would now be struggling to maintain food self-sufficiency.

17. USDA, FAS, "Reference Tables on Wheat, Corn and Total Coarse Grains Supply Distributions for Individual Countries," *Foreign Agriculture Circular* FG-19-83, Washington, D.C., July 1983.

18. A similar miscalculation in 1972, when record acreage of U.S. cropland was idled, contributed to the food shortages of the 1972-74 period. Figures on cropland idled in the United States are from Randy Weber, USDA, Agricultural Stabilization and Conservation Service, private communication, August 23, 1983.

19. For a discussion of manipulations of grain exports for political purposes, see Dan Morgan, *Merchants of Grain* (New York: Penguin Books, 1980).

20. India's grain stocks, typically ranging from 11-15 million tons, were drawn down to scarcely 5 million tons during 1973-75, a time of poor harvests. After this harrowing experience, India adopted a target stock level of 21-24 million tons as part of a beefed-up food security system. Bumper harvests in the late 1970s helped the government achieve its target, but after stocks were drawn down to 12-15 million tons during the 1979-80 drought, New Delhi has apparently decided for reasons of cost to maintain a more modest level of grain reserves.

21. Weber, private communication. The grain equivalent of idled cropland is calculated assuming a marginal grain yield of 3.1 metric tons per hectare.

22. For a discussion of the demographic consequences of famine in China, see H. Yuan Tien, "China: Demographic Billionaire," *Population Bulletin* (Washington, D.C.: Population Reference Bureau, 1983); U.S. food aid shipments to India are unpublished data on P.L. 480 from Dewain Rahe, USDA, FAS, private communication, August, 31, 1983.

23. Angola, Benin, Botswana, Cape Verde, Central African Republic, Chad, Ethiopia, Gambia, Ghana, Guinea, Lesotho, Mali, Mauritania, Mozambique,Sao Tome and Principe, Senegal, Somalia, Swaziland, Tanzania, Togo, Zambia, and Zimbabwe. The team of experts concluded that four million tons of emergency grain supplies would be needed to avoid starvation among the 145 million people living in these countries. Michael deCourcy Hinds, "U.S. Giving Peru and Bolivia Millions in Food Aid," *New York Times*, June 2, 1983; FAO and World Food Programme Special Task Force, "Exceptional International Assistance Required in Food Supplies, Agriculture and Animal Husbandry for African Countries in 1983/84," Rome, September 30, 1983.

24. Amartya Sen, "How Is India Doing?" *New York Review of Books*, December 16, 1982.

25. U.S. food aid shipments under P.L. 480 are from Rahe, private communication; International Monetary Fund, Washington, D.C., press release, May 21, 1981.

26. Mac Margolis, "Brazil 'dust bowl' 5 times size of Italy," *Christian Science Monitor*, August 26, 1983.

27. USDA, ERS, note 1 above.

CHAPTER 15

# Environmental Impacts of Early Societies and the Rise of Agriculture



CHARLES H. SOUTHWICK

With so much emphasis on our modern environmental problems, we should consider the environmental relationships of earlier societies. To what extent did preindustrial societies alter their environments? Is environmental destruction a product of our industrial age? These are important questions for a proper perspective on global ecology, and some of the answers are surprising.

The main part of this article is reprinted with permission from Chapter 6 of *Ecology and the Quality of Our Environment*, Second Edition, by Charles H. Southwick (1976), pp. 119–132. 

## HUNTER-GATHERERS AS ECOLOGISTS

Early man[1] was a successful practicing ecologist. He survived in a rich and competitive biotic community, and his relationship to this community was continually intimate. He was by no means the strongest, the swiftest, nor the hardiest among his congeners, but he possessed several distinct advantages. He had a close effective social organization, he had unusual manipulative ability, and he had, of course, an emerging intelligence. Thus, he developed tools and fire in early Paleolithic time (500,000 to 1,000,000 years ago), and he accumulated knowledge at a faster rate than other primates.

Much of this knowledge was ecological. It was knowledge of his environment and the most effective use of it. It was detailed knowledge of food and water resources, some of which can still be seen in the "primitive" peoples of today. The Kalahari bushmen, for example, can find water in a barren desert where other men would surely die of thirst. The Australian aborigine can locate grubs and lizards in the Australian deserts far better than a modern biologist. Elton (1933) pointed out that, "The Arawak of the South American equatorial forest knows where to find every kind of animal and catch it, and also the names of the trees and the uses to which they can be put." Bates (1960) found on the Micronesian atoll of Ifaluk that the native people had detailed knowledge of how the plants on the islands could be used for food, medicine, construction, and ornament. They also possessed de-

[1]The term "early man" refers here to pre-agricultural man—that is, human societies living entirely from hunting and gathering, without cultivating plants or domesticating animals.

tailed knowledge of the reefs and sea around them. In all societies of hunter-gatherers and throughout the several million years of Paleolithic man, survival depended on knowledge of the environment.

There is increasing evidence that early man was well adapted to his environment. The popular conception that he barely clung to life through a precarious and difficult struggle is definitely misleading (Lee and DeVore, 1968). Recent studies of both ancient and modern hunter-gatherers have shown that they frequently had an abundant life, with ample resources. Present-day primitive peoples are declining, of course, but usually as a result of modern forces—a deteriorating environment or competition from agricultural peoples. Other studies on former hunter-gatherers have shown that malnutrition was rare, starvation infrequent, and chronic diseases, as we know them today, of relatively low incidence—all evidence of sound ecologic balance in their way of life (Dunn, 1968; Neel, 1970). There was, however, high infant mortality, primarily through infectious disease, and high "social mortality" (infanticide, geronticide, warfare, etc.), and these served as primary mechanisms of population regulation.

Early man's practical knowledge of his environment does not necessarily mean that he was a good conservationist. He knew enough practical ecology to survive and even prosper, but he exploited his environment at every opportunity. He was a persistent forager and relentless hunter whose primary goal was survival. He was often nomadic "in part because prolonged habitation in any one area depleted game and firewood and accumulated wastes to the extent that the region was no longer habitable" (Guthrie, 1971). Some scholars have felt that the overzealous hunting of early man contributed to the widespread extinction of animals in the Pleistocene Epoch[2] (Martin and Wright, 1967). Although this is a controversial hypothesis, there is some evidence that early man was not always conservative in his hunting practices or environmental protection. For example, plains Indians of North America were known to have killed many more bison than they could utilize by driving them over cliffs.

As pointed out by Lee and DeVore (1968): "Cultural man has been on earth for some 2,000,000 years; for over 99 percent of this period he has lived as a hunter-gatherer. Only in the last 10,000 years has man begun to domesticate plants and animals, to use metals, and to harness energy other than the human body." The civilizations of agricultural

[2]The Pleistocene Epoch is also known as the Ice Ages, that period from 10,000 to 600,000 years ago, in which great populations of large mammals, including wooly mammoths, mastodons, giant pigs, royal bison, camels, horses, giant armadillos, and great ground sloths roamed the temperate regions of North America and Europe.

and industrial man have a long way to go before they can match the longevity of primitive man.

## PASTORAL MAN AND THE DOMESTICATION OF ANIMALS

With the rise of civilization and its elaborate divisions of labor, more niches for the nonecologist became available. The weaver, potter, and tool-maker did not require the same broad ecological knowledge as the hunter and gatherer. The most significant ecological achievement of civilized man was, of course, the domestication of plants and animals for greater productivity and control over the means of subsistence. Thus, the development of pastoral and agricultural life, in the Neolithic period of 10,000 B.C. to 6,000 B.C., altered the entire pattern of human existence. Permanent villages became established, intergroup cooperation and trade routes developed, and a demand arose for a new type of ecologic knowledge, that associated with the husbandry of plants and animals. Economics displaced ecology as the vital key to survival success. This period of history was dramatically portrayed in 1975 in the television series "The Ascent of Man," by Dr. Jacob Bronowski.

The beginning of recorded history, about 3,000 B.C., showed civilizations in Egypt and Mesopotamia with cities and a high degree of vocational specialization. The Bronze Age, shortly after 3,000 B.C., and the Iron Age, starting about 1,000 B.C., accelerated agriculture, exploration, and conquest, but did little for man's concern for his environment. Environmental exploitation took place on a much broader scale. Major land changes occurred. Forests were cut, fields cleared, pastures grazed and plowed, and the landscape was carved to fit the new economic demands of man. A general increase in aridity occurred throughout much of the world then occupied by civilized man, and some of the great land barrens of the modern world probably developed during this time (Marsh, 1864; Sauer, 1938; Sears, 1935; Thomas, 1955–1956).

There have been, in general, two theories to explain the development of these great land barrens, especially those in the Middle East and North Africa. The theories differ in their views on the role of man. One has viewed man as the victim of climatic change, and the other has viewed man as the perpetrator of climatic change. Thus, in one view, man had to adapt to desert conditions imposed upon him by great continental forces of climate and geography, and in the other view, man himself was the major desert-making force.

The "man-as-victim" view has probably prevailed in most academic circles until recent years, not so much because of positive evidence to support it, but because it seemed more logical and was kinder to man's ego. There is now increasing evidence from archeology and ecology that man's role as a desert-making force has been underestimated. Man has not simply been a passive victim of climatic change; he has often played a significant role in inducing this change. This view does not assert that man's activities have been the only forces in desert formation, but it does assert that they were influential in accelerating and intensifying desert formation.



The geographer Carl Sauer believed that there were three great periods of habitat destruction in the history of man: First, a period at least three to four thousand years ago, when the great herds of pastoral man and the early successes of agriculture caused extensive land scarring, erosion, and irreversible aridity in Africa, the Middle East, and Asia. Second, the latter days of Rome and the disorderly period immediately following, when many of the Mediterranean lands were despoiled. Third, the transatlantic expansion of European commerce and peoples into the New World, when rapid and disastrous land exploitation occurred throughout the Americas. Sauer documented these views in an important book first published in 1938 and reprinted in a more recent anthology (Leighly, 1967).

Does this mean that each major advance of man carries with it the possibility of some destructive force? Keeping in mind the domestication of hoofed animals, the expansion of the Roman empire, the invasion of the New World, the development of gunpowder, the rise of modern medicine, the internal combustion engine, and now the advent of the atomic age—all major technologic achievements which also contained destructive capabilities, one is tempted to believe the above statement may have some elements of truth.

## THE RISE OF AGRICULTURE

Agricultural man set into motion several major forces which had significant geographic and meteorologic consequences: deforestation, overgrazing, intensive burning, and land scarring. Geographic consequences came about through increased erosion, soil loss, and declining water tables; meteorologic consequences resulted from reductions in atmospheric humidity and cloud cover, increased heat reflectivity and a lowering of rainfall. Ecological studies have shown that forests help to maintain the level of rainfall necessary for their own existence, and that deforestation results not only in an immediate lowering of ground water levels, but also in long-term lowering of rainfall. Forests recycle moisture back into their immediate atmosphere by transpiration where it again falls as rain. Transpiration return from an acre of forest may reach 2,500 gallons per day (McCormick, 1959), and thus create a natural system of water reuse. If the forest is removed, this natural reuse cycle is broken, and water is lost through rapid runoff.

Grasslands perform the same function to a lesser degree, and operate with a smaller amount of available moisture. It is interesting to examine Middle Eastern history in light of this principle.

## CIVILIZATIONS OF THE MIDDLE EAST, NORTH AFRICA, AND THE MEDITERRANEAN



The span of history from 5,000 B.C. to 200 A.D., which we know primarily as the period of great civilizations—Sumeria, Babylonia, Assyria, Phoenicia, Egypt, Greece, and Rome—was also a period of unprecedented environmental disturbance. We tend to concentrate our attention on the superb achievements of these civilizations in literature, art, government, and science, while we virtually forget their incompetence in land management. These golden civilizations prospered at the expense of their environments. They left a landscape which has never recovered, and a legacy to future civilizations which ushered in a period of dark ages lasting for more than a thousand years.

As specific examples of the destructive nature of these civilizations, we can cite critical landscape changes in many areas of the Middle East, North Africa, and Mediterranean lands. As late as 7,000 B.C., the headwaters of the Tigris and Euphrates Rivers were covered with forests and grasslands (Saggs, 1962). In fact, most of the area now occupied by Iran and Iraq was productive and well-watered (Sauer, 1938). Domestic cattle appeared in the seventh millenium B.C., probably around 6,300 B.C. (Perkins, 1969). Herds of domestic cattle, sheep, and goats found very favorable pasture in these virgin grasslands. Their great success provided a major stimulus for the developing civilizations of Mesopotamia and Sumeria, but it also provided the first significant onslaught on these grasslands and their adjacent forests. The herdsmen prospered, utilizing the stored capital of thousands of years, and the upsurge of prosperity was followed by more people and larger herds. Forests were cut to provide additional pasture and more land was exposed to the increasing livestock populations. Residential communities developed around 4,000 B.C., and further deforestation occurred to provide timbers for developing cities. Elaborate agricultural practices ensued and irrigation canals extended the reach of the rivers. At first, the erosion and silt load of the canals was manageable, and the alluvial soil along the rivers was remarkably fertile. With passing generations, however, siltation increased and it became necessary to occupy great numbers of slaves and laborers with the job of keeping irrigation channels free of silt (Dasmann, 1968). After 3,000 B.C., the increased silt of the Tigris and Euphrates filled in the Persian Gulf 180 miles from its origin. All of these changes—loss of vegetation and soil, lowered water tables, declining agricultural productivity, diminishing rainfall, and the added economic burden of siltation in the irrigation canals—were significant factors in the fall of the great Babylonian empire (Saggs, 1962). Successive waves of invaders—Kassites, Elamites, Assyrians, and eventually Persians—conquered this tragic land and increased its devastation.

At the same time that the Sumerian and early Babylonian empires were flourishing in the Middle East, there was a thriving civilization in the Indus Valley of present-day Pakistan. A prosperous and advanced culture existed at the site of Mohenjo-Daro in the province of Sind (Wallbank, 1958). There is evidence of an urban civilization, with well-planned streets, dwellings, municipal halls, and palaces, and a well-engineered drainage system. A sophisticated governmental structure apparently existed. There was skillful use of bronze, copper, silver, and lead, and there was also beautifully glazed pottery, delicate jewelry, and carefully woven cotton textiles. This prosperous civilization came to an abrupt end about 1,500 B.C. No one knows the reason for its total collapse, but there was again a loss in the means of subsistence. The Sind today is mostly desert, except in those portions bordering the Indus River and its related irrigation canals.

Man, over a period of 3,000 years in the sixth to the third millenia B.C. had become a major force in environmental change. His powerful civilizations and glorious cities were intimately dependent upon the land which gave them birth. When the pastoral and agricultural riches of the land were destroyed, these civilizations could no longer be supported, decline began, and they became more vulnerable to invading armies.

In Africa, similar patterns have been traced. Even the Sahara within written history has not been as extensive as it is now. In arid parts of Egypt and the Sudan, Davidson (1959) found abundant evidence of productivity within historical times:

> Even as late as the third millenium large numbers of cattle are known to have found grazing in lower Nubia (formerly Ethiopia, now part of Sudan) where, as Arkell says, "desert conditions are so severe today that the owner of an ox-driven water-wheel has difficulty in keeping one or two beasts alive throughout the year." And anyone who has traveled in these dusty latitudes will have noticed how the wilderness of sand and rock that lies to the west of the Nile, far out upon the empty plains, is scored with ancient wadi beds which must once have carried a steady seasonal flow of water, but are now as dry as the desert air.

Throughout Saharan Africa, there are "lost cities" of former grandeur in areas now too arid to support life (Davidson, 1959). It would be extravagant and incorrect to claim that man's activities produced the Sahara desert, for it was largely the result of great climatic shifts over the course of geologic history. But the evidence is clear that man extended Saharan conditions and forced once productive and well-watered areas to become barren wastelands. The Sahara is still actively extending its arid conditions—in some areas at the rate of many miles

**Figure 1.** The remnants of a once-forested hillside in North Africa. Pollen deposits indicate that this range of hills was covered by coniferous forest before the time of Roman expansion. Now only the inorganic skeleton of a former ecosystem remains—the vegetation and soil have been lost.

per year (Wade, 1974). In recent centuries, over 250 million acres (390,000 square miles) in Africa have been converted from agricultural and pastoral production to desert, primarily through the agent of man and his livestock. The most recent developments in this process will be discussed later in this chapter.

According to Sauer (1938), the second great period of human destruction upon the landscape occurred during the latter days of Rome. Phoenicia, Greece, and Rome all prospered on the riches of the Mediterranean lands. One of the best accounts of this Roman exploitation is provided by Marsh (1864), who pointed out:

> The Roman Empire, at the period of its greatest expansion, comprised the regions of the earth most distinguished by a happy combination of physical advantages. The provinces bordering on the principal and secondary basins of the Mediterranean enjoyed a healthfulness and equability of climate, a fertility of soil, a variety of vegetable and mineral products, and natural facilities for the transportation and distribution of exchangeable commodities, which have not been possessed in an equal degree by any territory of like extent in the Old World or the New. The abundance of the land and of the waters adequately supplied every material want . . . the luxurious harvests of cereals that waved on every field from the shores of the Rhine to the banks of the Nile, the vines that festooned the hillsides of Syria, of Italy, and of Greece. . . .

Even much of North Africa was rich and productive, clothed in cedar forests which provided further wealth for Rome. But Marsh hastened to add a more recent description of these lands, as he observed them in the mid-nineteenth century during his tenure as United States Ambassador to Italy:

> If we compare the present physical conditions of the countries of which I am speaking, with the descriptions that ancient historians and geographers have given of their fertility and general capability of ministering to human uses, we shall find that more than one-half of their whole extent—including the provinces most celebrated for the profusion and variety of their spontaneous and their cultivated products, and for the wealth and social advancement of their inhabitants—is either deserted by civilized man and surrendered to hopeless desolation, or at least greatly reduced in both productiveness and population. Vast forests have disappeared from mountain spurs and ridges; the vegetable earth accumulated beneath the trees by the decay of leaves and fallen trunks, the soil of the alpine pastures which skirted and indented the woods, and the mould of the upland fields, are washed away; meadows, once fertilized by irrigation, are waste and unproductive....
>
> Besides the direct testimony of history to the ancient fertility of the regions to which I refer—Northern Africa, the greater Arabian peninsula, Syria, Mesopotamia, Armenia, and many other provinces of Asia Minor, Greece, Sicily, and parts of even Italy and Spain—the multitude and extent of yet remaining architectural ruins, and of decayed works of internal improvement, show that at former epochs a dense population inhabited those now lonely districts.
>
> It appears, then, that the fairest and fruitfulest provinces of the Roman Empire, precisely that portion of terrestrial surface, in short, which, about the commencement of the Christian era, was endowed with the greatest superiority of soil, climate and position, which had been carried to the highest pitch of physical improvement, and which thus combined the natural and artificial conditions best fitting it for the habitation and enjoyment of a dense and highly refined and cultivated population, is now completely exhausted of its fertility, or so diminished in productiveness, as, with the exception of a few favored oases which have escaped the general ruin, to be no longer capable of affording sustenance to civilized man.
>
> The decay of these once flourishing countries is partly due, no doubt, to that class of geological causes, whose action we can neither resist nor guide, and partly also to the direct violence of hostile human force; but it is, in far greater proportion, the result of man's ignorant disregard of the laws of nature....

## ENVIRONMENTAL DESTRUCTION IN OTHER PARTS OF THE WORLD

The same phenomenon of ecological deterioration has been discussed at length by Paul Sears (1935), Fairfield Osborn (1948), and Carl Sauer

(1967). They have shown that the environmental destructiveness of man has not been limited to the Middle East, Africa, and the Mediterranean, but has, in fact, occurred throughout China, India, sub-Saharan Africa and the New World. In India, for example, the original vegetation of Rajasthan and the southern borders of the Gangetic basin—areas which are now arid and semidesert—was deciduous forest and far richer in natural soil moisture than at the present time (Champion, 1936).

In northern India, the magnificent city of Fatehpur Sikri, just 23 miles west of Agra, was built in the sixteenth century, only to be abandoned within two decades due to a lack of water. It was overcome by the arid landscape its people helped to create.

The Great Thar desert of western India has increased its size by 60,000 acres in the last 100 years (Ehrlich and Ehrlich, 1970). Dr. A. Krishnan of the Central Arid Zone Research Institute of Jodhpur, India, said in 1973 that archeological evidence and recent carbon dating studies had established beyond doubt that the Rajasthan desert was largely caused by deforestation, overgrazing, and deterioration of the soil (*Calcutta Statesman*, Sept. 15, 1973, p. 6). This conclusion also held out some hope; since the desert was primarily man-made, Dr. Krishnan felt it could be reclaimed by adequate soil stabilization and large-scale establishment of grasses and trees. This requires, of course, high capital expenditures and relief from human and livestock population pressures—factors which are virtually impossible to achieve over large areas.

In North America, the present desolate shifting sand areas bordering the Colorado river were rich grassland pastures as late as the seventeenth century (Sauer, 1967). Mexico was extensively forested over vast areas now occupied by dry and rocky plains (Osborn, 1948). Of all these destructive changes wrought by man, Sears (1935) wrote:

> Wherever we turn, to Asia, Europe, or Africa, we shall find the same story repeated with an almost mechanical regularity. The net productiveness of the land has been decreased. Fertility has been consumed and soil destroyed at a rate far in excess of the capacity of either man or nature to replace.

Fairfield Osborn, in *Our Plundered Planet*, supported these views and pointed out that destructive land-use practices have not been confined to ancient peoples and former centuries, but have in fact been very recent phenomena as well. This can be seen in Australia, where forest destruction, overgrazing, and unlimited burning, have greatly accelerated wind erosion, and have thus extended the Australian deserts within the last 100 years.

The most dramatic and tragic desert formation of recent years has been the southern expansion of the Sahara desert into the countries of Mauritania, Mali, Upper Volta, Niger, Chad, the Sudan, and Ethiopia—the region of central Africa known as the Sahel. Although

**Figure 2.** Desertification in the Sahel. Several years of drought in sub-Saharan Africa, coupled with heavy grazing pressure, greatly extended the Sahara desert with a tragic loss of human and animal life. *(UPI photograph)*

drought conditions throughout the late 1960s and early 1970s triggered this great southern movement of the desert, several scientific studies have shown that human activities have been the primary cause of desert formation in the Sahel (Wade, 1974). Edward Fei, chief of the AID Special Task Force on the Sahel, said, "The desertification is man-caused, exacerbated by many years of lower rainfall." The French hydrologist Marcel Roche stated, "The phenomenon of desertification, if it exists at all, is perhaps due to the process of human and animal occupation, certainly not to climatic changes." These scientists concluded that the primary agents of desert expansion into the Sahel have been overgrazing, excessive cutting of trees for firewood, slash and burn agriculture, and excessive wind and water erosion during storms. Dramatic evidence of the effect of overgrazing was provided on the Ekrafane ranch in the Sahel where a 250,000 acre range was divided up into five sectors and cattle were allowed to graze on only one sector a year. Wade (1974) noted: "Although the ranch was started only 5 years ago, at the same time as the drought began, the simple protection afforded the land was enough to make the difference between pasture and desert." The Ekrafane ranch maintained green vegetation which

**Figure 3.** Devastated farmlands of the American dust bowl, converted from productive grassland to desert through human land misuse. *(Photograph by U.S. Soil Conservation Service, from Dasmann, 1968.)*

was visible from a satellite photo showing the surrounding desert.

The American dustbowl of the 1920s and 1930s is another well-documented story of land misuse. A once fertile grassland changed into a dusty pit of ecologic and human tragedy. Although the United States had the land resources and capital reserves to recover from the dustbowl, such circumstances could occur again in the Great Plains if agricultural pressure becomes too intense at a time of unfavorably dry weather.

Throughout much of present-day Latin America and in many tropical or subtropical regions of the world, the dangerous method of slash and burn agriculture is actively practiced. Forest is cut and burned and the cleared area is planted to bananas, manioc, or some other starchy tropical plant. This denudes forest tracts rapidly, permits only one or two years of productive agriculture, and leaves a wake of destruction (Figure 4). After a few years, the subsistence farmer must repeat the process in a new area. A forest that may have taken 1,000 years to develop can be destroyed in a few years. In the tropics, where natural soils are thin and very susceptible to rapid leaching and erosion, the entire ecosystem may be irreparably damaged. We have either ignored the lessons of ecologic history or are too blinded by the economic pressures of life to appreciate their importance to us.

**Figure 4.** Slash and burn agriculture in the Amazon Basin. The tropical forest is cut and burned to clear open space for marginal agricultural planting. One or two productive years may be obtained before the thin tropical soil is depleted and baked into an impervious hardpan.

Warfare, of course, has always had devastating and tragic effects not only on human populations, but on the environment as well. In Indochina in our own time, the widespread use of herbicides, the cratering effects of bombs, shells, and rockets, and the physical disturbances of heavy vehicles churning across the landscape have produced extensive ecological damage, some of which may persist for decades or even centuries.

Considerable progress was made in a more positive direction in the late 1960s and early 1970s by the well-publicized "Green Revolution." Many countries throughout Asia, Africa, and Latin America made dramatic advances in agricultural production through the development of irrigation, new varieties of seeds, and increased fertilizer production (Horsfall, 1970). India, for example, became self-sufficient in food grain production in 1971 for the first time in more than 20 years, although it again required grain imports in 1974. The Philippines became self-sufficient in rice production in 1970 (Athwal, 1971).

The Green Revolution offers some hope that the trends of environmental destruction which prevailed for many centuries can be reversed, at least in agricultural environments. This requires knowledge and skill in the development of new strains of crops, in the management of soil and water, in the control of pests, and in other areas of agricultural science. It also requires vast new supplies of water, fertilizer, pesticides, and petroleum, along with climatic good fortune. It is clear, however, that the productive success of the Green Revolution is highly vulnerable to economic factors, such as the price of petroleum and fertilizer, and to ecologic factors such as rainfall.

The progress of the Green Revolution suffered a setback in many countries in 1973 and 1974, however, due to shortages of water, fertilizer, pesticides, and petroleum. It became painfully evident that the methods used to modernize agriculture and increase production had also increased the farmer's vulnerability to economic and ecologic conditions.

## CONCLUSIONS

There is little doubt that many environments throughout the world have been rendered barren and inhospitable by excessive pressures from the axes, plows, hoofed animals, and military machines of man. The pioneer civilizations altered their own biotic and physical environments and displayed man's ability to trigger ecologic changes leading to his own downfall.

In the decade of the 1970s there is very little evidence that man recognizes this aspect of his own history or realizes its applicability to his present predicaments. In country after country around the world I have seen the conservation movement losing ground, either through neglect or ridicule. In many tropical lands, deforestation is occurring at increasing rates, and the people believe that their forests are unlimited. In India, overgrazing and hydroelectric power development schemes are invading the new national parks and forest reserves which still exist. In the Himalaya Mountains, the landscape is being carved within an inch of its life to support a burgeoning human population. In Africa and Latin America, slash and burn agriculture is encroaching upon natural areas which have existed in ecologic balance for thousands of years.



Much of this environmental destruction is done in the name of economic development. The pressures of increasing populations place a greater burden than ever on the finite blanket of life that covers this earth. That blanket is getting torn and shredded in thousands of places. Like a wound or burn on the skin, its ability to heal depends upon the extent of the lesion and the health of the surrounding tissue. A point can be reached where the healing capacity is lost.

Even the Green Revolution, although it offers some hope, concerns only the agricultural environment and does not involve natural environments, forests, mountains, urban areas, or many other habitats vital to man. These areas are still being exploited and are still deteriorating throughout much of the world, and the effects of these trends can be as profound and as destructive to our civilizations as they were to previous ones.

## References


Athwal, D. S. Semi-dwarf rice and wheat in global food needs. *Quart. Rev. of Biology*, 46:1–34 (1971).

Bates, M. *The Forest and the Sea. A Look at the Economy of Nature and the Ecology of Man.* New York: Mentor Books, 1960.

Champion, H. G. A preliminary survey of the forest types of India and Burma. *Indian Forest Records* (Silviculture Series), Vol. 1, No. 1, 1936.

Dasmann, R. *Environmental Conservation.* New York: Wiley, 1968.

Davidson, B. *The Lost Cities of Africa.* Boston: Little, Brown, 1959.

Dunn, F. L. Health and disease in hunter gatherers. Chap 23 in *Man the Hunter*, Lee and DeVore (eds.). Chicago: Aldine, 1968, pp. 221–228.

Ehrlich, P. R. and A. H. Ehrlich. *Population, Resources, Environment: Issues in Human Ecology* (2nd ed.). San Francisco: W.H. Freeman and Co., 1972.

Elton, C. S. *The Ecology of Animals.* London: Methuen, 1933.

Guthrie, D. A. Primitive man's relationship to nature. *Bioscience*, 21:721–723 (1971).

Horsfall, J. G. The green revolution: agriculture in the face of the population explosion. In *The Environmental Crisis*, H. W. Helfrich, Jr. New Haven: Yale University Press, 1970, pp. 85–98.

Lee, R. B. and I. DeVore (ed.). *Man the Hunter.* Chicago: Aldine, 1968.

Leighly, J. (ed.). *Land and Life. A selection from the writings of Carl Ortwin Sauer.* Berkeley: University of California Press, 1967.

Marsh, G. P. *Man and Nature.* D. Lowenthal (ed.). Cambridge: Harvard University Press. (Reprinted 1965. Originally published 1864.)

Martin, P. S. and H. E. Wright. Pleistocene Extinctions. Proc. 7th Cong. Int. Assoc. for Quarternary Res., Vol. 6. New Haven: Yale University Press, 1967.

McCormick, J. *Forest Ecology.* New York: Harper & Bros., 1959.

Neel, J. V. Lessons from a "primitive" people. *Science*, 170:815–822 (1970).

Osborn, F. *Our Plundered Planet.* Boston: Little, Brown, 1948.

Perkins, D., Jr. Fauna of Catal Huyuk: evidence of early cattle domestication in Anatolia. *Science*, 164:177–179 (1969).

Saggs, H. W. F. *The Greatness that was Babylon.* New York: Hawthorne Books, 1962.

Sauer, C. O. Theme of plant and animal destruction in economic history. *Journal of Farm Econom.* 20:765–775 (1938).

Sears, P. B. *Deserts on the March.* Norman: University of Oklahoma Press, 1935.

Thomas, W. L. (ed.). *Man's Role in Changing the Face of the Earth.* Chicago: University of Chicago Press, 1955–1956.

Wade, N. Sahelian drought: no victory for western aid. *Science*, 185:234–237 (1974).

CHAPTER 16

# Sahel Will Suffer Even if Rains Come



JOHN WALSH

The news from the Sahel region* of Africa is discouragingly familiar. As it did between 1968 and 1974, drought has caused the deaths of thousands of people, the decimation of livestock herds, and the loss of productive land to desert. This time the drought area is larger, extending through much of Africa south of the Sahara. And the emergency appears even more serious because it is occurring against the background of a crisis in development that afflicts most of Sub-Saharan Africa. Population growth is outstripping modest increases in food production. And a combination of high energy costs and a slump in world prices for exports from the region have devastated the vulnerable economies of those countries.

The last Sahel drought won world attention and caused a major mobilization of international relief, development assistance and research in behalf of the Sahel. But although assistance is now running at a level of nearly $1.7 billion a year, a decline in per capita food production has become an established trend. Concern about slow growth has prompted donor organizations to undertake intensive evaluations of development theory and practice in Africa.

In a speech in January, World Bank vice president for operations Ernest Stern struck a note of institutional self-criticism that is common today among development organizations. Assessing development efforts in Africa, Stern said, "We, I think it is fair to say, among all our achievements, have failed in Africa, along with everybody else. We have not fully understood the problems. We have not identified the priorities. We have not always designed our projects to fit the agroclimatic conditions of Africa and the social, cultural, and political frameworks of African countries. This is evidenced by the percentage of poorly performing projects in the agricultural portfolio and by the fact that we, and everybody else, are still unclear about what can be done in agriculture in Africa."

Stern went on to say that while he did not have the solutions, some of the elements were clear. Two main points were "we need to do very much more to support research," and "the designs of agricultural projects in Africa must be made more consistent with the implementation capacity of many of those nations."

The immediate crisis, of course, is caused by 2 years of severe drought. Among the Sahelian countries, Mauritania appears to be most seriously affected. The drought is reportedly responsible for 100,000 deaths in Mozambique. Food relief is being provided for the hardest hit areas, but these efforts are impeded in a number of regions by inadequate transportation and distribution systems and, in several countries, by warfare and civil conflict.

Beyond the present emergency, however, loom formidable long-term problems of development. There is growing recognition that recurrent droughts impose an element of unpredictability in the Sahel that limits agricultural options available in other regions. Soil is fragile in the fashion of arid regions and population growth has resulted in overcultivation and overgrazing with destructive consequences.

The countries of the Sahel are among the poorest in the world. When they gained independence, they were woefully short of capital resources as well as of technically trained manpower and able

*Sahel, an Arab word for coast or border, is normally used to denote six Sub-Saharan countries that won independence from France at the end of the 1950's—Mauritania, Mali, Niger, Senegal, Upper Volta, and Chad—plus the Cape Verde Islands and the Gambia. But similar conditions prevail in the countries that lie eastward along the tier such as the Sudan, Ethiopia, and Somalia. And drought conditions this year have seriously affected many countries in central and southern Africa, particularly Mozambique.

Reprinted from *Science 224:* 467–471 (1984).


John Walsh is a member of the editorial staff of *Science*.

# Desertification Defines Ordeal of the Sahel

The image of the sands of the Sahara advancing inexorably, engulfing everything in their path was a powerful symbol of the 1970's drought in the Sahel. Desertification, however, stands for a process of degradation of the environment that usually happens more slowly if no less surely than the dramatic stereotype. Desertification is a product of climate and human activity and in the Sahel it has been accelerated alarmingly by both.*

This interaction has been a main focus of research in the Sahel. In respect to climate, an earlier debate over whether the recent dry years were atypical seems to have given way to agreement among climate researchers that the dry period that began in 1967 in the Sahel falls into a pattern of sparse, highly variable, and unevenly distributed rainfall that has prevailed in the region for the last 2500 years.

What certainly differs is the pressures exerted on the ecosystem in the last 50 years by population growth and changing patterns of land use. Human activity, of course, has been significant for much longer than that. A millennium ago, the inhabitants of the Sahel were mainly nomads, living by herding and hunting. Major impacts on the ecosystem were made by bush fires set to improve grazing or aid hunting and by the caravans engaged in the trans-Sahara trade. A single caravan could include several thousand animals; the practice was to send out hundreds of people to cut trees for charcoal to be used as fuel on the desert-crossing and for emergency rations for the animals. The fires and the tree-cutting began the trends toward simplification of vegetation and degradation of the soil that plague the Sahel.

Arab chronicles testify that in the centuries before European nations began to vie for trading advantage a complex economy was maintained despite turmoil in the region. One feature was the warehouses in which supplies of food for several years were stored against crop failures. Famines seem to have occurred when drought coincided with local wars.

European colonization brought pacification resulting in the extension of settled agriculture northward into the Sahel and new stresses on the fragile soil. The colonial era also saw the enlargement of cattle herds and the displacement of camels by cattle as the basis of the economy and culture. Cattle, however, are poorly suited to the Sahel. They require large quantities of water, have a low conversion efficiency as feeders, and are vulnerable to stress. Many deep tube wells have been sunk in the region to water livestock. Especially in dry weather, the congregation of cattle around water sources causes wide areas to be denuded of vegetation and desertification accelerated.

Traditional agriculture in the Sahel represented a sophisticated adaptation to the conditions in the region. Land was relatively plentiful and farmers would cultivate a plot for 2 or 3 years then allow it to remain fallow for 10 years or more. Trees were left on cultivated land, contributing to soil stability and fertility, and providing browse for animals. Land tenure systems varied widely, but in general allowed multiple use and multiple users. Herdsmen and farmers cooperated, for example, with livestock often allowed to forage on harvested fields, their droppings, in turn, fertilizing the land.

Under the old system, the movement of people was relatively unrestricted and, in bad times, they were adept at finding water and filling out their diet by hunting and gathering. Limits on migration imposed by national boundaries, land use rules imposed by the new governments, and urbanization have nullified these survival skills.

Recognition that desertification vitiates action to increase food production has encouraged antidesertification efforts in the past decade. An evaluation of those efforts is contained in a report prepared as a follow-up to a 1977 U.N. conference on desertification. The report, "Assessment of Desertification in the Sudano-Sahel," is the product of a study sponsored by the Sudano-Sahelian office of the U.N. and headed by Leonard Berry of Clark University.

The report finds some causes for encouragement, particularly in a spreading awareness of the problem of desertification and the creation of new institutions and the strengthening of existing ones to combat it. Among antidesertification projects such as those aimed at sand dune fixation, reforestation, creation of woodlots, and provision of substitutes for fuelwood, some successes are described. The verdict, however, is that "Despite the efforts of all concerned the record of success in the battle against desertification is at best mixed." A majority of indicators developed to assess environmental change show moderate or severe deterioration in the region. "The reasons for this are numerous but high among them are still the lack of defined practical policies of resource management by the governments, inherent problems of marginal land development and the need for innovative approaches, the dearth of trained national personnel and the serious shortage of funds put to this end." Asked to sum up the difficulties, Berry said, "A shortage of money. A shortage of good concepts of what to do. And an inability to implement good concepts."

The report makes the point that evaluations of progress or lack of it on desertification are limited by still inadequate data and it urges vastly improved monitoring of land and resource conditions and climate. Although the debate on whether the current dry years were an anomaly seems to have quieted, not all the questions have been answered. A new discussion has been simmering since the mid-1970's over the possibility of what is termed biogeophysical feedback. In a 1982 paper, "Sahel: A Climatic Perspective," prepared for the Club du Sahel, Sharon E. Nicholson of Clark University wrote that, "Several researchers have proposed various mechanisms by which droughts can 'self-accelerate.' The changes of the land surface induced by a drought (removal of vegetation, increased reflectivity, reduced soil moisture) in turn influence the atmosphere in such a way as to strenghten conditions that first produced the drought. . . . The persistence of conditions of abnormal rainfall over one or two decades may be manifestations of such 'feedback' between the atmosphere and surface." The issue remains unsettled, but it would seem grimly consistent with the recent fortunes of the Sahel if drought proved to be self-perpetuating.

---

*Environmental history and baseline information on the Sahel are discussed in a report, *Environmental Change in the West African Sahel* (National Research Council, Washington, D.C., 1983).

managers. As new nations, they faced disputes with their neighbors and disunity among tribal and ethnic groups at home. Their governments generally reacted against the colonial past by following highly nationalistic, centralizing, socialist-oriented policies. These policies typically favored urban populations at the expense of rural people and agriculture. The result, in broad terms, was a decline in agriculture.

The experience of the 1970's convinced both donors and recipients that a long-term coordinated effort to increase food production was required and new strategies and organization needed to carry it out.

The eight Sahelian states in 1974 formed the Permanent Interstate Committee for Drought Control (CILSS) to provide a regional focus for action and make specific proposals for assistance. The principal donor countries followed in 1976 by forming the complementary Club du Sahel headquartered in the OECD (Organization for Economic Cooperation and Development) in Paris. The World Bank and its associated lending agencies sharply increased their programs for the Sahel and other regions of Africa. And the U.S. Agency for International Development (AID) in 1977 established a separate Sahel Development Program and upped the level of assistance.

The $1.7-billion funding level for 1983 represents a resumption of a general upward trend after a falling off in 1982. In the period 1975 to 1982 nearly $11 billion in assistance was provided with France, the largest donor, accounting for nearly 20 percent of the total. About 35 percent of the total during the period was not spent on development projects but on such things as food imports and debt service that did not contribute directly to solving the region's long-term problems. The U.S. contribution was about $135 million last year, some $85 million in development assistance and $37 million in food aid.

It is difficult to trace trends in research funding because research support is often included in development project grants and contracts. Funding is said to be increasing substantially for applied research on agricultural and environmental problems, however. Lack of such research is viewed as a partial answer to the obvious question of why the Green Revolution has largely missed Africa.

A wide-ranging 1981 report, *Food Problems and Prospects in Sub-Saharan Africa: The Decade of the 1980's*, done by the U.S. Department of Agriculture Economic Research Service for AID suggests several reasons why agriculture in the region has fared relatively poorly. In technology terms, Africa has lacked the packages of inputs that made for success elsewhere. Rice varieties raised so successfully in Asia proved highly vulnerable to disease in Africa. And plant breeders have so far not developed high-yielding crop varieties adapted to African conditions. In Asia, the Green Revolution flourished in irrigated conditions. Irrigated farming in Africa has not had the advantages of farmer experience and commercial marketing and transportation systems that existed in Asia. And both funds for fertilizer and other inputs and agricultural labor have been in short supply.

Similarly, limited research has been done on improving rainfed agriculture in African conditions. Only now is work beginning on the formidable task of producing plant varieties more resistant to drought, disease, and insects there.

Other items on the research agenda include studies in support of agroforestry to find ways to improve on the old practice of growing woody plants, food crops and livestock on the same land (see box). Research on the hydrography of the region is needed to make better use of limited water resources. And livestock research, pioneered by the French, is viewed as needing to be greatly expanded. But development experts agree that better understanding of underlying environmental and resource problems must be backed by new economic, political, and social policies.

Development strategy has evolved rapidly during the past decade. During the 1950's, Africa received little attention from the international development community. In the Sahel, rainfall had been greater than usual through the decade. Many African countries were emerging from colonial rule and the U.S. view at the time was that development aid in the area was the responsibility of

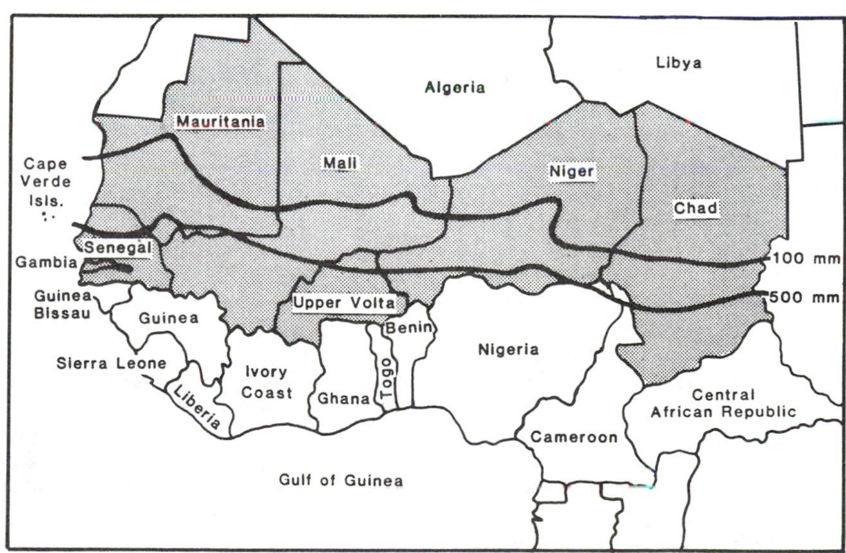


*Sahel countries (shaded) straddle rainfall zone indicated by 100- and 500-mm isohyets.*

the former colonial powers. The outlook changed in the 1960's both because the drought elicited a humanitarian response and because the countries of Sub-Saharan Africa registered a strong claim for development assistance.

Conditions in the Sahel contributed to a major shift in the focus of development programs in the early 1970's. This was signaled in a speech in Nairobi in 1973 by Robert S. McNamara, then president of the World Bank. McNamara indicated that the bank would reduce emphasis on support of major capital projects such as highways, railroads, dams, power plants, and port facilities designed to give a developing country an infrastructure required for a modern economy. A main objection to such projects was that development depended on a trickle-down effect. More bank resources were to go into projects to aid directly the poorest people, especially smallholders and the landless, in the poorest countries. The main vehicle for such assistance was to be integrated rural development projects designed to introduce and support new technologies that promised to increase agricultural productivity.

Earthscan

*Herds on the move in Niger.*

The McNamara model won wide acceptance among development planners, but encountered some difficulties in implementation. What is involved is a technology transfer process that requires project designers to work with local people to be sure that proposed technology is appropriate and can be absorbed. Despite mounting research activity, gaps in knowledge continued to hinder effective design efforts. In addition, cooperation of government officials in the recipient country at the national, regional, and local levels is necessary if a program is not to be an isolated pilot project with no real prospect of being adopted more widely. Such cooperation was frequently lacking.

Furthermore, in some ways, grand-scale infrastructure projects are easier for both donors and recipients to carry out. Building a highway or an airport requires the sort of concentration and deployment of resources and manpower that Western contractors excel at. In a developing country it may be more difficult to establish and operate a successful system of rural clinics than a central hospital or to make an irrigation project work than to construct a major dam. At the secondary level there is often a continuing need for the trained manpower and managerial expertise that is in short supply in many developing countries. Also a factor is the problem of recurrent costs. Unlike the one-shot capital investment that often suffices for big infrastructure projects, many agriculture, health, education, and transportation projects of direct benefit to rural people require continuing support which developing countries find difficult to finance.

Donor nations have increasingly blamed the difficulties of development on the domestic policies of the recipient nations. The practice of setting investment, trade, and exchange policies to protect nascent industries was seen as damaging to agriculture. A particular target of the critics was the so-called parastatal agencies which in many Sahelian countries were given control over consumer and producer prices and granted authority to market food and conduct trade relations. These agencies, typically, have kept farm gate prices low, subsidizing food for city dwellers, but depriving farmers of incentives to venture beyond subsistence agriculture.

An influential document in this discussion is a World Bank report, *Accelerated Development in Sub-Saharan Africa: An Agenda for Action*, known as the Berg Report for Elliot Berg who headed the staff group that wrote it. Published in 1981 with an update in 1983, the report is regarded as a major expression of the case for "conditionality," that is of international agencies making approval of projects conditional on agreement by recipients to make specific policy and institutional changes.

The report certainly does not advocate that the bank simply issue take-it-or-leave-it ultimatums to would-be clients. Rather it argues that conditions in Africa require that the bank make more money available and change its own policies, for example, to finance recurrent costs in projects that require it and to provide funds to support policy reforms.

From the viewpoint of the recipient governments, making policy changes involves decided difficulties and risks. After independence, authorities had little choice but to rely on a small educated elite and expatriate specialists. Advice to shift economic power to the private sector may be logical, but in many countries the private sector is feeble. And policies rigged to subsidize food and imported goods for urban populations may be economically unsound but politically prudent. Devaluation of the currency can be a rational step fiscally, but it may also be a recipe for a coup. Nevertheless, the Sahelian countries to varying degrees have adopted policies designed to bolster production of food for domestic use and revised pricing, marketing and tax policies to support the new priorities.

Many observers see external constraints—drought, low world commodity prices, energy costs—as the major current obstacle to development. Others see the reliance of Sahelian countries on food aid and financial support as creating a long-term dependence on such assistance. Probably most threatening to the Sahelian future, however, is population growth. Common estimates have population in the region growing at an annual 2.7 percent and food production at 1.5 percent. Continuation of the trend would negate even dim hopes for food sufficiency for the region. A 1983 report on development looking to the year 2008 by the U.N.'s Economic Commission for Africa says that the picture that emerges "under the historical trend scenario is almost a nightmare." Effects on physical resources and social services would be disastrous and "socioeconomic conditions would be characterized by a degradation of the very essence of human dignity."

Population control plays little part in current development programs or government policies in the region. African attitudes toward family planning have traditionally been conditioned by suspicion of donor country motives. French aid policy has not given family planning much place and current U.S. policy has prescribed provision of assistance for voluntary programs only when such aid is specifically requested. The issue has been recognized as a sensitive one politically for Sahelian governments, which do not lack for threats to stability. U.S. population experts, however, detect a willingness among Sahelian officials to discuss the population problem, which is seen as a hopeful development.

In a matter of weeks, the Sahel will begin to see whether the seasonal rains that make it possible to plant and grow the crops of the region will return or whether the drought will continue. But among the stern constraints that retard development in the Sahel, population growth is seen by many as the factor that will determine whether the region in the future will go from crisis to catastrophe.

CHAPTER 17

# Tropical Forests: What Will Be Their Fate?

HUGH H. ILTIS

The destruction of tropical forests in the world today is so extensive, so devastating, so irrevocable that humanity may soon lose its richest, most diverse, and most valuable biotic resource. As a consequence, life will lose forever much of its capability for continued evolution. Many groups of the larger vertebrates (mammals, birds, reptiles) will be especially affected, but countless other organisms—less spectacular insect, mollusk, and plant species—will be lost as well.[1] The economic, esthetic, and cultural losses to future generations will be incalculable.

The damage may not be visible to us here in the United States, for the tropics are faraway, hot and humid places about which we know very little and have been taught even less. To

Hugh H. Iltis

Reprinted by permission from *Environment 25(10):* 55–60 (1983). 

Hugh H. Iltis is professor of botany and director of the herbarium at the University of Wisconsin, Madison, WI.

most people the word "tropics" itself conjures up a picture of pristine nature, mysterious and marvelous; of great apes and elephants, of orchids and palms.

Precisely because the tropics *are* mysterious and wonderful, it has been hard to escape the nineteenth-century colonial mentality of a wild and green El Dorado just waiting to be explored in a dugout canoe: a tropical last frontier with unlimited riches still begging to be "developed" by enterprising pioneers (now, of course, ostensibly for the benefit of native populations).[2] The horrendous destruction of nature over the past half century—the critical reason for the now-widespread concern among biologists—was almost always conveniently ignored or belittled. The realistic ecological picture today, however, and its long-range implications are very grim indeed.

Those of us who take pride in being environmentally enlightened have been teaching pollution control and contour plowing, nature preservation and Aldo Leopold's land ethic, but mostly with reference to our own rather impoverished temperate biota.[3] What little concern we may have had for the tropics was rarely based on reality. In our comfortable ignorance, these lush lands and their wild animals seemed so safe from destruction that we rarely worried about their fate. The underdevelopment, the innocence of their illiterate populations, and the endemic tropical plagues like malaria and yellow fever that kept their populations well in check seemed sure to offer their ample protection.

Since 1945, however, all this has changed. We now have DDT and 2,4,5-T; the all-powerful (and greedy) multinational corporations with their woodchippers and jungle smashers (one acre an hour, as advertised by Le Tourneau); the vast and hungry army of the poor and the landless; and the devastating, self-serving, post-World War II development syndrome.

Inevitably, what has resulted from these activities has been the systematic, barbaric obliteration of nature for the "benefit of man." As a consequence, we are faced today with the greatest biological calamity this world has ever known—the imminent decimation and extermination of the world's tropical biota. As E.O. Wilson put it, such a great loss of genetic diversity would be worse than

*energy depletion, economic collapse, limited nuclear war or conquest by a totalitarian government. [In fact,] [a]s terrible as those catastrophes would be for us, they could be repaired within a few generations . . . . The one process ongoing in the 1980s that will take millions of years to correct is the loss of genetic and species diversity by the destruction of natural habitats. This is the folly our descendants are least likely to forgive us.*[4]

## Tropical Destruction

It is in the tropics, in particular in their inconceivably diverse and beautiful but fragile wet forests, that biological genocide is in full swing. Of the estimated 8 (to possibly 30) million species of plants and animals on Earth, a vast preponderance live in tropical ecosystems. The destruction of such habitats, therefore, would bring in its wake the extermination of literally millions of species—species of which, for the most part, we do not yet have a description, a picture, a life history, or even a name.

Indeed, the utter devastation that humanity is now bringing to tropical ecosystems—the vast and uncontrolled forest destruction in many regions of Asia, Australia, Africa, and Latin America—has to be seen to be believed. During 1962, at the foothills of the Andes in a Peruvian valley near San Ramon, I stood on a narrow hanging bridge suspended over a clear mountain stream and watched a troop of spider monkeys, a hundred feet up, jumping from one tree to the next and eating fruits from a gigantic fig tree. Iridescent blue, giant *Morpho* butterflies sailed erratically through a sun-flecked opening in the forest, while a pair of banana-billed toucans sat motionless on a branch silently watching—a scene seemingly straight out of Genesis.

Today, none of this exists, for later that year, with a development grant provided by the U.S.-sponsored "Alianza para el Progreso," an energetic man bought the land (and presumably all the Indians living on it), cut down the forest, and planted coffee and bananas on its 45-degree slopes. Such land clearing is now widespread all over the Andes, and has resulted in massive soil erosion, siltation of rivers, and even climatic changes. The dramatic and unprecedented fluctuations in recent Amazonian water levels[5] are among the serious and unexpected consequences. Loss of species is another.

With such extensive habitat destruction, it is small wonder that all around the world primates, along with many of the larger vertebrates, are facing extinction. In three months of South American field work in 1977, we saw only *one* spider monkey—a pet on a silver chain in a hotel lobby.

That was in Ecuador, where we visited the lowlands of the Pacific slope. The forests there, separated from those of the Amazon basin by the lofty peaks of the Andes, are the home of several thousand unique local species (termed endemics) that

**We are faced today with the greatest biological calamity this world has ever known—the imminent decimation and extermination of the world's tropical biota.**

evolved here in isolation, providing fascinating examples of geographic speciation.

Near Santo Domingo de los Colorados lies a small remnant of moist tropical forest, studied intensively by Calloway Dodson and Alwyn Gentry. As described in their *Flora of Rio Palenque*,[6] this tract of only 167 hectares (circa 420 acres) has over 1,100 species of plants in 123 families. Almost half of them are woody. Nearly 6 percent of them were new to science, and 4 percent are endemics, known to be from nowhere else on earth.

A small "sierra" only three miles away has strikingly different and also highly endemic flora, a diversity quite unexpected by any temperate-zone botanist. But in the tropics, whether in Panama, Mexico, Colombia, Borneo, or here in Ecuador, local floras are saturated with unique taxa: local endemism is the rule, widespread species the exception.

For example, in Ecuador— a country no bigger than Minnesota—there may well be as many as 20,000 different species of plants, over one-fifth of them endemic. On the other hand, there are only around 17,000 plant species in all of North America, and only 1,700 native species in all of Minnesota, which, at best, can boast of only one endemic—a semi-sterile Dogtooth Violet (*Erythronium propullans*) perhaps of hybrid origin.

Because of this overwhelming biotic diversity and the briefness of their acquaintance with it, biologists remain quite ignorant of most tropical species (especially insects) and their interrelations. How very much we have yet to learn is evidenced by the fact that one-fourth of the neotropical wet-forest plant species and more than nine-tenths of its animal species are still unknown to science.

Rio Palenque is a mere 400 acres of wild nature, yet it is the only preserved remnant of this particular type of tropical forest on the whole of the Pacific slope of Ecuador. It is now but a tiny island of virgin forest complexity set in a vast sea of sterile, cultivated uniformity—thousands of square miles of monocultured bananas, sugar cane, African oil palms, and corn.

What only 25 years ago was the most inaccessible and unknown of tropical forests—a dream for any botanical explorer, stretching uninterrupted from Quevedo to Esmeraldas—and a region from which hardly any biological specimens were available for scientific study, has now been almost totally destroyed. Only an occasional plank-rooted forest giant, uncut because of its immense size, stands alone in a field of bananas. It is a pathetic, grim reminder that now we shall never know what botanical or zoological riches these forests once held. In many ways it is already too late for their study.

## Spreading Development

This picture of tropical forest "conversion," of century-old giants being felled and mountains of bulldozed primeval wilderness being burned to give way to cow pastures, corn fields, and tree plantations to provide food and firewood for the hungry, is being repeated again and again—in Brazil, in Panama, in Indonesia, and in Mexico.[7] The U.N. Food and Agriculture Organization (FAO) estimates that 7.5 million hectares of closed forest and 3.8 million hectares of open forest are currently being destroyed *each year*.[8] At this rate, it will be all over in 20 years.

Even the seasonally dry forests now are getting the axe. I cannot tell you with what feelings of horror and hopelessness I witnessed the bulldozing of a hundred-thousand-acre plot of virgin tropical-dry forest in the coastal plain of southwestern Mexico. These arid lands supported 40-foot tall "candelabra" cacti and an army of small trees, yet they are essentially useless for agriculture because of very low rainfall. Does a miserable crop of sorghum for cattle feed every three or four years justify so mindless a destruction?[9]

Nearby, in the moist oak-fir-pine forests of the 10,000-foot Sierra de Manantlán, we discovered three years ago a wild perennial grass (*Zea diploperennis*) that is ancestral to corn and of immense economic potential.[10] Today, lumber trucks still continue to roar down these mountains every half-hour, hauling gigantic logs to be made into boards and into broom handles that are exported to the United States in order to gain badly needed foreign exchange. The trucks are witnesses to the steady, if slow, devastation of a biological treasure-house.[11] If the encroaching agricultural development had reached the minuscule habitats of *Zea diploperennis*, a dozen cows in a week's time could have obliterated this species, and with it the possibility of ever developing highly virus-resistant or even perennial corn.[12]

For many of us in North America, the destruction of these Latin American habitats has another, more special significance. The migratory birds we love to see in the summer depend on wild places in Latin America to survive the winter. Warblers in the Ohio Valley already are becoming rarer. Farmers and orchard owners depend upon these birds to a considerable degree for insect control. Many of our larger birds are moving closer to extinction year by year. From both the standpoint of the birds' survival and our own, this is indeed "one world."[13]

## Bioclimactic Paradox

Many well-meaning American advisors and humanitarians have been misled by the luxuriance of the tropical forests. How many times have they announced that the answer to world hunger lies in their sustained agricultural utilization? Yet, ignorance of ecology in this case is fatal.

As Professor J. Chang[14] of the University of Hawaii has explained, in the tropical regions, despite their lushness, annual grasses such as wheat, rye, barley, and rice have relatively low agricultural productivity compared to those species found in the cool temperate climates from whence they originally came. This is due to a simple bioclimatic fact: during the long and warm tropical nights, a plant's respiration burns up most of the surplus carbohydrates that it produces during the relatively short photosynthetic day. The 16-hour days followed by cool, 8-hour nights, as found in the Dakotas or the Ukraine, permit much greater accumulations of photosynthate in such plants, creating a bumper crop come harvest time.

In addition, high rainfall tends to leach the already nutrient-poor lateritic soils to sterile gravels in many parts of the lowland tropics, and there are no climatic controls (i.e., the freezing temperatures of winter) to knock back insect pests. Dreams of making bread baskets out of these regions evaporate into the fantasies that they are—editorials and lead articles in prominent journals not withstanding.[15]

Thus, the tropics—the wet tropics in particular—present a climatically determined paradox. Biologically, they are rich beyond belief, but in many significant ways they are agriculturally quite poor. That Iowa or South Dakota can never become a Mexico, Panama, or Amazonian Brazil in terms of biological diversity seems obvious. At the same time, tropical countries can never become an Iowa or a South Dakota in terms of agricultural productivity. This ecological fact will have serious political consequences and will be unpleasant to face.

There is another side to this great productivity-diversity-preservation paradox: the countries most desperately in need of money to create and maintain the gigantic national parks necessary to preserve their unique biological riches—both for themselves and for the whole world—are the very ones that are usually too poor to afford them. Statistics on park personnel bear this out. Compared to the tropical nations, the industrialized countries spend ten times the money and support ten times the staff per unit area of park.

Thus, the countries with so much to preserve can, in their poverty, preserve only a little. Just think of the incredible collection of large game animals that now crowd the remaining central African savannas and forests: rhinos, lions, gazelles, and giraffes—all residents of poor countries that have disastrous economies and exploding populations. Yet, the Serengeti is still a park. Worldwide, even despite great handicaps, countries small and large make valiant efforts (often with assistance from the World Wildlife Fund ) to preserve their patrimony of wildlife and diversity.

A bright light amid all this destruction is the massive effort by a variety of people in Latin America to preserve the tropical biota within their countries.[16] Costa Rica has a national system of 22 well-administered parks and reserves, which, considering the small size of the country, is unrivaled by any other in Latin America.[17] In fact, 8 percent of its area is under effective protection, and Costa Rica's per capita financial commitment to its parks is higher than that of the United States.

Significant advances have been made in the Amazonian regions of Bolivia, Brazil, Ecuador, and Venezuela, where, during the past 16 years, nearly 12 million hectares of forest have been placed under protection.[18] Admittedly, most of these are as yet unstaffed and exist "on paper only." Nevertheless, even this is a first step, and recognition must be given to the many forces in Latin America working for preservation. They should be encouraged to continue these efforts.

## An Ironic Lesson

In addition, it may be wise to consider the proposal by Ira Rubinoff, director of the Smithsonian Tropical Research Institute, for a global system of protected tropical forest preserves that would be supported by the international community through taxation of all countries whose people enjoy a per capita income greater than $1,500 a year. It should be remembered that developed nations have an economic stake in the survival of tropical forests at least as great as, or greater than, the nations in which the forests grow.[19] One percent of our bloated defense budget would go a long way toward setting aside 1,000 preserves of 600,000 acres (240,000 hectares) each, which would insure the preservation of approximately 10 percent of the rain forests (only 2 percent are now so protected).

For Americans, there is an ironic lesson in the loss of the tropical biota, for what Latin American countries in particular are unable or unwilling to preserve will haunt us in the centuries to come. It is not only the songbirds that we shall miss. The rich, overdeveloped nations have long used their economic clout to exploit mercilessly the poor, underdeveloped tropical nations, and directly—or indirectly—have added the insult to injury of abusing their biological treasures.

Politicans or businessmen are hardly ever interested in the preservation of rare Brazilian reptiles or Mexican birds, or in the percentage of endemics in the Colombian biota. The decisions they make concerning the fate of this bountiful life are almost always in terms of short-range profit, not long-range wisdom. Just ask the American lumber companies, or the gigantic Japanese concerns that are (literally) ripping off the forest resources of Southeast Asia and the "living museum" of New Guinea.

But this is one world, and the catastrophic loss of wildlife will affect us all. Our children and theirs may well wish to study the tropics, will want to see them for themselves, and, at the very least, will want access to the vast storehouse of economic plants and animals that these forests contain[20] and the opportunity to study the important ecological phenomena that they exhibit. If we are to keep faith with our children, the overdeveloped nations must learn, now and quickly, as an integral part of their foreign policy and foreign aid, to approach the problem of extinction seriously and in a new and much more farsighted, financially responsible way. Establishment of an enlightened foreign policy depends on efforts to subsidize the staffing and upkeep of preservation efforts in the tropics, to help train biologists, to build local museums of natural history, and to translate or otherwise make available the scientific literature so that people of these areas can become experts of their own biota.[21]

The trend, of course, has been the other way. We have persisted in our political and economic domination of the tropics. Furthermore, the biologically ignorant leaders of the overdeveloped nations have continually beaten the drum for rapid "sustainable development" of the tropics, unmindful that the tropical forests cannot be exploited on a permanent basis without destroying them.[22]

Try as these countries may, saddled as they are with their burgeoning populations, they will simply never be able to reach a high level of wealth unless they happen to own oil wells. But in the process of a biologically insensitive "development," they will surely destroy their own biotic wealth.

There is both hypocrisy and tragedy here: hypocrisy because these geographic-economic factors are well known but too inconvenient to be accepted, and tragedy because what all the underdeveloped countries need now more than anything else are freely available and medically safe population control measures on the one hand, and peace and freedom from economic manipulation and exploitation by the overdeveloped countries on the other. On both of these counts, the United States, the only country we ourselves can hope to influence directly, is unfortunately ambiguous, vacillating, and self-serving.

## Educating the Public

In the meantime, preservation falls by the wayside, exploitation of the tropical forests and their conversion to hamburger-producing cow pastures continues unabated,[23] and extermination and extinction of species are occurring at a wholesale rate, now claiming perhaps many thousands of species each year.[24] To a considerable extent, the tone has been set by rapacious multinational and national corporations based in the United States, Japan, and Europe—something that we need to be aware of, something that we must, with dedication and courage, try to correct within this decade. No small part of this will be the education of the American, European, and Japanese public.

That the forces opposing a rational solution to these ecological problems

**Developed nations have an economic stake in the survival of tropical forests at least as great or greater than the nations in which the forests grow.**

are all-powerful and influential need hardly be pointed out. In 1970, we had Earth Days from coast to coast, dedicating ourselves to a biologically sane world. Now, 13 years later, this profound intellectual revolution is all but dead, sabotaged by administrative officials and by heavily subsidized and carefully orchestrated campaigns of the media and their corporate allies, which have obscured the very real dangers in neglecting the world's ecology.

I need only to point out that the special anniversary issue of *Time* magazine, "The Most Amazing 60 Years in History," published October 5, 1983, does not mention or illustrate in its 168 pages of text and lavish pictures one *single* environmental event, fact, or problem: *nothing* whatever on Earth Day or Rachel Carson's *Silent Spring* or on DDT or 2,4,5-T, or on extinction of species or pollution, or soil erosion; in fact, absolutely not one word on *any* environmental issue. This indifference to reality reaches even to *Time*'s total silence on the population explosion. This is, on reflection, surely the most terrifying fact of the past 60 years: the near tripling of the world's population,[25] an increase of fully 2.95 billion additional people since 1923, the year when *Time* magazine came into existence.

This brings me to a final point. While corporations exploit and their political bedfellows run interference for dubious economic aims, let us not forget (well-meaning liberals included) that equally responsible for these biological extinctions are poverty, hunger, and ignorance—the chop-chop of a million axes, the cravings of a billion mouths.

Although the people of the tropics do need more protein and more firewood, they need birth control even more. For by now, any knowledgeable observer of the world's scene must come to the conclusion that the *food* vs. *population* race can never be won—or rather, can only be won by decreasing the birth rate by whatever means of birth control are available. It certainly will not be won by furthering the immaculate misconceptions of raising more food by cutting down more forests, by plowing up more prairies, and by draining more wetlands.[26] We are running out of all of these, and the population bomb keeps on ticking.

The world's net population increase in 1982 was 82 million people,[27] the highest *yearly* increase ever—and mostly in the tropics. (An estimated 40 to 50 million abortions allowed that many additional children *not* to be born.) Only in an ecologically educated world public, in nations self-restrained in both resource use and reproduction, is there any hope for the conjunction of a healthy and well-fed population, a biotically rich earth, and peaceful coexistance.

We must impress on our students these facts: that the world's carrying capacity is finite, that there are too many people now, that the actual, very real collision between resources (including food supply, clean living space, *and* tropical forests) and populations is the biggest, most fundamental, and most nearly insoluble problem that has ever confronted the human race. Whether wild ecosystems of any sort will long survive (even in the United States) will depend upon its resolution. The outlook is not hopeful. Men and women of good will have no alternative but to work for the preservation of a nature-rich good earth.


## ACKNOWLEDGMENTS

The experiences on which this article is based were partly supported by various National Science Foundation grants and the University of Wisconsin Herbarium's E.K. and O.N. Allen Herbarium Funds. I thank D.A. Kolterman, S.L. Solheim, and D.M. Waller for helpful criticisms of the manuscript. Thanks are due also to Cathie Beckwith for faithful typing under deadline pressure. I dedicate this paper to Paul and Anne Ehrlich, Norman Myers, Peter Raven, Ray Fosberg, Otto and Isa Degener, and Jack Sharp—all tireless advocates of a nature rich world.


## NOTES

1. Anonymous, **The World's Tropical Forests: A Policy, Strategy and Program for the United States** (Washington, D.C.: U.S. Government Printing Office, 1980); G.O. Barney (study director), **The Global 2000 Report to the President: Entering the Twenty-first Century** (Washington, D.C.: U.S. GPO, 1980); D.W. Ehrenfeld, **Conserving Life on Earth** (New York: Oxford University Press, 1972); P. Ehrlich and A. Ehrlich, **Extinction: The Causes and Consequences of the Disappearance of Species** (New York: Random House, 1981); A. Gomez-Pompa, C. Vasquez-Yanes, and S. Guevara, "The Tropical Rain Forest: A Nonrenewable Resource," **Science** 177(1972): 762–765; H. H. Iltis, "Shepherds Leading Sheep to Slaughter—The Biology Teacher and Man's Mad and Final War on Nature," **The American Biology Teacher** 34(1972):127–130, 137, 201–205, 221; H.H. Iltis, "To the Taxonomist and Ecologist: Whose Fight

is the Preservation of Nature?" **BioScience** 17(1967): 886–890; N. Myers, **The Sinking Ark** (Oxford: Pergamon Press, 1979); N. Myers, **Conversion of Tropical Moist Forests** (Washington, D.C.: National Academy of Sciences Press, 1980); National Research Council, Committee on Research Priorities in Tropical Biology, **Research Priorities in Tropical Biology** (Washington, D.C.: National Academy of Sciences Press, 1980); G.T. Prance and T.S. Elias eds., **Extinction Is Forever** (Bronx, N.Y.: The New York Botanical Garden, 1977); P.H. Raven, "Tropical Rain Forests: A Global Responsibility," **Natural History** 90(1981):28–32.

2. For pro-development appraisals of Amazonian colonization, see Tad Szulc, "Pioneers Carve a New Frontier—Will the Next Century Belong to Brazil?" **Parade Magazine**, September 4, 1983, pp. 4–6. This article contains an economic justification and humanistic glorification of biological destruction, and reached perhaps 30 million or more American households; see also P. H. Abelson's editorial, "Rain Forests of Amazonia," **Science** 221 (1983): 507. Equally uncritical is this tragic view of Amazonia by a prominent American businessman-diplomat: "The cause of this discouraging rate of development [of the Amazonian rain forest] is that the ground itself must first be cleared of jungle . . . and civilization itself introduced, before new farms can be laid out and made productive. . . . Whole new traditions and ways of life must be established . . . . Just to look at the geography is to see the formidable nature of the challenge. One huge belt of land . . . lies on the equator in the heart of the heat and fevers of the tropics. The Amazon River, unlike the Mississippi, flows through vast tracks of what are sill sodden, malaria-ridden, impenetrable jungle wastelands, its waters patrolled by alligators and man-eating snakes. In contrast, the gentle, traffic-moving rivers of Europe have been channels of trade for a thousand years." S.L. Linowitz, "The Future of the Americas," **Science** 181(1973): 916–920.

3. The former pre-occupation with the preservation of local plant communities is shown by M.L. Fernald of Harvard University in his famous pioneering essay, "Must all Rare Plants Suffer the Fate of Franklinia?" **Journal of the Franklin Institute** 226(1938): 383–397.

4. E.O. Wilson, as quoted in P. Schabecoff, "A Million Species Are Endangered," **New York Times** November 22, 1981; cf. **Proceedings of the U.S. Strategy Conference on Biological Diversity** (Washington, D.C.: Department of State, 1982).

5. A.H. Gentry and J. Lopez-Parodi, "Deforestation and Increased Flooding of the Upper Amazon," **Science** 210 (1980): 1354–1356; I. Friedman, "The Amazon Basin, Another Sahel?" **Science** 197(1977): 7.

6. C.H. Dodson and A.H. Gentry, "Flora of the Rio Palenque Science Center," **Selbyana** 4(1978):1–628.

7. J.D. Nations and D.I. Komer, "Rainforests and the Hamburger Society," **Environment** 25(1983):12–20; see also note 1 above.

8. P.M. Fearnside, "Deforestation in the Brazilian Amazon: How Fast Is It Occurring?" **Interciencia** 7(1982): 82–88; the utilitarian, anti-preservation opposition creates the impression that there are no hard data on tropical deforestation, that environmentalists (such as N. Myers) exaggerate the extent of damage [e.g., the Lugo-Brown critique of Myers' book in **Interciencia** 7(1982): 89–93], and that, since there is nothing really to worry about, scientists and preservationists are misleading the public. But "it is irrelevant in the long range, whether the proportion of forests destroyed is 0.6% or 2% of the biome per year" [N. Myers, **Interciencia** (1982)7:358], whether 60,000 $km^2$ or 200,000 $km^2$ of primary virgin forest are converted to *permanent* cultivation *each year*, because even the lower figure is an incredibly large area—1/3 as large as the state of Wisconsin. In either case, it represents ecological insanity. Sad to note, in the eyes of the world's power brokers, nature destruction is always justified, if by doing so people get fed and hunger is alleviated. The crucial, ultimate question, "what are we going to do then?" after three or four decades, once everything is gone and the world will be even fuller with people than now, is conveniently neglected.

9. D. Poore, "Deforestation and the Population Factor," **IUCN Bulletin**, January-February-March 1983; reprinted in **Parks** 8(1983):11–12.

10. H.H. Iltis et al., "*Zea diploperennis* (Gramineae): A New Teosinte from Mexico," **Science** 203(1979): 186–187; N.D. Vietmeyer, "A Wild Relative May Give Corn Perennial Genes," **Smithsonian** 10(1979): 68–75; L.R. Nault et al., "Response of Annual and Perennial Teosintes (*Zea*) to Six Maize Viruses," **Plant Disease** 66(1982): 61–62; and L.R Nault and W.R. Findley, "*Zea diploperennis*: A Primitive Relative Offers New Traits to Improve Corn," **Desert Plants** 3(1982): 203–205.

11. There are currently attempts being made by the Universidad de Guadalajara and the Instituto Nacional de Investigaciones Sobre Recursos Bioticos (INIREB), Xalapa, to set aside part of this magnificent mountain range as a scientific preserve.

12. See note 10 above.

13. B. Webster, "Songbirds Decline in America," **New York Times**, August. 12, 1980; J.W. Fitzpatrick, "Northern Birds at Home in the Neotropics," **Natural History** 91(1982): 40–47.

14. J. Chang, "Potential Photosynthesis and Crop Productivity," **Annals of the Assn. of Amer. Geographers** 60(1970): 92–101; D.M. Gates, "The Flow of Energy in the Biosphere," **Scientific American** 224(1971): 88–100. At the same time, the quite effective agricultural methodologies evolved by primitive or indigenous peoples in the Amazon and elsewhere are also in need of deliberate protection. They can teach us a great deal about how forests can be utilized to some extent and with minimum impact on ecosystem function. But, just like the tropical forests themselves, the life, knowledge, and culture of these forest farmers are being destroyed.

15. See note 2 above.

16. W. M. Denevan, "Latin America," in G.A. Kless, ed., **World Systems of Traditional Resource Management** (N.Y.: Halstead Press, 1980), pp. 217–244; P.M. Fearnside, "Development Alternatives in the Brazilian Amazon: An Ecological Evaluation," **Interciencia** 8(1983): 65–78.

17. D.H. Janzen, ed., **Costa Rican Natural History** (Chicago: University of Chicago Press, 1983); M.A. Boza and R. Mendoza, **The National Parks of Costa Rica**, published under the auspices of the Costa Rican Institute of Tourism, the National University, the National Park Service, and the National Open University (Madrid: INCAFO, 1981); and personal communication with Alvaro Ugalde, director of the Costa Rica National Parks Service and executive director of the Costa Rica National Parks Foundation, November 1983.

18. G.B. Wetterberg, G.T. Prance, and T.E. Lovejoy, "Conservation Progress in Amazonia: A Structural Review," **Parks** 6(1981): 5–10; A. Gentry, "Extinction and Conservation of Plant Species in Tropical America, A Phytogeographical Perspective," in I. Hedberg ed., **Systematic Botany, Plant Utlization and Biosphere Conservation** (Stockholm: Almquist and Wiksell, 1979).

19. J.S. Denslow and T.C. Moermond, "Why We Must Save the Rain Forests," **Capital Times** (Madison, Wisconsin), August 28, 1982; I. Rubinoff, "A Strategy for Preserving Tropical Rainforests," **AMBIO** 12, no. 5(1983):255-258; and J. D. Nations and D. I. Komer, "Central America's Tropical Rainforests: Positive Steps for Survival," **AMBIO** 12, no. 5 (1983): 232–238.

20. N. Myers, **A Wealth of Wild Species** (Boulder, Colorado: Westview Press, 1983); G. Wilkes, "The World's Crop Plant Germplasm: An Endangered Resource," **Bulletin of the Atomic Scientists** 33 (1977): 8–16; J.V Neel, "Lessons from a Primitive People," **Science** 170(1970):815–822; and H. H. Iltis, "Discovery of No. 832: An Essay in Defense of the National Science Foundation," **Desert Plants** 3, no. 4 (June 1982): 175–192.

21. H.H. Iltis and D.A. Kolterman, "Botanical Translations: Needs and Responsibilities," **BioScience** 33(1983): 613.

22. M. Jacobs, "The Spirits of Bali," **IUCN Bulletin** 14(1983): 64–65.

23. Nations and Komer, note 7 above.

24. See note 1 above.

25. Anonymous, "World Growth-Rate Breaks Record," **Wisconsin State Journal**, August 31, 1983.

26. L.R. Brown, "World Population Growth, Soil Erosion, and Food Security," **Science** 214(1981): 995–1002. "We have now squarely to face this paradox . . . . We have increased human hunger by feeding the hungry. We have increased human suffering by healing the sick. We have increased human want by giving to the needy. It is almost impossible for us to face the fact that this is so. The truth comes as a shocking discovery, for we have all been brought up in the Christian tradition in which caring for the least of our brethren has been counted the highest virtue." Rev. Duncan Howlett, All Souls Church, Washington, D.C., December 6, 1969. [Quoted in **The Other Side**, The Environmental Fund Newsletter, Washington, D.C., September 1979].

27. "World Growth-Rate," note 25 above.

# SECTION IV
# HUMANISTIC CONSIDERATIONS

CHAPTER 18

# Stabilizing Population

LESTER R. BROWN

Progress at halting world population growth has been extraordinarily uneven during the last decade. Some countries have been highly successful, others complete failures. Some have reached zero population growth without trying. At one end of the spectrum are several European countries that have completed the demographic transition and whose population growth has ceased. At the other end are some 34 Third World countries whose populations are expanding at 3 percent or more annually.[1]

Assessing national population policies is complicated by the changing criteria of success. Until recently, progress was defined simply in terms of slowing population growth, so as to widen the margin of economic growth over that of population. During the seventies, however, the inadequacy of this goal became evident. Forests were shrinking, grasslands deteriorating, and soils eroding. Biological thresholds of sustainable yield were crossed in scores of countries. When a biological system is being taxed, even a modest increase in human numbers can be destructive. Failing to halt population growth before such a critical threshold is crossed leads to food shortages, fuel scarcities, and a declining standard of living.

The relationship between population and economic growth is also changing. When the world economy was expanding at 4 percent or more per year, the average rate of national economic growth was well above that of even the most rapidly expanding population. But the much slower economic growth of the eighties means national rates are beginning to fall below population growth in many of the countries with high birth rates. The failure of these governments to introduce effective population policies is leading to a broad-based decline in living standards. In these circumstances rapid population growth does not merely slow improvements in income, it precludes them.

Reprinted from *State of the World—1984* by Lester R. Brown, William U. Chandler, Christopher Flavin, Sandra Postel, Linda Starke and Edward Wolf. By permission of W.W. Norton Company, Inc. 

Lester R. Brown is President and Senior Researcher, Worldwatch Institute.

## Population Arithmetic

Over the past decade many people concerned about rapid population growth have cited with some relief the gradual decline in the growth rate. Peaking at something like 1.9 percent around 1970, annual growth in world population declined to 1.7 percent in 1983. Although this is encouraging, the rate is not falling fast enough to reduce the number of people added each year. In 1970 world population increased by 70 million; in 1983 the addition was 79 million.[2] By the criterion of sheer numbers, the worldwide effort to get the brakes on population growth is falling short.

There are few areas of public policy in which a lack of understanding of basic arithmetic has hindered effective policy-making as much as it has in population. Few national political leaders seem to understand what an annual population growth of 3 percent, relatively innocuous in the near term, will lead to over a century. Political leaders and economic planners often think of a 3 percent growth rate as simply three times as much as a 1 percent rate. Although this is true for one year, it is not true over the longer term, as the rates compound. A population growing 1 percent annually will not even triple in a century, but one growing 3 percent annually will increase nineteenfold. (See Table 1.)

Stopping world population growth requires bringing the number of births and deaths roughly into balance, as they already are in several countries. How far the world has to go in this task can be seen in the data for 1983, when an estimated 131 million births and 52 million deaths yielded the net increase of some 79 million. A clearer idea of the

**Table 1. Relationship of Population Growth Per Year and Per Century**

| Population Growth Per Year | Population Growth Per Century |
|---|---|
| (percent) | (percent) |
| 1 | 270 |
| 2 | 724 |
| 3 | 1,922 |

SOURCE: Worldwatch Institute.

dynamics of this growth can be gained from analyzing the annual additions of various countries. Leading the list is India, where there are now over 15 million more births than deaths each year. (See Table 2.) China, which reached the one billion mark in 1981, adds 13.3 million people annually, fewer than India because of its much lower growth rate. Together, these two demographic giants account for over one-third of the earth's new inhabitants each year. The five leading national contributors to world population growth—these two, plus Brazil, Bangladesh, and Nigeria—constitute nearly half the 79 million annual increment.

Many relatively small Third World countries have an astounding number of new citizens to feed, clothe, and shelter each year because their growth rates are so high. Nigeria, for example, has 84 million people, less than one-third as many as the Soviet Union; yet each year there are 2.8 million more Nigerians, compared with 2.2 million more Soviets. In the Western Hemisphere, the U.S. population is four times that of Mexico's but the latter grows by nearly 2 million people a year, whereas the U.S. natural increase is 1.6 million. A similar situation exists in Asia: Burma, with only one-third the population of Japan, adds more people each year.

## Countries with Zero Population Growth

The first country in the modern era to bring births and deaths into equilibrium was East Germany, where population growth came to a halt in 1969. It was closely followed by West Germany, whose population stopped growing in 1972. During the decade since, several other countries—most recently Italy, Switzerland, and Norway—have joined the ranks of the demographic no-growth countries.

**By 1983, there were 12 countries, all in Europe, where births and deaths were in equilibrium.**

By 1983 there were 12 countries, all in Europe, where births and deaths were in equilibrium. (See Table 3.) For them, in contrast to the rest of the world, a year of zero economic growth does not automatically lead to a decline in living standards. With the exception of East Germany and Hungary, all are in Western Europe. They range from tiny Luxembourg to three of the four largest countries in Western Europe—Italy, the United Kingdom, and West Germany. With the recent addition of Switzerland to the list, Europe's German-speaking population has stabilized. Together, these dozen countries contain some 244 million people. Although this represents only 5.2 percent of the world total, it is at least a beginning toward the eventual stabilization of world population, a prerequisite of a sustainable society.

Population stabilization was not an explicit national goal in any of these 12 countries. Declines in fertility flowed from economic gains and social improvements. As incomes rose and as employment opportunities for women expanded, couples chose to have fewer children. The improved availability of family planning services and the liberalization of abortion laws gave couples the means to achieve this. Population stabilization in these countries has been the result, therefore, of individual preferences, the product of converging economic, social, and demographic forces.

Several other European countries could reestablish equilibrium between births and deaths in a matter of years. Among these in Western Europe are Finland, France, Greece, the Netherlands, Portugal, and Spain. In Eastern Europe, Bulgaria and Czechoslovakia are soon likely to follow East Germany and Hungary along the path of population stabilization.

Encouraging though it will be when these countries reach population stability, major gains in raising the share of people in the world who live in a similar situation awaits further fertility declines in the three largest industrial countries —Japan, the Soviet Union, and the United States—which now have annual population growth rates of 0.7, 0.8, and 0.7 percent, respectively. The postwar decline in the Soviet crude birth rate leveled off in the early seventies, hovering around 18 since then. The U.S. birth rate, after declining rather steadily from 1957 until the late seventies, picked up slightly during the early eighties. Japan's birth rate, however, has been falling steadily for over a decade and, at 13, is now the lowest of these three.[3]

With the number of young people entering their reproductive years now falling steadily in both the United States and Japan, it is possible to envisage a time in the not-too-distant future when births and deaths will come into balance, as they have in so much of Europe. In both countries this will be hastened by the aging of the population and the consequent increase in death rates. The situation is somewhat less clear for the So-

**Table 2. Principal Sources of World Population Growth, by Country, 1983**

| Country | Population | Annual Growth Rate | Annual Increase |
|---|---|---|---|
| | (million) | (percent) | (million) |
| India | 730.0 | 2.1 | 15.33 |
| China | 1,023.3 | 1.3 | 13.30 |
| Brazil | 131.3 | 2.3 | 3.02 |
| Bangladesh | 96.5 | 3.1 | 2.99 |
| Nigeria | 84.2 | 3.3 | 2.78 |
| Pakistan | 95.7 | 2.8 | 2.68 |
| Indonesia | 155.6 | 1.7 | 2.65 |
| Soviet Union | 272.0 | 0.8 | 2.18 |
| Mexico | 75.7 | 2.6 | 1.97 |
| United States | 234.2 | 0.7 | 1.64 |

SOURCE: Population Reference Bureau, *1983 World Population Data Sheet* (Washington, D.C.: 1983); China's growth rate is author's estimate based on 1982 census by Chinese Government.

Table 3. **Countries With Zero Population Growth, 1983**[1]

| Country | Crude Birth Rate | Crude Death Rate | Annual Rate of Increase (+) or Decrease (−) | Population |
|---|---|---|---|---|
| | (per 1,000 population) | | (percent) | (million) |
| Austria | 12 | 12 | 0.0 | 7.6 |
| Belgium | 12 | 11 | +0.2 | 9.9 |
| Denmark | 10 | 11 | −0.1 | 5.1 |
| East Germany | 14 | 14 | 0.0 | 16.9 |
| Hungary | 13 | 14 | −0.1 | 10.7 |
| Italy | 11 | 9 | +0.2 | 56.6 |
| Luxembourg | 12 | 11 | +0.1 | 0.4 |
| Norway | 12 | 10 | +0.2 | 4.1 |
| Sweden | 11 | 11 | 0.0 | 8.3 |
| Switzerland | 11 | 9 | +0.2 | 6.5 |
| United Kingdom | 13 | 12 | +0.1 | 56.0 |
| West Germany | 10 | 12 | −0.2 | 61.6 |
| Total Population | | | | 243.5 |

[1]Zero population growth is here defined as within a range of plus or minus 0.2 percent change in population size per year.
SOURCE: Worldwatch Institute estimates, based on data in United Nations, *Monthly Bulletin of Statistics*, New York, monthly.

viet Union, which has a highly fertile, rapidly expanding Asian minority. The moderate overall Soviet growth rate embraces both European republics where population growth has ceased and Asian republics where it expands at typical Third World rates. Indeed, some of the Soviet republics in Asia have higher birth rates than some demographically progressive states in India.[4] Overall, nevertheless, barring any unexpected rises in fertility, it is only a matter of time until natural population growth comes to a halt in these three countries, since each has a net reproductive rate well below the replacement rate of 2.1.

## Rapid National Fertility Declines

The demographic transition, the shift from high mortality and high fertility to low rates of both, has historically been a slow process. In Western Europe death rates and birth rates fell gradually over a few centuries. In countries now in the early stages of modernization, the start of the transition has been much faster. The precipitous decline in mortality of roughly a generation ago was often compressed into a decade or so and usually occurred before the decline in fertility had begun. This produced record population increases against a backdrop of insufficient, inequitably distributed resources. The result was historically unprecedented growth in human numbers, commonly exceeding 3 percent per year. These rates underline the urgency of completing the last phase of the demographic transition—the decline in fertility—as rapidly as possible. But reducing births is a much more complex undertaking than reducing deaths. Mortality rates can be brought down quickly through public health measures and childhood vaccination programs, whereas the reduction in births requires changes in values that must then translate into changes in reproductive behavior.

As of the mid-eighties hope on this front springs not so much from the broad global trends as from the sharp reduction in fertility in countries that represent a wide cross section of cultures, religions, and political systems. This diversity is evident in the four countries in the Western Hemisphere that have the lowest crude birth rates—Cuba (14), Canada (15), the United States (16), and Barbados (17).[5] It would be hard to find two national cultures more similar than the United States and Canada, so their similar fertility levels are not surprising, but the contrasts between them and the two Caribbean island cultures are striking.

The four developing countries in Asia, on the other hand, that were among the first to lower their birth rates—Hong Kong (17), Singapore (17), Taiwan (23), and South Korea (25)—are rather homogenous in that all are predominantly Sinitic (Chinese-based or -related) cultures.[6] In addition to their common or similar ethnic backgrounds, all four are economically vigorous societies moving rapidly along the path to modernization. And they all have had well-designed family planning programs, with contraceptive services widely available, for at least a decade and a half.

**In 1979 China became the first country to launch a one-child family program.**

Perhaps the most impressive family planning achievement in the developing world has occurred in China. Family planning was periodically caught up in the ideological crosscurrents of the Chinese communist party from 1949 until the early seventies, but since that time a sustained national effort has been under way to reduce births. By the mid-seventies the Chinese leadership was urging all couples to stop at two children, and in 1979 China became the first country to launch a one-child family program.

By 1980 China had lowered its crude birth rate to 20, an achievement many thought impossible in such a large country still at an early stage of economic development. Although the 1982 census showed a slightly higher birth rate of 21, this does not diminish the success embodied in China's remarkable reduction in fertility from 34 to 20 in only a decade. This precipitous decline closely parallels that in Japan from 1948 to 1958, when the birth rate fell 47 percent, from 34 to 18.[7]

China's family planning program is distinctive in several ways. To begin with, the national leadership has been deeply involved in designing and supporting the program. More than most countries, China has fostered public discussion of the population problem and particularly the effect that continuing growth has on future living standards. As population policy was integrated into overall economic planning, it led in the late seventies to the establishment of birth quotas. The overall plan aimed to ensure that targets for raising living standards were met; couples who had no children were given priority in the quota allocation. Once births were allocated to people at the production-team level, peer pressure played an important role in assuring compliance. To even consider such an ambitious birth control program requires, of course, the ready availability of family planning services, including abortion.

From birth planning it was a relatively small step to the provision of economic incentives to couples who would agree to have only one child. The one-child family campaign was not introduced because China's leaders were enamored with the concept per se, but because the buildup in population pressure left few alternatives. Gaining social acceptance of this concept is not easy in a society where large families are traditional and where there is still a strong preference for sons, particularly in the countryside. This unprecedented policy initiative puts China one step ahead of other Third World countries in trying to halt population growth.

On a smaller scale, neighboring Thailand's reduction of fertility has also been impressive. Its estimated crude birth rate of 26 is not yet as low as China's but it apparently started from a higher level in 1970. Data on annual birth rates are not available in Thailand, but various fertility surveys taken from 1970 onward have measured contraceptive usage. The proportion of married Thai women of reproductive age who were using contraceptives went from 14 percent in 1970 to nearly 60 percent in 1981, a level approaching that in industrial societies.[8] (See Table 4.)

In 1983, Thailand's birth rate was 26 —only one point higher than that of South Korea, which launched its family planning program many years earlier. Indeed, it was roughly the same as the U.S. rate immediately after World War II. Thailand's achievement is all the more laudable because its reproductive revolution has preceded widespread economic development. Central to its success is a public education program that emphasizes both the economic and social advantages of small families and the ready availability of contraceptive services. Its vigorous family planning program is credited with perhaps 80 percent of the national fertility decline since 1970. Behind this imaginative program is Mechai Viravaidya, the innovative and charismatic head of the family planning group.[9]

Analysts of the Thai fertility decline associate the rapid change in reproductive attitudes and behavior with the high degree of social and economic independence of women. The influence of Buddhism, which does not restrict contraception and is not particularly pronatalist, may also contribute to the ready acceptance of family planning. And Buddhism emphasizes individual responsibility, which may have created a social environment particularly receptive to a progressive family planning program.

One of the most rapid fertility declines on record has occurred in Cuba since the mid-sixties. The country's 1983 crude birth rate of 14 per 1,000, lower than that of the United States, is all the more remarkable because it was not the result of a concerted national program to lower fertility and curb population growth. This recent birth rate decline should be seen, however, in historical perspective, since Cuba, along with Uruguay and Argentina, experienced a gradual decline in birth rates in the early decades of the twentieth century. At the

**Table 4. Thailand: Share of Married Women Practicing Contraception, 1970–81**

| Year | Percent |
|---|---|
| 1970 | 14 |
| 1975 | 37 |
| 1979 | 50 |
| 1981 | 58 |

SOURCE: *International Family Planning Perspectives*, September 1980 and June 1982.

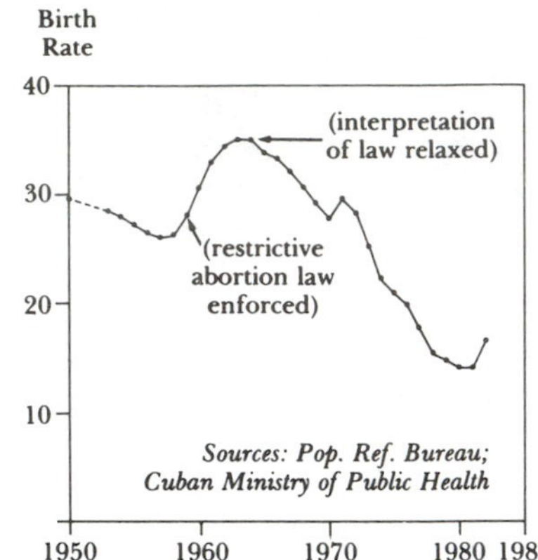


**Figure 1. Cuba's Crude Birth Rate, 1950-82**

time of the Castro takeover in 1959 the birth rate was 28, already well below those of most of Latin America.

One of the early actions of the new government was to begin enforcing the rather restrictive abortion law that was already on the books. This pushed birth rates up sharply, to over 35 in 1963. (See Figure 1.) In 1964, the interpretation of the restrictive abortion law was relaxed and, abetted by widespread social gains for women and broad-based improvements in health care services, the birth rate began a precipitous decline that continued for the next 16 years. By 1980 it had fallen to 14. Although the birth rate had been more than halved, the population growth rate had been cut by some two-thirds, to less than 1 percent per year.[10]

After China, the most populous Third World country to reduce its fertility sharply is Indonesia. Until the late sixties Indonesia was a pronatalist country. Its only population program was one of transmigration from densely populated Java to the sparsely populated outer islands. After years of struggling at great cost to move Javanese to these territories, and discovering that the resettlement flow was a mere trickle compared with the torrent of births, the Indonesian Government shifted its attention to family planning.

Indonesia's national family planning program was launched in 1969. By 1976, when the World Fertility Survey conducted a study there, fertility had fallen substantially in some regions—as much as 12 percent in West Java to 34 percent in Bali. Indonesia's family planning program has been distinguished by strong government support and a highly acclaimed local approach that goes far beyond the more traditional clinic system. The direct involvement of the local village leaders as motivators, field workers, and even contraceptive distributors has been central to the program's success. By 1983 Indonesia's birth rate had fallen to 32 and its population growth rate was reportedly 1.7 percent per year, well below the nearly 3 percent that existed before the program began.[11]

Another Latin American country—Mexico—is the site of one of the most recent fertility reductions in a major Third World country. Prior to 1972 the official Mexican policy was like Indonesia's—the more people the better. President Echeverria's announcement in 1972 that Mexico would launch a family planning program came as a surprise. At that time Mexico's crude birth rate was 44 and its overall population growth rate was 3.5 percent annually—one of the highest in the world. Within a decade the birth rate dropped to 32. Combined with a death rate of 7, this still resulted in a population growth rate of 2.5 percent, which was far too rapid. And there are remote rural areas where family planning services are not yet available. Nonetheless the progress of a decade indicates that Mexico has at least set the stage for a continuing decline in fertility. The challenge now will be limiting births over the rest of this century among the enormous number of young people entering their reproductive years as a result of the extraordinarily high birth rate before the mid-seventies.[12]

The countries that have achieved rapid national fertility declines represent a wide variety of cultures. The common denominators are a committed leadership and locally designed programs. Experience to date shows a broad popular interest in planning families and a willingness, in some cases an eagerness, to take advantage of services when they become available. Each country desiring to reduce fertility must of course design its own program, one that is responsive to its values, traditions, and needs.

## Population Versus Economic Growth

Overall, global economic growth during the third quarter of this century was more than double that of population. As will be noted in Chapter 3, this growth was closely tied to an abundance of cheap oil. With the 1973 oil price hike, economic growth began to slow. With the 1979 increase it slowed further. The following four years the world economy expanded less than 2 percent per year. Although these four years are not in themselves enough to constitute a trend, the less favorable resource conditions that contributed to the slower growth during this time may also lead to slow growth over the long term.

The *Wall Street Journal,* in a survey of European economic analysts at the end of 1982, reported that many expect the prolonged recession of the early eighties to become permanent. In contrast to earlier recessions, when recovery translated into a 4–6 percent growth rate, the *Journal* noted that these analysts saw European recovery from the current recession in terms of a 1–2 percent rate.[13]

For parts of the Third World, prospects are even grimmer. A 1983 assessment of Africa's future by the Economic Commission for Africa reported that "the historical trend scenario is almost a nightmare." It continued: "The potential population explosion would have tremendous repercussions on the region's physical resources such as land. . . . At the national level, the socio-economic conditions would be characterized by a degradation of the very essence of human dignity. The rural population . . . will face an almost disastrous situation of land scarcity whereby whole families would have to subsist on a mere hectare of land." A World Bank analysis of Africa's economic outlook described the Commission's assessment as "graphic but realistic."[14]

Given the emerging constraints on growth, particularly in basic sectors such as food and energy, the world is going to have great difficulty resuming the rapid economic growth of the 1950–73 period. Expanding global economic output even by an average of 2 percent per year may tax the skills of economic policymakers. Countries with rapid population growth may thus face declines in living standards—unless they quickly reduce their birth rates.

Slower economic growth will have an impact everywhere, but the effect will vary widely according to national population growth rates. For West Germany or Belgium, which have attained zero population growth, a 2 percent rate of annual economic growth will still raise incomes 2 percent each year. For countries such as Kenya and Ecuador, whose numbers expand at more than 3 percent per year, a 2 percent economic growth rate will produce steady declines in incomes and living standards.

If an economic growth rate of 2 percent per year becomes the new norm, then nearly half the world's people—those living in countries with annual population growth rates of 2 percent or more—face possible income stagnation or decline. Countries where the threat of falling income is greatest are those whose numbers grow at 3 percent or more per year. These 34 countries, almost all in Africa, the Middle East, and Central America, have a combined population of 394 million.[15]

Within this group, countries with exportable surpluses of oil may be tempted to neglect population policy, with the result that their populations will continue to multiply rapidly, sustained by the imported resources that petroleum buys. Over the long term, however, as oil reserves dwindle and the exportable surplus disappears, they may find themselves with populations that far exceed the carrying capacity of local resources. Countries such as Iran and Nigeria, where oil production and exports have already peaked, illustrate well the risks that oil-producing countries with rapidly expanding populations face over the long term. Situations like these have the greatest long-term potential for massive human suffering.

Unfortunately, economic growth has already fallen behind population growth in many countries. The World Bank reports that per capita GNP declined in 18 countries from 1970 to 1979. (See Table 5.) In a few cases this was caused by disruption of economic activity associated with political conflicts and instability, but in the great majority population growth simply outstripped that of economic output.

**Table 5. Countries Experiencing a Decline in Per Capita Income, 1970–79**

| Country | Population | Annual Rate of Decline |
|---|---|---|
| | (million) | (percent) |
| Angola | 6.9 | −9.6 |
| Bhutan | 1.3 | −0.1 |
| Chad | 4.4 | −2.4 |
| Congo | 1.5 | −0.2 |
| Ghana | 11.3 | −3.0 |
| Jamaica | 2.2 | −3.7 |
| Libya | 2.9 | −1.6 |
| Madagascar | 8.5 | −2.5 |
| Mauritania | 1.6 | −0.7 |
| Mozambique | 10.2 | −5.3 |
| Nicaragua | 2.6 | −1.6 |
| Niger | 5.2 | −1.2 |
| Sierra Leone | 3.4 | −1.2 |
| Uganda | 12.8 | −3.5 |
| Upper Volta | 5.6 | −1.2 |
| Zaire | 27.5 | −2.6 |
| Zambia | 5.6 | −1.9 |
| Zimbabwe | 7.1 | −1.7 |
| Total | 120.6 | |

SOURCE: World Bank, *1981 World Bank Atlas* (Washington, D.C.: 1982).

Central to the decline in per capita income during the seventies in many of the countries in Africa, a largely rural continent, was the decade-long decline in per capita food production. The growth in Africa's food supply compares favorably with that for the world as a whole, but its increase in human numbers is far more rapid. Plagued by the fastest population growth of any continent in history, as well as by widespread soil erosion and desertification, Africa's food production per person has fallen 11 percent since 1970.[16]

Given the prevailing economic conditions of the early eighties, the ranks of the 18 countries with declining per capita GNP are likely to swell dramatically. In a report published in late 1981 the World Bank projected a decline in the average income of 187 million people in 24 low-income countries in sub-Saharan Africa. This is the first time the World Bank has projected a decline in living standards for a major region of the world.[17]

As the growth in the production of material goods and services slows, the distribution issue must be viewed against a new backdrop, one unfamiliar to this generation. With the changing growth prospect, pressure to reformulate economic and social policies with basic human needs in mind will no doubt increase. This task will be far more difficult than it was in the era of rapid economic growth, with its underlying belief that a rising tide raises all ships. As economist Herman Daly has perceptively observed, turning our focus to meeting basic human needs will "make fewer demands on our environmental resources, but much greater demands on our moral resources."[18]

## Population/Resource Projections

Existing projections of population and resources do not provide a solid foundation for formulating intelligent population policies. U.N. projections show world population continuing to grow until it reaches 11 billion before leveling off a century or so hence.[19] These population figures are the product of two sets of assumptions—one explicit and one implicit. The explicit assumptions are demographic. They include country-by-country assumptions about future fertility levels, sex ratios, life expectancies, and numerous other demographic variables. If these explicit assumptions hold, the projected increases in world population will materialize.

**Africa's food production per person has fallen 11 percent since 1970.**

But population growth does not occur in a vacuum. Current projections of world population are based on the implicit assumption that the energy, food, and other natural resources required to support human life are going to be as readily available in the future as they have been in the past. They assume that the production of the major biological support systems—fisheries, forests, grasslands, and croplands—that satisfy basic human needs for food, shelter, and clothing will continue to expand along with population. It is perhaps unfortunate that population projections are done by demographers on their own. Projections by an interdisciplinary team of analysts including, for example, agronomists, economists, and ecologists would be more realistic since population growth cannot be divorced from the carrying capacity of local biological systems. Continuing increases in human population on the one hand and the progressive deterioration of life-support systems on the other are not compatible over the long term.

The unreality of current population projections can be seen in the numbers put forth for individual countries. According to the World Bank, which uses slightly lower projections than the U.N. medium-level ones, the growth ahead for some countries can only be described as phenomenal. India is projected to add more than another billion people to its 1983 population of 730 million before stabilizing at 1.84 billion, while neighboring Bangladesh and Pakistan could increase from 96 million each at present to 430 million and 411 million respectively. If these three countries grow as projected, the Indian subcontinent would be home to 2.7 billion people, more than the entire world population in 1950. The 84 million Nigerians of today are projected to increase to 623 million, more people than now live in all of Africa. Mexico would grow from 76 million to 215 million people, roughly the size of the current U.S. population. And the populations of several Central American countries—El Salvador, Guatemala, and Nicaragua—would triple before stabilizing.[20] Clearly, national and global resources would be stretched to the breaking point well before increases of this size materialize.

Many of the economic stresses afflicting the world economy during the mid-eighties have their origins in changing population/resource trends. When world population reached the three billion mark in 1960, the per capita production of almost every major commodity was still increasing. But when world population moved toward four billion, demands began to approach the sustainable yield of many biological support systems: The world harvest of forest products began to fall behind population growth; 20 years of rapid growth in the world fish catch came to an end; and per capita growth in world beef output halted in 1976. World production of oil, the leading energy resource, peaked in 1979. The growth in world grain production, which had been steadily outdistancing population since mid-century, has barely kept pace with population growth since 1973. In North America food production has continued to outstrip population growth. But in Africa, as mentioned earlier, the reverse is true. Since 1970 per capita food production there has been falling by some 1 percent per year.[21]

Faced with this emerging picture of a new population/resource balance, President Carter was frustrated by the inability of U.S. Government agencies to provide coherent sets of projections in important areas such as energy, water, and food. This lack of global foresight capability led him to launch the Global 2000 study in 1977. Undertaken by the U.S. Council on Environmental Quality and the Department of State in cooperation with 17 specialized government departments and agencies, the study examined, among other things, the consequences of the projected increases in world population.

The Global 2000 study team discovered numerous inconsistencies among government projections and policies. In assembling sector forecasts from the various agencies, overlapping resource claims came to light. The Department of Agriculture, for example, was incorporating in its projections water that the Department of Energy was expecting to use to launch synfuels projects. The risk of such a lack of coherency in planning is that it can lead to resource scarcities, as in the case of water, or to excess capacity, as in the case of electrical generation. In either case resources are wasted.

The *Global 2000 Report,* the end product of this first effort to project global economic, environment, and resource trends, observed in an oft-quoted conclusion that "barring revolutionary advances in technology, life for most people on earth will be more precarious in 2000 than it is now—unless the nations of the world act decisively to alter current trends." Five years have passed since the study went to press. Unfortunately, life is more difficult now for many of the world's people than it was then.[22]

The only other comprehensive attempt to project world trends to the end of the century was undertaken in the mid-seventies by the United Nations. Directed by economist Wassily Leontief, it foresaw a rosy economic future for the world. The weakness of this undertaking was that it was done almost entirely by economists and lacked a solid interdisciplinary foundation. The result was pie-in-the-sky projections and misleading conclusions about where the world was headed. For example, the U.N. study postulated for the remainder of this century an average agricultural growth rate in the developing world of 5.3 percent, a rate with little foundation in either the historical experience of developed countries or in current Third World realities. The authors noted that soil erosion would affect future food production, but because data was sketchy they chose simply to ignore it. If agronomists had been involved in the project they could at least have estimated erosion's impact on future productivity, thus making the projections more realistic and useful.[23]

Concerned with the declines in per capita production of the basic resources that underpin the global economy, several individual countries decided to examine the implications of the Global 2000 study for their own national policies. Countries as varied as Japan, Mexico, and West Germany have expressed interest in the "Global 2000" approach. China, well aware of indigenous resource constraints, is undertaking an ambitious China 2000 study. Japan has organized a Year 2000 Committee to focus on the implications of the changing global resource outlook for the Japanese economy.[24] The United States, unfortunately, has failed to follow through on *Global 2000.* The result has been an international vacuum as the dominant world economy and traditional world leader on such matters sits on the sidelines, failing to provide leadership on this critical complex of issues.

One of the first countries to examine systematically the long-term population/resource balance was China. As part of the post-Mao reassessment, Chinese leaders projected future population size based on the assumption that couples would have only two children. Even under this scenario, given the country's youthful age structure China would add another 300 or 400 million people before population growth ceased. After relating these projections to the availability of land, water, energy, and other basic resources and to the capacity of the economy to provide jobs, the leadership concluded that they had no choice but to press for a one-child family lest they jeopardize their hard-earned gains in living standards.[25]

Elizabeth Croll, China scholar at Oxford University, observes that Beijing had concluded that "unless the population to be fed, housed and clothed is reduced, the goals of any development strategy in China are bound to fail."[26] The principal difference between China and other densely populated developing countries such as Bangladesh, India, Egypt, Nigeria, and Mexico may be that the Chinese have had the courage and the foresight to make these projections and to translate their findings into public policy. If others took a serious look at future population/resource balances, they too might well decide that they should press for a one-child family.

## New Population Initiatives

As world population moves toward five billion, the per capita production of many basic commodities is falling. The effort to raise incomes and living standards is faltering in many countries, particularly where population is growing most rapidly. The fall in global economic growth from over 4 percent annually to 2 percent is dividing the world into two groups: those where economic growth exceeds population growth, and those where it does not. For one group, living standards are rising. For the other they are falling. One group can hope that the future will be better than the present. In the other, hope is turning to despair.

The fall in per capita incomes is occurring, almost without exception, in countries that have given little attention to the human side of the population/resources equation. The attention of political leaders and the allocation of budgetary resources have both focused almost entirely on the supply side, on expanding output. One consequence of this imbalance is that more and more of the growth in output is required to satisfy the increase in population, with little remaining for improvements in per capita consumption.

Where population growth is rapid, changing economic circumstances call for new population policies. Many developing countries are stalled in the middle of the demographic transition discussed earlier. If economic growth remains slow, barely keeping pace with population growth or even falling behind it, then high-fertility developing countries cannot count on economic improvements to reduce births as they did in the industrial countries. Fortunately, recent experience has shown that countries with broad-based but inexpensive health care systems and well-designed family planning programs that encourage small families can bring fertility down even without the widespread economic gains that characterized the demographic transition in the industrial societies.[27]

Making family planning services universally available is the very first step that all countries should take. Beyond helping to curb population by preventing unwanted births, it makes people aware that they can control their fertility, in effect creating its own demand. And because family planning leads to better spacing of births, it lowers infant mortality, which in turn fosters lower fertility. Although each country represented at the U.N. Conference on World Population in Bucharest in 1974 agreed that access to family planning services was a basic human right, not all governments have followed through. As a result, an estimated one-third of all couples still go to bed unprotected from unplanned pregnancy.[28]

Essential though family planning services are, they are not in themselves sufficient. Data from the World Fertility Survey show the desired family size in many countries is still four or five children. Although this is lower than it was just a few years ago, in some countries having this many children would lead to vast population gains that would steadily reduce living standards.[29] Reducing fertility to the level circumstances call for requires public education programs that inform people about the future relationship between population and resources and about the economic consequences of continuing on the current demographic path. More projections like those done in Beijing are needed. In traditional countries such as China, childbearing decisions are still influenced by a parental desire to be looked after in old age. By emphasizing future population/resource relationships, government officials can shift parents' considerations of childbearing decisions from their own well-being to that of their children.

For most Third World countries, however, the provision of family planning services and public education programs based on projections will not slow population growth quickly enough to avoid a decline in living standards. Governments in these countries will need to reorient economic and social policies to lower fertility further. Traditionally, government policies that affected population growth encouraged large families. In most countries, for example, income tax deductions and maternity leaves are available without restriction. Now some countries are beginning to alter these long-standing policies. For example, South Korea and Pakistan limit income tax deductions to two children. Tanzania, Sri Lanka, and Nepal have gone even further and entirely eliminated tax deductions for dependent children.[30]

Several governments restrict maternity benefits. In the Philippines, they are limited to the first four births; in Ghana, Hong Kong, and Malaysia, to three; and in South Korea, even more stringent, to only two. Tanzania, adapting this general approach to African family planning programs' emphasis on child spacing, provides paid maternity leave to employed women only once every three years.[31]

Some governments use access to education and to public housing or to low-interest loans for the purchase of housing as a carrot to encourage small families. In China, the certificate awarded to couples who pledge to have only one child entitles them to preferred access to schooling for that child. South Korean government employees who stop at two can deduct their educational expenditures from income tax. In Singapore, having no more than two children gives couples preferred access to housing, much of which is government-constructed.[32]

Other countries have designed reward systems to encourage sterilization or the use of contraceptives. India, for example, was the first government to provide small, one-time payments to men and women who were sterilized. In some countries such payments are regarded more as compensation for lost wages and travel costs than as incentives per se. The Indian payments in 1983 ranged from $11–13, for example, roughly two weeks' wages for rural laborers. In Bangladesh, those sterilized receive new clothing—a sari for women and a lungi for men—plus travel-cost reimbursement.[33]

Experience shows that such payments do make sterilization more popular. In a 1983 Worldwatch Paper, Judith Jacobsen noted that "in Sri Lanka, the number of sterilizations performed in government family planning programs increased when payments were first introduced in January 1980. Nine months later, when payments were increased fivefold to match those offered on private estates, sterilization clinics were swamped. The number of sterilizations performed at one clinic increased from 6 to 35 a day after the introduction of the payment and rose to 150 a day after the increase."[34]

Some countries have begun to experiment with community incentives. A community that achieves certain family planning goals, either measured in terms of contraceptive usage or in reduction in births, becomes eligible for a new school, a village well, a community irrigation pump, or even a television set. This approach, being tried in Thailand and Indonesia, has the advantage of mobilizing peer pressure to limit family size.[35]

There is an urgent need for data to be gathered regularly in this vital area, to permit continuing evaluations of progress and reports to the public. The collection and monthly publication of birth rates would help measure progress on this critical front, much as is now the case with employment, inflation, or the balance of payments. It would also contribute to public awareness of the need to reduce family size and halt population growth.

In many countries, reducing the birth rate rapidly enough to avoid a decline in living standards will require a Herculean effort and the constant attention of national political leaders. Administratively, successful implementation of population programs may require that responsibility for them be escalated from a division in the health ministry, where it commonly resides, to a cabinet-level committee that regularly reviews the situation. Unorthodox though this may be, it is the level of attention befitting the gravity of the issue.

Among the Third World countries that have successfully reduced fertility, no two approaches are identical. But all have involved national leaders' commitment to reduce fertility, the widespread availability of family planning services, and a public education program that links population growth to the long-term social interest as well as to benefits for individual families. Several of the more successful countries have often used some combination of economic incentives or disincentives to encourage small families.

Governments will be forced to settle differences between private interests, sometimes better served by larger families, and the public or social interest, which is invariably better served by smaller families. Reconciling these differences can be extraordinarily complex and politically costly. Failure, however, could be catastrophic. The issue is how —not whether—population growth will eventually be slowed. Will it be humanely, through foresight and leadership, or will living standards deteriorate until death rates begin to rise? The latter would certainly reestablish a population equilibrium of sorts, but it would be through high birth and death rates, rather than the low birth and death rates characteristic of countries that have completed the demographic transition.

In an age of slower economic growth, improvements in living standards may depend more on the skills of family planners than on those of economic planners. As the outlines of the new economic era become more visible, population policy seems assured a high place on national political agendas. Too many governments have delayed facing the issue for too long. When they belatedly do so, they may discover, as China has, that circumstances force them to press for a one-child family.

## References


1. Information on growth rates, referred to throughout this chapter, and other demographic parameters for international comparison can be found in Population Reference Bureau, *World Population Data Sheet* (Washington, D.C.: annual).

2. Population growth in recent years is reviewed in the United Nations, Department of International Economic and Social Affairs (DIESA), *World Population Trends and Policies—1981 Monitoring Report, Volume 1*, Population Studies, No. 79 (New York: 1982); 1983 estimates are from Population Reference Bureau, *World Population Data Sheet.*

3. Population Reference Bureau, *World Population Data Sheet.*

4. Stephen Rapawy and Godfrey Baldwin, "Demographic Trends in the Soviet Union: 1950–2000," in Joint Economic Committee, U.S. Congress, *Soviet Economy in the 1980s: Problems and Prospects* (Washington, D.C.: U.S. Government Printing Office, 1983); Centre for Monitoring Indian Economy, *Basic Statistics Relating to the Indian Economy Vol. 2: States* (Bombay: 1982).

5. Population Reference Bureau, *World Population Data Sheet.*

6. Ibid.

7. H. Yuan Tien, "China: Demographic Billionaire," *Population Bulletin* (Washington, D.C.: Population Reference Bureau, April 1983); United Nations DIESA, *Demographic Yearbook Special Issue: Historical Supplement* (New York: 1979).

8. John Knodel, Nibhon Debavalya, and Peerasit Kamnuansilpa, "Thailand's Continuing Reproductive Revolution," *International Family Planning Perspectives,* September 1980; Peerasit Kamnuansilpa, Aphichat Chamratrithirong, and John Knodel, "Thailand's Reproductive Revolution: An Update," *International Family Planning Perspectives,* June 1982.

9. Henry P. David, "Mechai's Way," *People* (London), Vol. 9, No. 4, 1982.

10. Sergio Diaz-Briquets and Lisandro Perez, "Cuba: The Demography of Revolution," *Population Bulletin* (Washington, D.C.: Population Reference Bureau, April 1981); Henry P. David, "Cuba: Low Fertility, Relatively High Abortion," *Intercom,* Population Reference Bureau, Washington, D.C., July/August 1983.

11. Terence H. Hull, V.J. Hull, and M. Singarimbun, "Indonesia's Family Planning Story: Success and Challenge," *Population Bulletin* (Washington, D.C.: Population Reference Bureau, November 1977); Geoffrey McNicoll and Masri Singarimbun, "Fertility Decline in Indonesia I: Background and Proximate Determinants," and "Fertility Decline in Indonesia II: Analysis and Interpretation," Center for Policy Studies Working Papers, Nos. 92 and 93 (New York: Population Council, 1982).

12. John S. Nagel, "Mexico's Population Policy Turnaround," *Population Bulletin* (Washington, D.C.: Population Reference Bureau, December 1978); Jorge Martinez Manautou, ed., *The Demographic Revolution in Mexico 1970–1980* (Mexico City: Mexican Institute of Social Security, 1982).

13. Debbie C. Tennison, "Europe is Adjusting to a Long Recession That Some Economists See as Permanent," *Wall Street Journal,* December 2, 1982.

14. Economic Commission for Africa quote and World Bank assessment both from World Bank, *Sub-Saharan Africa: Progress Report on Development Prospects and Programs* (Washington, D.C.: 1983)

15. Population Reference Bureau, *World Population Data Sheet.*

16. U.S. Department of Agriculture (USDA), Economic Research Service (ERS), *World Indices of Agricultural and Food Production, 1950–82* (unpublished printout) (Washington, D.C.: 1983).

17. World Bank, *Accelerated Development in Sub-Saharan Africa: An Agenda for Action* (Washington, D.C.: 1983).

18. Quoted in John Harte and Robert Socolow, eds., *The Patient Earth* (New York: Holt, Rinehart and Winston, 1971).

19. United Nations, *Prospects of Population: Methodology and Assumptions,* Population Studies No. 67 (New York: 1979).

20. Estimates of hypothetical stationary populations are from World Bank, *World Development Report 1983* (New York: Oxford University Press, 1983); current-year populations are from Population Reference Bureau, *World Population Data Sheet.*

21. See, for example, Lester R. Brown and Pamela Shaw, *Six Steps to a Sustainable Society* (Washington, D.C.: Worldwatch Institute, March 1982). For oil production trends, see Chapter 3 of this report; grain production trend is from USDA, ERS, *World Indices.*

22. Council on Environmental Quality and U.S. Department of State, *The Global 2000 Report to the President, Volume I* (Washington, D.C.: U.S. Government Printing Office, 1980).

23. Wassily Leontief et al., *The Future of the World Economy* (New York: Oxford University Press, 1977).

24. Gerald O. Barney and Associates, Arlington, Va., private communication, October 26, 1983.

25. Tien, "China: Demographic Billionaire."

26. Elizabeth J. Croll, "Production Versus Reproduction: A Threat to China's Development Strategy," *World Development,* June 1983.

27. See, for example, John Ratcliffe, "Kerala: Testbed for Transition Theory," *Populi,* U.N. Fund for Population Activities, New York, Vol. 5, No. 2, 1978, David Winder, "Literacy—the third world's beacon of hope," *Christian Science Monitor,* May 12, 1983, and W. Parker Maudlin, "The Determinants of Fertility Decline in Developing Countries: An Overview of the Available Evidence," *International Family Planning Perspectives,* September 1982.

28. Author's estimate, based on Bruce Stokes, *Filling the Family Planning Gap* (Washington, D.C.: Worldwatch Institute, May 1977) and subsequent progress in family planning programs in China and other countries.

29. Robert Lightbourne, Jr., and Susheela Singh, with Cynthia P. Green, "The World Fertility Survey: Charting Global Childbearing," *Population Bulletin* (Washington, D.C.: Population Reference Bureau, March 1982); Mary Mederios Kent and Ann Larson, "Family Size Preferences: Evidence From the World Fertility Surveys," *Reports on the World Fertility Survey 4* (Washington, D.C.: Population Reference Bureau, April 1982).

30. Nuray Fincancioglu, "Carrots and Sticks," *People* (London), Vol. 9, No. 4, 1982.

31. Ibid.

32. Pi-chao Chen, "11 M Chinese Opt for Only One Child Glory Certificate," *People* (London), Vol. 9, No. 4, 1982; Planned Parenthood Federation of Korea, *New Population Policy in Korea: Social and Legal Support for Small Families* (Seoul: 1982); Saw Swee-Hock, *Population Control for Zero Growth in Singapore* (Singapore: Oxford University Press, 1980).

33. Figures on current payments in India are from Pravin Visaria and Leela Visaria, "India's Population: Second and Growing," *Population Bulletin* (Washington, D.C.: Population Reference Bureau, October 1981); Indian wage data are from Centre for Monitoring Indian Economy, *Basic Statistics;* Bangladesh information is from Michael Jordan, U.S. Department of State, Washington, D.C., private communication, January 5, 1983.

34. Judith Jacobsen, *Promoting Population Stabilization: Incentives for Small Families* (Washington, D.C.: Worldwatch Institute, June 1983).

35. For an early review of community incentives, see Lenni W. Kangas, "Integrated Incentives for Fertility Control," *Science,* September 25, 1970; efforts in Thailand are described in Henry P. David, "Incentives, Reproductive Behavior, and Integrated Community Development in Asia," *Studies in Family Planning,* May 1982; Indonesian efforts are described in Hull, Hull, and Singarimbun, "Indonesia's Family Planning Story."

CHAPTER 19

# The Human Condition: Economics and Health

CHARLES H. SOUTHWICK

It is possible today to tour the globe and come away convinced that the world is increasingly prosperous. All it takes is money, first class travel, and overnight stops in five-star hotels. The experience is impressive; even in the crowded cities of Cairo, Nairobi, New Delhi, Bangkok, Singapore, Hong Kong, and Rio de Janeiro, one can emerge with the conviction of elegance, wealth, and great economic progress. It is equally possible to tour the world visiting refugee camps, urban slums, and areas of drought, poverty, famine, and armed conflict, and to emerge with a totally depressing view of the human condition.

Such disparities, which have existed throughout history, are now more apparent than ever before. They are more visible, more widely known, and involve more people on both sides of the equation than ever before. Extremes of wealth and poverty can be seen daily on television and readily documented in both personal experience and statistical fact. Possibly this is the origin of the totally different perceptions that come from Global 2000 (Chapter 2) and the ideas of Julian Simon and Herman Kahn (Chapters 3 and 4). The purpose of this chapter is to consider some of the facts relating to economics and health, the two most obvious indicators of the quality of life and the human condition.

## Economic Disparities

A common way to compare the relative economic status of nations is to use the gross national product (GNP) per capita per year. Although GNP per capita is not the same as income per capita, it is generally accepted as a comparative measure of the economic status of a nation's peoples and is used to separate nations into different income groups (World Tables, Vol. I, 3rd Ed., Economic Data, 1983). Gross national product per capita is certainly not an infallible index of the quality of life, and attempts to relate the two have been criticized on the basis that the many items that enter into GNP, such as excessive military production and burgeoning costs of medical technology, do not necessarily translate into a higher quality of life for a nation's citizens (Miller, 1982). Nonetheless, many other aspects of living standards are related to GNP, and this chapter will explore a number of these relationships.

In 1980, the GNP per capita of different nations varied from a high of $30,070 (United Arab Emirates) to a low of $80 (Bhutan). The worldwide average in 1980 was $2,430 and the worldwide median was $1,340 (Kurian, 1984). Out of 171 nations for which data are available, 20 had GNPs per capita over $10,000, and these were all nations with economies dominated by industry, technology, or oil export. The United States ranked fourteenth with a GNP per capita of $11,360.

At the other end of the economic scale, 54 nations had GNPs per capita in 1980 of less than $600. Table 1 provides a sample of representative nations in five different categories of GNP per capita: high, upper middle, middle, lower, and very low. These categories obviously contain nations of very diverse economics, geography, and culture; they are based solely on the criterion of GNP per capita and are not intended to imply any other similarities. They demonstrate, however, the very great disparities of economic wealth that exist between different nations.

The argument is sometimes made that per capita GNP or per capita income figures do not mean much when compared internationally because basic commodity prices vary so much.

**Table 1.** Economic indicators of selected nations in five different categories of gross national product per capita.

| Nation | GNP/capita per year (U.S. dollars)[a] | Energy consumption per capita per year (kg coal equiv.)[b] | Food consumption per capita per day (calories per day)[b] |
|---|---|---|---|
| | *High GNP/capita* | | |
| United Arab Emirates | 30,070 | 9,289 | n.a.[c] |
| Kuwait | 22,840 | 11,977 | n.a. |
| Switzerland | 16,440 | 5,223 | 3570 |
| West Germany | 13,590 | 6,053 | 3561 |
| United States | 11,360 | 11,626 | 3658 |
| | *Upper middle GNP/capita* | | |
| Japan | 9,890 | 4,649 | 2912 |
| United Kingdom | 7,920 | 5,363 | 3338 |
| Italy | 6,480 | 3,725 | 3778 |
| Spain | 5,350 | 2,944 | 3319 |
| Soviet Union | 4,550 | 6,422 | 3372 |
| | *Middle GNP/capita* | | |
| Venezuela | 3,630 | 3,039 | 2764 |
| Iraq | 3,020 | 1,220 | 2677 |
| Portugal | 2,350 | 1,822 | 3155 |
| Mexico | 2,130 | 1,684 | 2819 |
| Brazil | 2,050 | 1,101 | 2614 |
| | *Low GNP/capita* | | |
| Malaysia | 1,670 | 881 | 2700 |
| Syria | 1,340 | 964 | 2907 |
| Ecuador | 1,220 | 692 | 2010 |
| Jamaica | 1,030 | 1,440 | 2662 |
| Nicaragua | 720 | 362 | 2234 |
| | *Very low GNP/capita* | | |
| Bolivia | 570 | 452 | 2084 |
| Kenya | 420 | 208 | 2034 |
| China, Peoples Republic of | 290 | 618 | 2527 |
| India | 240 | 210 | 1933 |
| Ethiopia | 140 | 24.5 | 1781 |
| Bangladesh | 120 | 49 | 1944 |

[a]Data from Kurian (1984).
[b]Data for 1981, from World Tables (1983).
[c]n.a., Not available.

This is only partially true, and it does not really alter the fact of tremendous economic inequality. Frequently basic commodities are more expensive in middle or lower income countries than in high income countries. In 1982, for example, the consumer price indices for food and clothing were, respectively, 81.2 and 209.8% higher in Italy than in the United States, even though Italy's GNP per capita was 43% lower than that of the United States (Statistical Abstract of the U.S., 1984). There are frequent occasions when rice and potatoes are more expensive per pound in India than in the United States.

Although the GNP per capita is a somewhat abstract statistic, it is interesting to note how many other indicators are related to this basic economic measurement. Tables 1 through 3 present other numerical facts broadly related to the quality of life and health: energy use, food consumption, living space, percentage of dwellings with piped water, percentage with electricity, percentage of literacy, life expectancy, infant mortality, hospital beds, and physicians. These are certainly not a comprehensive measure of living standards, and many intangibles enter into the elusive phrase "quality of life," but there would probably be general agreement that these represent some relative measures of the living conditions. We can obviously identify some characteristics of poverty and low living standards: low food consumption per capita, crowded living conditions, a high percentage of dwellings without water and electricity, low literacy, low life expectancy, high infant mortality, and few hospitals and physicans. Other measures are also of obvious importance, such as employment rates, educational levels, and disease rates; but these basic parameters are especially difficult to measure, and comparative data on national scales are lacking.

### Energy Use and Food Consumption

Energy use per capita shows some broad relationships to national GNP. In general, the high income countries exhibit high energy use, usually on the order of 5000 to 12,000 kilograms of coal equivalent per person per year (Stat. Abs. of U.S., 1984). By world standards, these are exorbitant levels of energy consumption, since the average of the developing economies was only 587 kilograms of coal equivalent per person per year in 1981 (World Tables, 1983). The energy uses per person per year for some nations were considerably lower than this: India (210), Bangladesh (49), and Ethiopia (24.5). Although these are primarily agricultural nations, India has been making great progress in industrialization and would certainly be higher now in per capita energy use than in 1981. Even agricultural enterprise requires increasing energy to be productive, however, so there are strong ties between energy use and crop productivity when viewed in these extremes. In Ethiopia, for example, where 75% of the population is registered as living in absolute poverty (Kurian, 1984), the relationships between low GNP, low energy use, poverty, and famine are too strong to ignore. Ecologically, one must conclude that this is not only a problem of economic management, it is also a problem of land misuse and drought. In 1984 and 1985, we have had tragic pictures of a degraded and exhausted landscape, devoid of vegetation, in which the human population has had no ability or opportunity to withstand a serious drought—a drought exacerbated by decades and even centuries of land abuse and further aggravated by several years of civil war.

Food consumption patterns also show relationships to GNP and energy use. Most of the high and middle income nations have daily food consumption on the order of 3000 to 3600 calories per person per day (Table 1). Japan is an exception, with only 2900 In some nations, such as the United States, both food consumption and energy use are too high to be healthy. Our pollution would be less and our national health would be better if both were lowered. Several other nations, including Sweden, Switzerland, and Australia all show greater longevity with less energy consumption and less food consumption per capita than the United States.

The other extremes are even more obvious, however. Many of the low and very low income countries with low energy use have very low food consumption per capita, usually fewer than 2500 calories per person per day, and sometimes fewer than 2000. India, for example, had a daily per capita food consumption of only 1933 calories in 1981, Bangladesh, 1944 calories, and Ethiopia, one of the lowest, with only 1781 calories. In 1985, the Ethiopian data will probably be even worse with extensive famine conditions. The Food and Nutrition Board of the National Research Council considers 2100 to 2700 calories per day as a normal daily per capita diet necessary to maintain reasonable health and function (Goodhart and Shils, 1980).

### Living Conditions

Other objective measures of living standards on which international data are available include living space, dwellings with piped water, and dwellings with electricity. Many other statistics can be obtained, including data on items such as telephones, refrigerators, radios, television sets, and automobiles per 1000 people, but most of these show the same general trends as space, water, and electricity.

With a few exceptions, living space, as defined by the numbers of individuals per room, shows an inverse relationship to income pattern. The high income and upper middle income countries show less than 1 person per room. Japan and the oil-rich nations are exceptions. Most of the low income nations have 1 to 2.5 individuals per room, and the very low often have over 2.5 persons per room. There are, of course, very significant cultural differences, such as the tendencies toward extended families, as well as differences in the sizes of rooms and even what constitutes a room in tropical houses with open space. Hence, this measure of crowding cannot be considered simply an economic variable. Nonetheless, crowding does have ecological components, especially in cities, where, for example, both ecological and economic constraints may impose severe crowding. In Calcutta, for example, urban population densities exceed 100,000 persons per square mile, even in low-rise housing areas, with an average of over 3 persons per room and less than 40 square feet of living space per person (Moorhouse, 1983; Kaylor, 1984).

Data on water and electricity are quite incomplete, but they tend to show that the high and upper middle income nations have over 90% of the dwellings with both piped water and electricity. In West Germany, for example, over 99% of the dwellings have both. The oil-exporting nations are an exception, and both Kuwait and the United Arab Emirates have water in less than one-third of their dwellings. Many of the middle and low income nations tend to run around the 50% figure for both water and electricity, although again there are a number of exceptions. Data are not available for many of the very low income nations, although Bolivia, with water and electricity in only

**Table 2.** Selected indicators of physical standards of living.[a]

| Nation | Living space: persons/room | Percentage dwellings without piped water | Percentage dwellings without electricity | Physical quality of life index[b] |
|---|---|---|---|---|
| United Arab Emirates | — | 69.2 | 75.8 | 65 |
| Kuwait | 2.1 | 82.5 | — | 77 |
| Switzerland | 0.6 | 3.9 | — | 97 |
| West Germany | — | 0.8 | 0.1 | 94 |
| United States | 0.6 | 2.5 | — | 96 |
| Japan | 1.1 | 1.7 | — | 98 |
| United Kingdom | 0.6 | 6.4 | — | 95 |
| Italy | 0.9 | 10.6 | 1.0 | 95 |
| Spain | 0.9 | 22.2 | — | 92 |
| Soviet Union | — | — | — | 90 |
| Venezuela | — | 27.6 | 21.6 | 81 |
| Iraq | — | 79.2 | 82.9 | 47 |
| Portugal | 0.8 | — | 35.8 | 81 |
| Mexico | 2.5 | 50.6 | 41.1 | 78 |
| Brazil | — | 59.8 | 44.3 | 74 |
| Malaysia | 2.6 | 52.5 | 56.6 | 69 |
| Syria | 0.4 | 59.8 | 58.3 | 64 |
| Ecuador | 2.3 | 66.6 | 58.8 | 71 |
| Jamaica | 1.9 | 54.0 | — | 88 |
| Nicaragua | — | 66.8 | 59.1 | 66 |
| Bolivia | — | 62.1 | 67.0 | 49 |
| Kenya | 2.5 | — | — | 53 |
| China, Peoples Republic of | — | — | — | 76 |
| India | 2.8 | — | — | 44 |
| Ethiopia | 2.7 | 25.7 | 41.3 | 20 |
| Bangladesh | — | — | — | 37 |

[a]Data from Kurian (1984)

[b]The PQLI is a concept developed by the Overseas Development Council and is calculated by averaging life expectancy, infant mortality, and literacy, each rated on a scale from 1 to 100 from worst to best. Life expectancy and literacy are directly proportional; infant mortality is inversely proportional to the PQLI. The scores are not weighted, and the PQLI is considered free of biases and distortions that affect other measures weighted to the GNP.

about one-third of its dwellings, may be typical. Ethiopia seems to be an enigma because it reports piped water in nearly 75% and electricity in almost 60% of its dwellings. This may represent national optimism, or sampling restricted to towns and cities, and it highlights problems common to many of these data. As a general rule, very low income nations have fewer than 50% of their dwellings with water and electricity.

## Education and Health

The quality of life is very much related to education and health, but these are difficult to measure because standards differ so much and primary data are often lacking. Some data are available, however, that relate to levels of education and health (Table 3). Literacy is one measure of educational status, even though national definitions vary considerably. Some nations define literacy as the ability to read and write; other nations define it as the ability to read only; and still other nations, as the attendance of school at any time in one's life. Even with these differences of definition, data on literacy show striking relationships to economic status. Most of the high and upper middle income nations show literacy rates above 90%, and many, such as the United States, West Germany, and the United Kingdom, have 99% literacy. Switzerland and the Soviet Union report 100% literacy, although the figure for the Soviet Union may be questioned in a nation of such great size and diversity. The exceptions to the relationship between high income and high literacy are some of the oil exporting states, such as the United Arab Emirates, which report only 18% literacy.

Many of the middle and low income nations report literacy rates from 25 to 86%, with averages around 50 to 60% being most common. Another problem in literacy data that give a single figure for any nation, is the significant difference that occurs between the sexes. In Syria, for example, male literacy is reported as 60% while female literacy is only 20%. In a few rare cases, female literacy is higher than male literacy; for example, in Jamaica female literacy is listed as 85% and male 79%. In the lowest income countries, literacy exists at only low levels: 20% in Kenya, 21% in Bangladesh, and 6% in Ethiopia.

**Table 3.** Selected data on health and education, 1980–1984.

| Nation | Life expectancy at birth[a] | Infant mortality[b] | Hospital facilities: population per bed[c] | Medical care: population per doctor | Percentage literacy | | |
|---|---|---|---|---|---|---|---|
| | | | | | Adult males | Adult females | Average |
| United Arab Emirates | 61.5 | 57.0 | 342 | 243 | 27 | 9 | 18 |
| Kuwait | 68.9 | 24.1 | 257 | 783 | 68 | 48 | 58 |
| Switzerland | 75.3 | 8.6 | 87 | 498 | 100 | 100 | 100 |
| West Germany | 73.3 | 14.7 | 85 | 490 | 99 | 99 | 99 |
| United States | 73.8 | 13.0 | 159 | 595 | 99 | 99 | 99 |
| Japan | 76.5 | 8.0 | 94 | 845 | 99 | 97 | 98 |
| United Kingdom | 73.5 | 12.6 | 120 | 659 | 99 | 99 | 99 |
| Italy | 72.8 | 15.3 | 97 | 485 | 95 | 93 | 94 |
| Spain | 72.3 | 15.1 | 185 | 557 | 94 | 86 | 90 |
| Soviet Union | 69.0 | 8.5 | 82 | 289 | 100 | 100 | 100 |
| Venezuela | 67.3 | 33.7 | 429 | 874 | 80 | 73 | 76.5 |
| Iraq | 55.1 | 30.0 | 496 | 2,208 | 36 | 13 | 24.5 |
| Portugal | 69.0 | 38.9 | 189 | 704 | 78 | 65 | 71.5 |
| Mexico | 65.4 | 60.2 | 863 | 1,251 | 78 | 70 | 74 |
| Brazil | 59.3 | 84.0 | 245 | 1,647 | 69 | 63 | 66 |
| Malaysia | 69.8 | 31.8 | 273 | 7,302 | 59 | 48 | 53.5 |
| Syria | 64.2 | 67.2 | 956 | 2,514 | 60 | 20 | 40 |
| Ecuador | 60.7 | 72.1 | 495 | 1,621 | 78 | 70 | 74 |
| Jamaica | 64.6 | 16.2 | 260 | 3,505 | 79 | 85 | 72 |
| Nicaragua | 55.3 | 42.9 | 474 | 1,465 | 58 | 57 | 57.5 |
| Bolivia | 48.7 | 138.2 | 526 | 2,118 | 75 | 51 | 63 |
| Kenya | 49.0 | 92.0 | 773 | 11,417 | 30 | 10 | 20 |
| China (PRC) | 67.3 | 48.7 | — | 2,602 | — | — | — |
| India | 45.5 | 122.0 | 1465 | 3,652 | 47 | 19 | 33 |
| Ethiopia | 39.0 | 84.2 | 3314 | 73,191 | 8 | 4 | 6 |
| Bangladesh | 46.2 | 139.0 | 4505 | 12,378 | 33 | 9 | 21 |
| Zambia | 48.3 | 259.0 | 273 | 10,373 | — | — | — |

[a]Life expectancy data calculated as average of both sexes, taken from 1982 United Nations Demographic Yearbook, 34th Issue, published 1984, New York, N.Y. 10017.

[b]Infant mortality data expressed as number of deaths of infants under one year of age per 1000 live births. Data from Kurian (1984), and 1982 United Nations Demographic Yearbook, 34th Issue, 1984.

[c]Data on hospital facilities, medical care, and literacy from Kurian (1984).

In the area of public health, international statistics on disease rates and the causes of death are published annually by the World Health Organization in Geneva, but in many countries it is difficult to evaluate the completeness of the data. Certainly most illnesses are not recorded officially and the causes of many deaths are undiagnosed. The reasons for this are a shortage of health personnel and diagnostic facilities in many developing countries, a lack of consistent record keeping, and finally the problem of multiple morbidity and mortality factors. This simply means that individuals often suffer from several illnesses at the same time. A child in a tropical nation, for example, may have a respiratory tract infection (such as flu or bronchitis) along with an intestinal infection (such as dysentery or gastritis), various parasitic infections, and perhaps some aspects of malnutrition. If the child dies at 18 months of age, the probable cause may be an interaction of pneumonia or measles, along with diarrhea, malaria, and malnutrition. Hence, it is often impossible to assign any single disease or cause of mortality. Furthermore, many developing nations have more than 10,000 people per physican, and most of these doctors have quite inadequate diagnostic facilities.

For these reasons, it is better to look at the broader indices of health—life expectancy, infant mortality, and medical facilities, hospital beds and physicians—available to the population. None of these necessarily reflects the complete health status of a population, but together they provide an indication of public health. When coupled with other facts, they help to portray both national and global conditions.

Life expectancy at birth usually varies directly with the economic status of a country. The high income and upper middle income nations typically have life expectancies from 72 to 76 years. The major exceptions are some of the Gulf oil-exporting nations and the Soviet Union, which have life expectancies from 60 to 69. Most of the middle and lower income nations have life expectancies from 55 to 69 years, whereas the very low income nations range primarily from 40- to 50-year life expectancies (Table 3). The Peoples Republic of China is an exception to the latter category, reporting an average longevity of 67 years. Ethiopia, Nigeria, Togo, Upper Volta, and Kampuchea (Cambodia) have the unfortunate distinction of having some of the lowest life expectancies in the world, respectively recorded at 39, 37, 35, 32, and 30 years of age in the 1982 United Nations Demographic Yearbook (published in 1984). Sexual differences in longevity exist, and with only a few national exceptions, males have shorter life expectancies than females. Bhutan, Liberia, Mongolia, Iran, and Nigeria are among the few exceptions in which males have longer life expectancies than women.

Infant mortality is often used as one of the best indicators of general health conditions in a population because it integrates several aspects of sanitation, nutrition, maternal health, and medical care (World Health Statistics Annual, WHO, Geneva, 1983). Infant mortality rates are usually inversely related to income status. Infant mortality (deaths in the first year of life) in the high and upper middle income nations is generally lower than 15 deaths per 1000 live births (Table 3). The Gulf oil-exporting nations are again an exception, with the United Arab Emirates at 57 and Kuwait at 24.1. The middle and lower income nations usually have infant mortalities in the range of 30 to 80, with a few exceptions at both extremes. Jamaica, for example, has a low infant mortality of only 16.2, whereas Turkey, with a comparable GNP per capita, showed an infant mortality rate of 153 in recent years.

The lowest income nations frequently have infant mortality rates of 100 to 250 under typical conditions, and these certainly increase in times of famine or epidemic disease. It is not uncommon in the world's poorest nations for 20 to 25% of the infants to die in their first year.

There are some notable exceptions to the general relationship between low income status and high infant mortality. Hong Kong, Singapore, Spain, and the United Kingdom, all with GNPs per capita considerably less than that of the United States, also have lower infant mortality rates. Our own national record on infant mortality is not exemplary despite one of the highest states of medical technology in the world. Some of the European nations with a GNP per capita equal to or exceeding our own, such as Denmark, Norway, and Sweden, have infant mortality rates approximately one-half that of the United States.

Health facilities and medical personnel also provide a general indicator of the quality of medical care and health status. The population per hospital bed in the high and upper middle income nations usually runs from 80 to 200 individuals. In the middle and low income nations, the same statistic is typically 200 to 1000, and in the lowest income nations, it is often over 1000, with some nations as high as 4500 persons per hospital bed. Exceptions occur, of course, but again the general picture is one of a strong link between economic status and hospital facilities.

The association between economic status and availability of medical personnel is even stronger. Most of the high and upper middle income nations have fewer than 800 persons per physician. The Soviet Union and United Arab Emirates have two of the lowest population to physician ratios—fewer than 300 persons per physician, compared to 595 in the United States and 845 in Japan.

Most of the middle income nations have 1000 to 2000 persons per physician, and most of the low income nations have 1000 to 8000. The dubious record is probably held by Ethiopia, with 73,191 persons per physician (Kurian, 1984). Similar relationships between economic status and health personnel exist in other professional categories—dentists, nurses, medical technicians. The numbers of health personnel needed in various professions is a complicated matter, influenced by the health profiles of a given country, its age distribution, and its conditions of sanitation, housing, nutrition, employment, and so forth; but in general terms, it is difficult for any nation to provide adequate health care with more than 1000 persons per physician. The global average in a recent year was 1237 persons per physician, with one-half the world's nations showing a population of 2000 or more persons per physician (Kurian, 1984).

### Global Health Patterns

Two disparate patterns exist in world health patterns. The industrial and technological nations of the world have chronic, noninfectious diseases as their most serious health problems and most frequent causes of death. In the United States, more than two-thirds of all adult mortality is attributable to various kinds of cancer and to cardiovascular diseases (Statistical Abstract of the U.S., 1984). Other primary causes of illness and death are accidents and metabolic diseases such as diabetes. Our most serious medical problems stem from behavior patterns and lifestyle: smoking, alcoholism, dietary habits, drug abuse, automobile use, crime, and pollution. These patterns are common in most of the industrialized nations of the world.

In the developing nations, especially those of low income status, the primary health problems involve infectious disease and nutrition. Tuberculosis, common respiratory diseases, malaria, schistosomiasis and other parasitic infections, diarrhea, dysentery, and protein–calorie–vitamin malnutrition are among the most common forms of illness and death. In malnourished children, measles, chickenpox, influenza, and pneumonia are frequent causes of death. This is not to say that cancers, cardiovascular disease, diabetes, accidents, alcoholism, and crime do not exist in nations of lower income status—they are often common—but they are frequently overridden by infectious and parasitic disease. Nor is it accurate to say that the industrialized world has eliminated infectious disease—the most advanced nations economically and industrially have not conquered tuberculosis and influenza, nor have they controlled sexually transmitted infectious diseases. Nonetheless, the most pressing medical needs of the high income nations differ markedly from those of the low income nations.

In 1967, the editors of *Johns Hopkins Magazine* produced an issue on world health and poverty built around the theme "The War Against Suffering," and it contained these stark facts (Neville, 1967):

> Two-thirds of the human race live in the underdeveloped areas of the world.
>
> Most of them drink unsafe water, prepare food dangerously, dispose of wastes recklessly, and live in unfit dwellings.
>
> In some parts of the developing world, every living individual is afflicted with intestinal parasites.
>
> In some parts of the developing world, half of the children born never reach their fifth birthday.
>
> Within the developing world at least 400 million people suffer from trachoma, a curable disease that causes progressive loss of sight.
>
> Within the developing world, despite a massive campaign by the World Health Organization to eradicate malaria, 400 million people live in malarious areas where no eradication program is under way. In some of these areas, most people have the disease.
>
> Schistosomiasis, a debilitating parasitic disease, afflicts 200 million people—more than the population of the United States [in the 1960 census, the population of the United States was 179,323,175; estimated to be 196,759,000 in 1966]. In some areas, the chances of escaping the disease are only one in five.
>
> Leprosy afflicts more than 10 million people; fewer than one in five are receiving any kind of treatment. In the countries where leprosy is prevalent, 740 million people are exposed to the risk of infection.
>
> Cholera, a disease on the decline in the 1950s, is spreading. The number of reported cases tripled between 1960 and 1964.
>
> Venereal diseases are spreading into the underdeveloped areas of the world, to an extent not yet measured.
>
> This is, in part, the health situation amid two-thirds of the world's people. By the year 2000, if present demographic trends continue, it could be the story of four-fifths of the world's people.

These statistics were compiled almost 20 years ago and conveyed concern about the status of the world's people in the year 2000. It is reasonable to ask how we are doing now.

Current statistics give cause for both encouragement and discouragement. The world has made progress in a number of areas of public health. Most notable is the eradication of smallpox. A viral disease once responsible for the disfiguration and death of millions of people around the world, smallpox is the first and only infectious disease to be completely eradicated from the world. The job was undertaken by the World Health Organization in the mid-1960s, at a relatively modest cost of $10 million dollars per year. Within 15 years, the last verifiable case of smallpox occurred. This phenomenal achievement was accomplished by a combination of vaccination, quarantine, and surveillance, and it now saves between $1 and $2 billion per year in medical costs (D. A. Henderson, personal communication)—a very cost-effective public health measure in economic terms alone, not to mention the savings in human misery.

The world has also made progress in the control of cholera. A bacterial infection spread by contaminated food and water, and formerly a disease of tragic epidemic proportions, cholera is now relatively rare. When it does occur, cholera is treated fairly easily by fluid replacement and antibiotic therapy. Other advancements in global medicine and public health have been made—better nutrition for many of Asia's 2 billion people, reductions in infant mortality in most of the world's nations, and a significant extension of life expectancies in many countries.

The world health picture is not all encouraging, however. The famines of Ethiopia, Chad, and other African nations in 1984 and 1985 are stark reminders that world food supplies are still not adequate nor properly distributed. The current situation is described by King (1983) as follows: "In spite of the 'green revolution' more people are hungry, even starving, than at any time in the past. A billion people are said to be malnourished, and 400 million on the brink of starvation. The mortality rate of children between one and five, which is perhaps the best indicator of nutrition, is 10 to 40 times higher in parts of Asia, Africa, and Latin America than it is in Europe or the United States. In Africa alone, 30 percent of children are estimated to be clinically underweight for their age, and 4 percent of them seriously so, with either kwashiorkor or marasmus. Nevertheless, the population continues to increase, often at 3 percent or more, in countries least able to increase food production."

Data on parasitic and zoonotic diseases are also alarming. Between 1962 and 1980, worldwide reported cases of malaria increased from 3,289,777 to 14,013,045 (World Health Statistics Annual, 1983). Still, the numbers of reported cases are a small percentage of total cases, and most experts feel that there are over 200 million cases of malaria in the world, with 1.5 million deaths annually (Katz et al., 1982). Nearly 2.5 billion people live in areas of active malaria transmission. In sub-Saharan Africa alone, 270 million people are exposed to malaria without any medical protection, and nearly a million children die each year from malaria (Weatherall et al., 1983).

Between 200 and 300 million people in the world have schistosomiasis, and more than 400 million are infected with filariasis (Katz et al., 1982), a parasitic disease transmitted by mosquitos. A billion people are known to suffer from helminthic infections, including schistosomiasis and filariasis. In Sri Lanka, Bangladesh, and Venezuela, recent studies show that over 90% of six-year-old children are infected. About 700 million people are thought to have hookworm and probably more have *Ascaris* infections, both parasitic worms of the gastrointestinal tract (King, 1983). These discouraging data provide an alarming credibility to the concern that the world may face 3 billion cases of human helminthic infections by the year 2000, a concern first expressed by Norman Stoll of the Rockefeller Foundation in 1947 and more recently reiterated by Andrew Davis (1984).

The prevalences of intestinal protozoan infections, such as amoebiasis and giardiasis, are unknown, as are those for the enteric bacterial and viral infections, such as shigellosis, hepatitis, and viral gastroenteritis, but they certainly number in the tens of millions annually. In 1975, 500 million cases of diarrhea occurred in the children of Asia, Africa, and Latin America, causing between 5 and 18 million deaths, "a situation comparable to that in the industrial world at the end of the last century" (King, 1983).

The prevalence of injury and disability is also shocking. More than 42 million people of the world are blind (almost 1% of the world's population), and 80% of these cases are preventable (WHO, 1982). On a global basis, blindness is caused primarily by trachoma (an intracellular bacterial disease of the eye), onchocerciasis (caused by a parasitic worm transmitted by the black fly), and vitamin A deficiencies.

These discouraging facts prompted Professor Norman Scotch of Boston University School of Public Health to state in 1984 that the three major breakthroughs in medical care are going to be "prevention, prevention and prevention." He was referring primarily to cardiovascular disease, but the concept is also applicable to worldwide public health.

### Trends in the Development Gap

The striking disparities between nations in economic status, standards of living, and public health raise the questions of whether the gaps between the rich and poor are widening or narrowing. Recent books have focused on this question (Cole, 1982; Weatherall et al., 1983). The development gap has certainly been closed in a most convincing way in a number of cases: Japan, Taiwan, Hong Kong, and Singapore are all nations that have made remarkable economic gains in the past 20 years. In many other nations, the picture is not at all encouraging.

On the issue of food consumption per capita, a sample of five high and upper middle income nations (United States, Japan, West Germany, Spain, and Italy) had an average increase of food consumption per capita of 12.8% between 1960 and 1980. A sample of five lower income nations (Syria, Iraq, Brazil, Bolivia, and

China), achieved the same, an increase of 12.7% in food consumption per capita. A sample of the lowest income nations, however, lost 13.3% in food consumption per capita in this same period (Chad, Ethiopia, Bangladesh, Kenya, and India).

Other economic indicators show a similar problem. From 1960 to 1980, the GNP per capita in the United States increased 306% (from an index level of 2979 to 11,363), whereas that of India increased only 207% (from 69 to 226). Hence, as great as the economic disparity between India and the United States was in 1960, it was even greater in 1980.

In regional terms, the GNP per capita of 53 African nations increased 454% from 1960 to 1979, but during the same period, the GNP per capita of 20 European nations increased 726%. Again, in relative terms, Africa continued to fall behind Europe in the past two decades; in other words, the development gap widened.

The most remarkable economic gains have been made by some of the Far Eastern nations, and these represent the greatest exceptions to the widening development gap. From 1960 to 1980, the GNP per capita of Singapore increased 969%; Hong Kong increased 1125%; and Japan increased a phenomenal 1821% (Statistical Yearbook of the United Nations, 1983).

The World Bank has projected trends in per capita GNP as shown in Figure 1, and they concluded that even if optimistic rates of economic growth are achieved in the developing world, there will still be 600 million people living in "absolute poverty" by the year 2000. This is defined by the world bank as "a condition of life so characterized by malnutrition, illiteracy, disease, high infant mortality, and low life expectancy as to be beneath any reasonable definition of human decency" (King, 1983).

## Summary and Conclusions

Global comparisons of economic status, standards of living, and basic health patterns show striking disparities between nations. High and upper middle income nations tend to have the highest energy consumption, highest food consumption per capita, and highest standards of living in physical terms. In education and health, these same nations also show the highest levels of literacy and life expectancy, and the lowest rates of infant mortality and infectious disease. Their health problems are characterized more by chronic diseases and accidents related to lifestyles. National exceptions to most of these generalizations can be found. The oil-exporting nations of the Gulf, for example, have very high GNPs per capita, but otherwise share many characteristics with low income developing nations such as moderately high illiteracy and infant mortality rates.

Although many nations of all economic conditions have made progress in the past 20 years, the greatest advancements by far have been made in those nations with economies dominated by industry and technology. In fact, in many discouraging cases, the development gap is widening; that is, the richest nations of the world continue to increase their economic status more rapidly than the poorest nations.

Particularly difficult situations occur in the lowest income nations with the highest rates of population growth. These nations also have the most rapid deterioration of their environments. With their basic resource bases in nonproductive and often exhausted conditions, especially soil, water, and agriculture, some of these nations represent tragic circumstances. Their economic, ecologic, and humanistic resources are often depleted, or undeveloped, and in a costly state of maintenance. Popula-

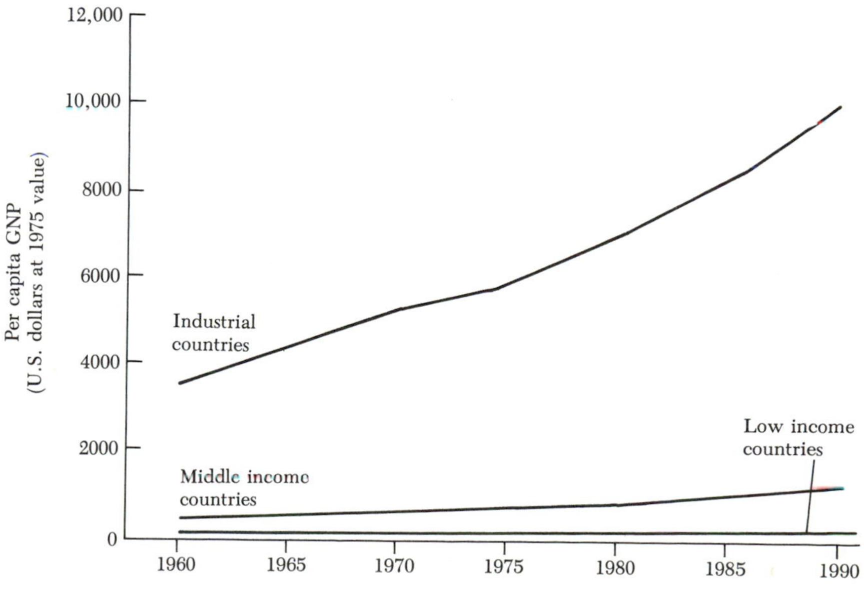

Per capita GNP (U.S. dollars at 1975 value)
12,000
10,000
8000
6000
4000
2000
0
1960 1965 1970 1975 1980 1985 1990
Industrial countries
Middle income countries
Low income countries

**Figure 1** Trends in per capita gross national product as projected by the World Bank (1979).

tions in many of these countries lack the human resources in terms of literacy, education, and health to solve the problems they face. It is certainly more expensive to restore soil fertility, agricultural productivity, and water resources—as it is to cure advanced disease states—than it would have been to prevent these problems in the first place. Yet these are the costs and the problems that the world now faces. The international community must find new techniques of ecologic and economic restoration that will eventually narrow the development gap—that will permit the low income nations of the world to achieve more of the productive benefits of the global economic community.

### References

Cole, J. P. 1982. The Development Gap: A Spatial Analysis of World Poverty and Inequality. New York: John Wiley. 454 pp.

Davis, Andrew. 1984. This wormy world. World Health. Geneva, Switzerland: WHO. March 1984, pp. 2–3.

Goodhart, R. S. and M. E. Shils (editors). 1980. Modern Nutrition in Health and Disease. Philadelphia: Lea and Febiger. p. 817.

Katz, M., D. D. Despommier, and R. W. Gwadz. 1982. Parasitic Diseases. New York: Springer-Verlag. 264 pp.

Kaylor, Robert. 1984. Calcutta. U.S. News and World Report, July 9, 1984, pp. 45–46.

King, M. H. 1983. Medicine in an Unjust World. *In* Oxford Textbook of Medicine. D. J. Weatherall, J. G. G. Ledingham, and D. A. Warrell (editors). Oxford: Oxford University Press. pp. 3.3–3.8.

Kurian, G. T. 1984. New Book of World Rankings. New York: Facts on File Publications. 490 pp.

Mahler, H. 1983. The Marathon of Health for All. WHO Chronicle 37(6): 187–191.

Miller, W. T. 1982. Living in the Environment, 3rd Ed. Belmont, Calif.: Wadsworth Publishing Company. 500 pp.

Moorhouse, G. 1983. Calcutta—The City Revealed. New York: Penguin Books. 393 pp.

Neville, A. 1967. The War against Suffering. Johns Hopkins Magazine, Summer 1967.

Parkin, D. B., J. Stjernsward, and C. S. Muir. 1984. Estimates of the Worldwide Frequency of 12 Major Cancers. Bulletin WHO 62(2): 163–182.

Statistical Abstract of the United States. 1984. U. S. Dept. of Commerce. Washington, D.C.: U. S. Government Printing Office, pp. 79, 853–885.

Statistical Yearbook of the United Nations. 1983. 32nd Issue, for the year 1981. New York: U. N. Publishing Division. 1070 pp.

United Nations Demographic Yearbook. 1984. 34th Issue for the year 1982. New York: U. N. Publishing Division. 1043 pp.

Weatherall, D. J., J. G. G. Ledingham, and D. A. Warrell. 1983. Oxford Textbook of Medicine, Vol. 1. Oxford and New York: Oxford University Press.

World Bank. 1979. World Development Report. Washington, D.C.

World Bank. 1983. Data Files. Baltimore: Johns Hopkins University Press.

World Health Organization. 1982. Seventh General Programme of Work—covering the Period 1984–1989. WHO: Geneva, Switzerland. 153 pp.

World Health Statistics Annual. 1983. WHO: Geneva, Switzerland. 807 pp.

World Tables. 1983. Social and Economic Data, 3rd Ed., Vol. 1. Baltimore and London: Johns Hopkins University Press. 585 pp.

CHAPTER 20

# Improving World Health: A Least Cost Strategy

WILLIAM U. CHANDLER

## Introduction

World health leaders have set a goal of "health for all by the year 2000," a step that has initiated a global effort to define health and to devise ways to achieve it.[1] One natural place to begin is with the world's principal causes of death. A large number of early deaths are preventable, and many more lives can be extended into old age. These lives can be saved for a surprisingly low cost.

Death, as characterized by Euripides in his tragedy *Alcestis*, said, "Those who could afford a late death would buy it." Nations that could afford it have invested heavily in medicine and sanitation, and as a result, their children are ten times more likely to survive to adulthood than the children of the least affluent.[2] Adults in richer lands also live longer, although, ironically, many now die prematurely of diseases associated with affluence.

Though their health care needs differ drastically, the rich and the poor do have one thing in common: both die unnecessarily. The rich die of heart disease and cancer, the poor die of diarrhea, pneumonia, and measles. (See Table 1.) Scientific medicine could vastly reduce the mortality caused by these illnesses. Yet, half the developing world lacks medical care of any kind.[3] The rich, though infinitely better cared for, have had to begin rationing health care in the face of rising costs. A policy of health for all can succeed, therefore, only if limited resources are used in the most efficient way possible.

I would like to thank Brian Brown and Susan Norris for assistance in the preparation of this paper, and George Alleyne, David Banta, Clyde Behney, Peter Bourne, John Foggle, Hellen Gelband, Davidson Gwatkin, Holly Gwin, James Heiby, Judith Jacobsen, Michael Jacobson, and Larry Miike for reviewing the manuscript.



William U. Chandler is Senior Researcher, Worldwatch Institute.

**Table 1: Causes of Death In Rich and Poor Countries, 1980**

| Disease or Cause of Death | Developed Countries[1] | Developing Countries[2] |
|---|---|---|
| | (percent) | |
| Diarrheal Infections and Parasites | 1 | 17 |
| Respiratory Infections | 8 | 18 |
| Heart Disease and Strokes | 48 | 18 |
| Cancer | 21 | 8 |
| Accidents | 7 | 7 |
| Other | 15 | 32 |
| Total | 100 | 100 |

[1] Seventy-five percent of all developed countries. [2] Chile, Peru, Mexico, Iran, and Philippines, unweighted average.

**Source:** World Health Organization Data Bank.

Death is inevitable, of course, but its occurence in either infancy or middle age has importance beyond the obvious meaning for the individual. High infant mortality rates portend a generation with reduced capacity for learning and work—for human development. Reduced mortality rates indicate a general improvement in health.

Likewise, reducing catastrophic heart attack in middle age not only prolongs and improves the lives of individuals, but saves for society its most productive members.

Impressive improvements in human health have been made recently in rich and poor countries alike. Since 1960, average life expectancy in 34 of the world's poorest countries has increased from 41 to 50 years. Much of this improvement is due to reduced infant mortality, brought about by adoption of modern medicine and improving economic conditions. With the worldwide economic recession, these improvements have slowed somewhat.

In developed nations, life expectancy was increased from an average of 70 to more than 75 years.[4] Several countries, including the United States, France, and the Netherlands, now spend over 8 percent of their entire economic output on health care. Developed nations, in fact, now spend more money on health care than the poorest half of the world spends on all items. The poor countries, many of which already allot half of their health care budgets to hospital-based care that reaches only a few, cannot hope to provide high levels of care in the near future. Waiting for incomes to rise will work far too slowly to help the present generation, because half the world cannot reasonably expect to earn more than $1,500 per capita by the year 2010.[5] At the same time, China, Costa Rica, Cuba, Chile, Sri Lanka and others have shown that major improvements in health can be accomplished at low income levels. Statistical analysis shows a low correlation between rates of infant death and income levels up to $2,000 per capita.[6]

**"Developed nations now spend more money on health care than the poorest half of the world spends on all items."**



Fortunately, major improvements in world health can be made with cost-effective preventive and primary care measures. The most important of these are providing maternal and child care for the world's poorest people, clean drinking water and sanitation facilities to the third of the world's population that lacks them, diet education for populations at high risk of heart disease and cancer, control of tobacco products, and basic research for low-cost cures.[7]

But the largest opportunities for reducing unnecessary death today go untaken. In developing countries, simple diarrhea will kill more people this decade than the Bubonic Plague throughout the Middle Ages.[8] Pneumonia will take a comparable toll. Most of the victims of these diseases will be young children. In India, Kenya, and Guatemala, for example, half of all deaths occur in children under five years of age. Respiratory and diarrheal diseases account for more than two-thirds of all childhood deaths in many developing countries. For those who survive childhood, life spans average six to eight years less than in developed countries; one month in ten is seriously disrupted by sickness. Trachoma, preventable with hygiene and curable with inexpensive antibiotics, has blinded 2 million people. Sleeping sickness retards rural economic development in a wide area of Africa by killing both animals and humans. Malaria kills people or deprives them of the energy for work throughout Africa, Asia, and South America. The list of other tropical ailments is long.[9]

When children die in the developed lands, it usually is in an automobile accident, fire, or fall. Measles, whooping cough, and diphtheria have been virtually eliminated as causes of death. Birth defects still take their toll, but even so, only 3 percent of all infant and child deaths occur in the developed world.[10] Many adults in industrialized countries, however, die prematurely as a result of avoidable effects of heart disease and cancer. About 40 percent of all cancer and 25 percent of all heart disease deaths in the United States, for example, occur in persons younger than sixty-five.[11] Although these degenerative diseases may be inevitable in late old age, perhaps half of the deaths in persons younger than sixty-five could be avoided with preventive medicine. The quality of life for many others could be vastly improved.[12]

Many people still do not understand the health risks they face. Sudanese women, for example, believe that during pregnancy it is harmful to their unborn children to eat eggs or fish, their major sources of protein. This attitude accounts in part for the low birth weights and high infant mortality in that part of Africa.[13] Similarly, many Americans do not yet understand the role of cholesterol and fat in producing heart disease and cancer. Most people in developed countries, in fact, have been taught that dairy products and meat are nearly perfect foods, and that eating more protein and less carbohydrates is good for them. Both groups can benefit most from that basic function of public health care, teaching people to help themselves. As David Banta of the Pan American Health Organization advises, "Never begin with the premise that people will not act on information."[14]

## Primary Health Care

Lowering infant and child mortality in developing lands offers the greatest opportunity in the world today for saving lives at a low cost. One-quarter of all deaths, totaling 15 million per year, occur among children under the age of five, and of these, two-thirds are infants, that is, less than a year old. Ninety-seven percent of these deaths befall developing countries, primarily nations where population growth is rapid. Three million infants die annually in India alone. Infant deaths in Bangladesh, Nigeria, Indonesia, and Pakistan together total some two million each year.[15]

Children of the Third World die of diseases usually not considered lethal elsewhere. Diarrhea, complicated or brought on by malnutrition, causes about a third of all child and infant deaths. Pneumonia vies with diarrheal diseases as the leading taker of young life. Measles, one of the most infectious diseases known, makes children more susceptible to pneumonia, though it is preventable by vaccination. Tetanus, whooping cough, diphtheria, and tuberculosis, also preventable by vaccinations, continue to take a heavy toll.[16]

Comparison with the developed world indicates the magnitude of this disaster. The worst incidence of infant mortality in the United States in 1982 was in Washington, D.C., where 2 percent of all babies alive at birth died before their first birthdays. This high rate, almost double the U.S. average of 1.1 percent, was due in part to low birthweights and poor prenatal care in very young, often impoverished women. In Upper Volta, however, 21 percent of all infants die. More than 76 countries today endure infant mortality rates greater than 10 percent, and in regions of India and within some countries in Africa the rate exceeds 50 percent. These areas not only fare catastrophically worse than developed nations, but worse than several developing countries as well. China and Sri Lanka, despite income levels among the lowest in the world, have infant mortality rates of "only" 4 to 5 percent.[17] (See Table 2.) Low death rates have also been achieved in parts of India, Thailand, and Haiti, where primary health care procedures—midwifery, maternal education on breastfeeding and weaning, vaccinations, oral rehydration of victims of diarrhea, and antibiotics against respiratory infections—have been implemented.

Improved levels of female education also play a critical role in reducing infant mortality. Mothers with elementary training are taught how to avoid the causes of infant death, and they are better able to understand health workers' instructions. Primary health care can be fashioned to fit their needs and to help minimize the risks of catastrophic illness. Primary care, in a sense, can be considered an educational service.

**Table 2: Infant Mortality and Female Literacy in Selected Countries**

| Country | Infant Mortality, 1981 | Female Literacy, 1980 |
|---|---|---|
| | (percent) | |
| Upper Volta | 21 | 5 |
| Afghanistan | 20 | 6 |
| Ethiopia | 15 | 5 |
| Bolivia | 13 | 58 |
| Nigeria | 13 | 23 |
| India | 12 | 29 |
| Pakistan | 12 | 18 |
| Saudi Arabia | 11 | 12 |
| Tanzania | 10 | 23 |
| Honduras | 9 | 62 |
| Brazil | 8 | 73 |
| Mexico | 5 | 80 |
| Philippines | 5 | 88 |
| Thailand | 5 | 83 |
| China | 4.1 | 66 |
| Yugoslavia | 3.1 | 81 |
| Costa Rica | 2.7 | 92 |
| Soviet Union | 2.6 | 98 |
| United States | 1.2 | 99 |
| Japan | 0.7 | 99 |

**Sources:** United Nations Population Division; World Bank; and Ruth Leger Sivard, *World Military and Social Expenditures 1983* (Washington, D.C.: World Priorities, 1983).

Simply providing these most basic health care measures would save between five and ten million young lives each year. These services are most effective when delivered through a system of workers who can give shots year after year at the required intervals, continuously teach breastfeeding and proper weaning, and provide birth control devices. Health workers with only limited training have cut infant mortality by half in demonstration projects, and for as little as $2 per person served per year. Extending primary health care to all of the world's peoples would cost an additional $10 billion annually, one twenty-fifth as much as the world spends on cigarettes. Nevertheless, for many developing countries, this increase would require a doubling of health expenditures.[18]

Women and children in developing countries usually lack the most basic advantages. The Third World infant is disadvantaged even be-

**"In the Gambia, weight gain in pregnancy may average only six pounds."**



fore birth because risk of death increases with low birth weight.[19] A Third World child's mother may gain only half as much weight in pregnancy as a woman in the developed world. During harvest seasons, when women must work exceptionally hard, or during rainy seasons, when food supplies are low, a childbearing woman may gain only a quarter as much weight as would be expected in the West. In the Gambia, weight gain in pregnancy during such difficult times may average only six pounds.[20] Women in food-short countries face not only scarcity along with everyone else, but usually receive low priority for food. A United Nations' survey showed that pregnant or lactating women throughout the Third World consume on average only 1750 calories per day, at least one-third fewer than recommended and one-third fewer than the men.[21] Often women in the developing world will be pregnant again after having lost a child only a few months earlier. Even when the last child has survived, a subsequent birth within 18 months and perhaps, as some research shows, even within 36 months, creates a significantly higher risk of infant mortality. The mother's system may not have fully recovered from the rigors of childbearing and she will be less able to gain weight and provide adequate milk.[22]

Half of all births in the Third World are delivered without any assistance from a trained midwife or doctor.[23] When midwives are available, a breech or otherwise complicated birth may be beyond their competence. In some areas, they may even cut the umbilical cord with a rusty razor and salve it with cow manure, a practice that contributes to the neonatal mortality caused by tetanus and infection in developing countries.[24]

If the child survives birth, it faces a treacherous year. The major threats include malnutrition due to poor weaning practices, diarrheal infection from contaminated water, and infectious diseases that prosper in malnourished children. Parents themselves frequently cause or permit malnutrition in their children. Children often become malnourished because infection depresses their appetites while it consumes additional calories. Common sense prompts parents to withhold food from a child infected with diarrhea because feeding increases stool volumes. The word diarrhea, in fact, means "to flow through." But the body still can use most of the food ingested during diarrhea, and so failure to feed children starves them unnecessarily. The resulting malnutrition then suppresses the body's immune responses. Repeated episodes of diarrhea, which are common where clean drinking water is unavailable, lead to further malnutrition, infection, and even death.

Dehydration caused by advanced diarrhea kills more dramatically. The large intestine normally absorbs salts and sugars through wall membranes using "pumps" in the cell membrane. Microorganisms such as Shigella dysentery produce toxins, shut down the pumps, or

**"Family planning services increase the chances of survival of all children."**



cause the membranes to excrete large quantities of fluids. Normally, the small intestine reabsorbs ten quarts of fluid per day, but diarrhea can reduce reabsorption to half this level. Dehydration follows, along with a high risk of death.[25]

Still another major disadvantage for children of the developing world is lack of access to medicine's most elegant solutions: vaccinations. Only one-third of these children are vaccinated against measles, for example. Inexpensive inoculations against the highly infectious and often deadly childhood diseases of whooping cough, diphtheria, and polio have been available for decades, but few Third World children receive them. Immunization campaigns have been attempted, but often yield poor results, in part because shots must be given to children over various intervals and to new infants year after year. Inadequate recordkeeping, refrigeration, and transportation each contribute to low participation in immunization campaigns.[26]

Different solutions for the health problems of the Third World are needed at each stage of life. Family planning services increase the chances of survival of all children. Parents in lands with high infant mortality rates typically produce more children than they desire because they have little access to contraception or because they want to ensure the survival of a minimum number. Surveys have consistently shown that large numbers of Third World women desire birth control devices but do not have access to them.[27] Women fitted with IUD's or provided with birth control pills can space their children and limit their number.

Supplementing the food supply of pregnant women can provide cost-effective benefits. United Nations University (UNU) studies have shown that increasing calories, protein, iron, and other nutrients in the diets of pregnant women greatly enhances the survival of infants at birth by increasing birthweights. The UNU found these supplements far more beneficial than providing food to mothers for the purpose of increasing breastmilk production. Case studies of supplemental feeding of pregnant women, lactating mothers, infants, and children in Chile, India, Haiti, and Thailand also argue for giving pregnant women highest priority in any nutrition intervention efforts.[28] Health workers can monitor the weight of pregnant women and urge them—and urge their families to let them—eat more. Where heavy scales cannot be transported, simple tape measures for measuring arm circumference can help identify women with a high risk of giving birth to a low-weight child.[29]

Workers trained in midwifery can perform deliveries without infecting mother or child. Simply teaching traditional midwives to sterilize the razor blade used to sever the umbilical cord would help avoid neonatal infection. Similarly, inoculating pregnant women against tetanus would also provide immunity for the infant. Traditional midwives can be readily trained in delivery and vaccination methods. Training can be given on the job, thus avoiding a long and difficult absence from the village.

All primary care workers should be trained to promote breastfeeding; its advantages have been abundantly documented. Studies around the world have shown that artificially fed infants are several times more likely to contract diarrheal diseases and die. Healthy mothers can satisfy an infant's nutritional requirements through breastfeeding for at least six months and thus avoid the risk of infection carried by contaminated formulas or water. The colostrum, or foremilk, that flows from the mother's breasts immediately after birth is rich in antibodies that will protect the baby from diseases against which it otherwise would have little resistance. The milk itself is more nutritious than any substitute, and it is free—the money saved can be used to assure the nourishment of the mother. The importance of breastfeeding is obviously much more critical when clean water and sterile containers are unavailable for preparing formula.[30]

Failures in breastfeeding emphasize the need for promoting good weaning practices. Exclusive breastfeeding for many women is not practical for more than a few weeks. A survey in Barbados showed that 90 percent of all mothers believed breastfeeding to be superior, but only 45 percent breastfed fully at three months, and only 17 percent at six months. The gap between belief and behavior probably is due to the mother's need to work outside the home and a lack of daycare facilities near work. Neither problem can be readily or cheaply solved. In such circumstances, adequate weaning foods and clean water are essential.[31]

Weaning exposes infants to contaminated water and to unnourishing, indigestible foods. Unhealthy weaning practices throughout developing countries may explain why infant mortality is so high in nations where breastfeeding for long periods is common. Many Third World mothers may breastfeed even up to 18 months but supplement diets only with watery gruels or starches. Infants in the Third World are usually not fed adult foods, which are more nourishing, until they are five years old.[32] Health workers, as part of their routine visits to homes or in patients' visits to health centers, can teach families how to prepare foods that children can digest.

Simple tools are available to health workers to assure that infants and young children are being fed properly. Most malnourished children do not display the swollen stomach or wasted appearance that characterize severe malnutrition, and their parents may not know they are underfed. They may simply appear sickly or small for their age. By the time the condition is properly diagnosed, the child will be chronically ill and may soon die. Growth charts make obvious when additional food should be given. The charts, which cost 10 to 15 cents each, help a parent or a health worker compare the weight of children against "normal" weight for their age. If a child gains no weight for three months, or falls below 65 percent of the norm, the chart will show that the child should receive additional food and perhaps other curative measures.[33]

**"Breast milk is more nutritious than any substitute."**



Health workers are essential in treating diarrhea because they can show parents how and when to use oral rehydration therapy (ORT), a new technique that reduces dehydration due to diarrhea quickly and inexpensively. ORT has been successfully demonstrated among thousands of families in Guatemala, Honduras, Egypt, India, Bangladesh, and elsewhere, and has been credited with cutting infant mortality by half or more in these projects. Administering glucose in a given proportion with salts enhances intestinal absorption of fluids and halts their hyper-secretion. ORT packets can be manufactured for pennies per dose, or when packets are unavailable, workers can teach parents how to mix simple salt and sugar in correct proportions. Songs or inexpensive posters can help convey instructions to the illiterate. An essential part of the delivery of this service is instruction in early use, for delay can be fatal. Health workers can again make use of posters to display symptoms that call for administration of the tablets or mixture.[34]

The key to the delivery of these services is workers who are quickly trained, accepted by the community, willing to live in the area, and who serve without the great expense of highly trained doctors. The use of such workers—paramedics, "barefoot doctors," or whatever they may be called around the world—has been demonstrated by a few model primary health centers. Health worker systems in India, for example, have saved lives at a cost of about $60 to $75 per life saved.[35] The low cost reflects the use of workers with levels of education far below a physician's. Even though the cost of saving young lives may be several times higher in most of the Third World, primary health centers still are comparatively inexpensive.

A typical primary health care system works like the one in Dhende Mao, India, where a nurse with two years of government-sponsored training provides prenatal care, nutritional advice, vitamin supplements, inoculations, and antibiotics. The nurse is the middle level of a three-tier "health pyramid," and receives patients who cannot be helped by a first-aid nurse, usually a man who runs a fever treatment center in his home. If the first-aid nurse cannot help the people who come to him, he sends them to the nurse in the clinic. If she cannot help them either, she refers them to a doctor some 12 miles away.[36] This system of referrals facilitates treatment of the more difficult cases and enhances the system's credibility by also making the primary care workers providers of access to more sophisticated care.

A test of primary health care during the early seventies in Narangwal, India, has produced astonishing results. Three different experimental projects were conducted providing either medical care, nutrition supplements, or combined medical care and nutrition supplements. Similar villages nearby without significant primary health care services were studied for comparison and used as scientific "controls." The combined medical care and nutrition project reduced infant mortality to 7 percent compared to 13 percent in the control areas (70 per thousand versus 130 per thousand). The nutrition campaigns only reduced the infant death rate to 10 percent. The medical care only project, however, was cheapest and most effective in reducing child death and disease. Nutrition care cost $4.80 (1983 dollars) per person per year, compared to about $2.40 per person per year for medical care. Nutrition intervention targeted exclusively at pregnant women halved the stillbirth rate compared to the control group. The workers used in each of the Narangwal variations were mainly illiterate.[37] (See Table 3.)

Another project at Miraj, India, also conducted during the seventies, yielded similar results. With a per capita annual cost of only 85¢ per year, the results were impressive: Infant mortality was reduced in three years from 6.8 percent to 2.3 percent. The number of children immunized against the basic childhood diseases increased from about 5 percent to 85 percent for some vaccines. Ninety-seven percent of mothers received prenatal care. The birth rate was simultaneously reduced.[38]

Comparable results were obtained in a program in Haiti that used community health workers to expand the services available at the Albert Schweitzer Memorial Hospital. At the outset of the project, infant mortality was slightly lower in the countryside surrounding the hospital than the national average. But for less than $3.85 per person per year, rates of mortality were cut to one-sixth the national levels. Eighty-five percent of the children were inoculated for diphtheria, tetanus, and whooping cough, compared with only 15 percent nationally. Though literate, the workers employed did not have extensive education.[39]

**Table 3: Primary Health Care Projects: Success Rates and Cost**

| Project | Infant Mortality Rate | | Annual Cost Per Capita |
|---|---|---|---|
| | Project Area | Control Area | |
| | (percent) | | (1983 dollars) |
| Guatemala, rural (1970-72) | 5.5 | 8.5 | $8.90 |
| Haiti, rural (1968-72) | 3.4 | 15.0 | $3.85 |
| India, Jamkhed (1971-76) | 3.9 | 9.0 | $1.55 |
| India, Narangwal (1970-73) | | | |
| Medical care | 7.0 | 12.8 | $2.40 |
| Nutrition Intervention | 9.7 | 12.8 | $4.80 |
| Combined | 8.1 | 12.8 | $3.80 |
| Nigeria, Imesi (1966-67) | 4.8 | 9.1 | $3.60 |

**Sources:** Davidson R. Gwatkin et al., *Can Health and Nutrition Interventions Make A Difference?* (Washington, D.C.: Overseas Development Council, 1980); Warren L. Berggren et al., "Reduction of Mortality in Rural Haiti Through A Primary-Health-Care Program," *The New England Journal of Medicine*, May 28, 1981; Rashid Faruqee and Ethna Johnson, "Health, Nutrition, and Family Planning in India," World Bank, February, 1982.

These primary health care demonstration projects were also encouraging because local communities were involved in their operation and the projects became substantially self-supporting. Residents donated labor and materials. Charges assessed for curative services and medicines recovered up to 75 percent of operating costs. Those most in need of help, the very poor, were provided care free of charge, and were even sought out with home visits. These special efforts, as much as anything, led to the success of the projects. The combined practice of curative and preventive care—for male adults as well as women and children—thus made the system attractive and acceptable. Curative care filled a special need and made "community participation" meaningful.

A major achievement of primary health care projects has been a large increase in families practicing family planning. In Jamkhed, India, family planning participation rates increased from 10 percent in the surrounding area to 50 percent in the project. In the Miraj project, the rate increased from less than a third to almost 90 percent of the "eligible" couples. A couple's willingness to practice birth control strongly depended upon their past history of infant and child losses. As health care reduced the risk of having too few children survive, the number of parents using contraceptives increased quickly. Family planning was stimulated even in areas where sterilization campaigns had caused intense resentment.[40]

Findings from some prototype primary health care projects are discouraging. Government sponsored health-care projects in India, for example, did not achieve nearly the success rate as nongovernment projects, partly because the Indian government typically spent only one-fourth as much per primary health care center. A survey by the U.S. Agency for International Development of 52 primary care projects it funded showed that common problems include poor administration, lack of minimally educated personnel, lack of medicines and supplies, poor communication, poor transportation, and inadequate follow-up training—all formidable problems.[41]

Ghana and Thailand have experienced difficulties with their primary health care efforts. Lack of training in Ghana was blamed for the failure of nurse practitioners to even look inside childrens' throats for infection, to listen to their breathing, or to pinch the skin—a common test for dehydration in children with diarrhea. One doctor in Kenya bitterly denounced the concept of integrated primary services when adequate resources are not provided. "In theory integration provides everything," he wrote. "In reality it throws dust in the eyes of the public."[42] These experiences underscore the importance of individual initiative, good organization, flexibility, and planning in the success of primary health care. Primitive primary health care may only be an emergency measure taken in the tradition of triage. But it is the only practical prescription for hundreds of millions of children. To reduce infant mortality from 20 percent to 5 percent would represent a major improvement, even though this level would be considered intolerable in developed nations.

To achieve "health for all," the World Health Organization (WHO) has set as a goal the expenditure of 5 percent of each country's GNP on health care.[43] This may seem an odd target, for countries with low levels of income will require a far higher proportion of GNP to attain equivalent levels of health care as richer countries. But even the achievement of the WHO goal will require a major reorientation of development priorities. More than 76 countries—the most impoverished and the least healthy—spend less than 2 percent of their GNP on public health. Altogether, about 100 countries spend less than the WHO goal. This means that the average per capita expenditure on health for more than two billion people is only $1 to $2 per year, although private expenditures on traditional medicine may also equal this amount.[44] To meet the WHO goal by providing primary health care to 1.5 billion people without adequate services, an additional annual expenditure—or shift in expenditures—of about $7 per capita would be needed. This extra cost would total just over $10 billion per year. Moreover, half the budgets of the Ministries of Health in developing countries now go to hospitals that provide intensive care, mainly for upper-income urban dwellers. These nations will have to shift priorities within their current health care systems, unless they double or triple total health care spending.

Nations and multilateral aid institutions such as the World Bank have long spent more on energy, industrial, and transportation development than health. The World Bank's International Development Association, despite recent efforts in health investment, has invested over ten times as much money in energy as in health, nutrition, and family planning. The Agency for International Development, which funded primary health care and oral rehydration therapy in Thailand long before the 1978 international conference on primary health care in Alma-Ata, recently cut funding for its health activities by 20 percent. (This cut was prompted by the Reagan Administration's opposition to the family planning aspects of primary care.) The entire budget of the World Bank, in fact, is less than the $10 billion or so required to bring world health care up to even the most minimal levels.[45] Additional transfers—aid—from the rich countries are clearly needed. Meanwhile, developing nations themselves can elevate health on their investment agendas and use available resources more efficiently. Many Third World doctors cite the waste of resources in expensive hospitals that fail to provide the type of care that is most needed and perform large numbers of operations that may not be necessary.[46]

Health can become a central effort of nations only at the insistence of the needy citizens. Unfortunately, dissent is tolerated in only one-third of the world's nations. The importance of constructive criticism is plainly visible in all free lands, and the absence of civil liberties to promote such an exchange of ideas and the expression of political will is debilitating. The recent example of Somalia is indicative. Doctors and nurses there critical of the government's failure to accelerate the implementation of primary health care—they merely handed out leaflets to passers-by—were imprisoned without trial for four years.[47] Human health development requires the struggle for human rights. To the extent that powerful governments support oppressive regimes, there is a far greater cost in human life than just the lives lost in armed struggle.

**"Average per capita expenditure on health for more than 2 billion people is only \$1 to \$2 per year."**



## Drinking Water and Toilets

Peter Bourne, president of Global Water, Inc., an organization formed to help implement the goals of the International Drinking Water Supply and Sanitation Decade, relates two stories that capture the meaning water carries for human health. The first comes from an African woman asked whether she understood the importance of encouraging her children to wash their hands after defecation, particularly before eating. She replied, "I have to carry our water seven miles every day. If I caught anyone wasting water by washing their hands, I would kill them." The second comes from another African woman asked how having water taps installed in her village had changed village life. Her immediate response was, "The babies no longer die."[48]

One-quarter of the world's people lack clean drinking water and sanitary human waste disposal. As a result, diarrheal diseases are endemic throughout the Third World and are the world's major cause of infant mortality. Cholera, typhoid fever, Guinea worm, schistosomiasis, and intestinal parasites also infect hundreds of millions. Many people, because they must visit rivers and swampy areas to obtain water, risk contracting malaria, river blindness, and sleeping sickness. Experts estimate that a sanitary water supply would eliminate half the diarrhea, including 90 percent of all cholera, 80 percent of sleeping sickness, and 100 percent of Guinea worm infestation, as well as smaller fractions of several other serious tropical diseases.[49] (See Table 4.)

Some observers have argued that water and sanitation systems should receive higher priority than other investments, including major reservoir projects, because they fundamentally improve the human condition, while some reservoirs have caused serous problems such as the tripling of cases of schistosomiasis. Clean drinking water, unfortunately, has not been a high priority for many countries. Four-fifths of the rural population of 73 African and Asian countries, where the populations are mostly rural, do not have access to clean drinking water. Most have no toilet or latrine. (See Table 5.) Worldwide, 1.3 billion people lack clean water and 1.7 billion lack adequate sanitation.[50]

## Conclusion

Least-cost health strategies will accord priority to preventive and primary methods designed to attack the world's leading causes of unnecessary death. The toll of childhood infections can be sharply cut

**Table 4: Principal Sources of Disease**

| Disease (common name) | Persons Infected | Controllable with Clean Water Supply and Basic Sanitation |
|---|---|---|
| | (millions) | (percent) |
| **Acquired In Drinking or Water Contact:** | | |
| Cholera | na | 90 |
| Typhoid Fever | na | 80 |
| Diarrhea | 500[1] | 50 |
| Guinea Worms | na | 100 |
| Schistosomiasis | 200 | 10 |
| **Acquired in Collecting Water:** | | |
| Malaria | 300 | na |
| Sleeping Sickness[2] | na | 80 |
| River Blindness | 20-30 | 20 |
| Elephantiasis[3] | 270 | na |
| **Acquired by Contact with Excreta:** | | |
| Roundworm | 650 | 40 |
| Whipworms | 350 | na |
| Hookworms | 450 | na |

[1] Estimate is for annual cases in children in developing countries. [2] Gambian Trypanosomiasis. [3] All Filariasis infection.

**Sources:** *Safe Water and Waste Disposal For Human Health: A Program Guide,* U.S. Agency for International Development, 1982; *Zoonoses And Communicable Diseases Common To Man and Animals* (Washington, D.C.: Pan American Health Organization, 1980).

by extending primary health care to the world's poor women and children. The cost would total only $10 billion per year—less than two-tenths of a percent of the world's annual economic output—and some five to ten million lives would be saved annually. Where the incidence of diarrheal, tropical, and parasitic diseases are highest, tens of billions of dollars worth of investments in wells and toilets will be necessary, but cost-effective. The toll of heart disease and cancer in middle age can probably be halved with diet modification and the control of smoking. Educational campaigns for reducing fat and cholesterol consumption, coupled with taxes on tobacco and restrictions on public smoking, can help extend millions of lives into old age, and at a favorable cost, compared with the half-way treatments available once the diseases have been acquired. But the best hope of low-cost cures to high-cost diseases such as malaria, sleeping sickness, and the cancers and diseases of the heart not yet understood lies in basic science. Fortunately, the additional cost of research and development would be comparatively low.

"Behavioral science is a vastly underfunded field."



**Table 5: Availability of Clean Drinking Water and Human Waste Disposal in Selected Countries**

| Country | Infant Mortality | Share of Population with Service: Clean Drinking Water Supply | Human Waste Disposal |
|---|---|---|---|
| | (percent) | (percent) | |
| Upper Volta | 21 | 31 | na |
| Afghanistan | 20 | 11 | na |
| Angola | 15 | 27 | na |
| Ethiopia | 15 | 16 | 14 |
| Bolivia | 13 | 37 | 24 |
| India | 12 | 42 | 20 |
| Pakistan | 12 | 34 | 6 |
| Turkey | 12 | 78 | 8 |
| Indonesia | 10 | 22 | 15 |
| Tanzania | 10 | 46 | 10 |
| Honduras | 9 | 44 | 20 |
| Brazil | 8 | 55 | 25 |
| Mexico | 5 | 57 | 28 |
| Philippines | 5 | 51 | 56 |
| Chile | 4.1 | 85 | 32 |
| Costa Rica | 2.7 | 72 | 97 |
| Portugal | 2.6 | 73 | na |
| Soviet Union | 2.6 | 76 | na |
| Cuba | 1.9 | 62 | 36 |
| United States | 1.2 | 99 | 99 |

**Sources:** *The State of the World's Children, 1984* (New York: Oxford University Press and the United Nations Children's Fund, 1983); Ruth Leger Sivard, "World Military and Social Expenditures, 1983," World Priorities, 1983; *Health Conditions in the Americas* (Washington, D.C.: Pan American Health Organization, 1982).

These elements of health provision deserve a high priority for public funds and human resources because they will effectively and cheaply save the largest number of lives. They also have a special urgency because their implementation is long overdue.

Emphasizing these five program elements does not diminish the importance of treating other acute ills. Reducing the toll from birth defects, highway accidents, falls, suicide, and homicide remains important. When a health investment such as air bags for reducing automobile fatalities is cost-effective, it can be made profitably, for it will pay for itself both in human and economic terms. Alcohol abuse, at least in the United States, exacts a high human and economic cost, and may not be generally preventable without new developments in behavioral science. Neither economic nor equity policy can justify neglect of these problems simply because their costs are outweighed by greater ones.



But even the largest opportunities for improving the human condition continue to be neglected. Primary health care is tragically underfunded, both in the training of appropriate personnel and in creation of delivery systems. Little funding is provided for appropriate sanitation alternatives. Diet education is haphazard at best, and anti-smoking efforts are sporadic. The policy failure common across these categories is a failure in the field of public health. Government health agencies around the world have been content to foster and develop private, intensive health care, and have allowed preventive and effective primary measures to languish. Preventive medicine comprises less than 2 percent of the training of U.S.-educated doctors.

Governments have an ultimate responsibility for the health and welfare of their people. Many observers believe that sufficient resources exist in developing countries to extend primary health care to all by the year 2000. Recurrent costs will ultimately have to be covered by the beneficiaries of the health services. Fortunately, some curative services can be provided cheaply enough for most to afford. Charging small fees for such services actually enhances their credibility, and the provision of curative services makes the preventive services more acceptable. The initial costs of health-workers training, facilities, equipment, and drugs will almost always have to come from an investment by the governments themselves or through external assistance. Charitable and religious organizations have made large contributions here, but their resources are far too limited to accomplish the task at hand. Governments can reallocate funds to higher priority primary care by saving on wasteful practices in tertiary care systems. Only by reallocating funds from other sectors of their economies, however, can they avoid diminishing the services provided in these centers.

Governments would be more inclined to take action if leaders were judged on their country's state of health. Poverty, inequity, and inefficiency may be covered up in short-run economic statistics, but infant mortality statistics reveal these clearly. Unfortunately, two-thirds of the world is ruled dictatorally, and the leaders thus are not accountable to their people. In these nations, pressure must come from the outside. The World Health Organization, the United Nations Childrens' Fund, the International Red Cross, and other world health leaders already serve in this capacity, and their efforts should be strongly supported and extended.

Multilateral and bilateral aid can supplement the health resources of poor nations. Sweden, for example, has long contributed about one percent of its gross economic product for aid, and usually requires

**"Sufficient resources exist to extend primary health care to all."**



that its funds go to the neediest in the promotion of equity and democracy. Aid can contribute best to "investing in people," in their health and education. When children are nourished, mentally sound, and strong, they can better take advantage of the educational opportunities available to them, and help themselves. The developed world as a whole, however, contributes only three-tenths of a percent of its total economic product to development aid. If the richest 20 percent of the world followed Sweden's example, development aid would be tripled, and more than $85 billion would be available each year, part of which could support basic primary health care and sanitation needs. Unfortunately, most aid—even Sweden's—has strings attached. Even if political concessions are not exacted, it usually is required that the beneficiaries spend much of the money buying the products and services of the donor. Such restrictions severely limit the benefit of aid.

In developed countries, the economic benefits of labeling foods, training doctors and teachers in nutrition, and providing dietary education will be enormous, and the costs almost trivial, compared with the alternative.

Tobacco control will require many decisions on many governmental levels. Strong health warnings and the prohibition of advertising are critical first steps. A tax on cigarettes of $2 per pack would discourage smoking and place the burden of smoking's costs to society on the smoker. Restricting public smoking would protect the health of nonsmokers. Progress will be painfully slow and difficult in the best of circumstances, and, for this reason, leadership at the highest levels of government will be critical. Parties with vested interests, the medical insurance industry for pecuniary reasons, and the health care industry for ethical reasons, can take the lead in pressuring governments to implement these public health priorities. Citizens and health promotion organizations will need broad support, both financial and political, to be effective.

Closing with a call for more research is a timeworn practice, but one that is justified both by the enormous contribution basic research has made to human health this century and by the equally great potential that biomedical science promises. Recent advances in biotechnology, biochemistry, and genetics may revolutionize medicine around the world. Applying this potential for the good of all humanity, however, is a challenge that may not be met. For along with the maldistribution of wealth goes the maldistribution of science. Rich countries—and their scientists—have an ethical responsibility to allocate a share of this good fortune to solving the problems of the poor.

## Notes

**1.** World Health Organization and United Nations Children's Fund, *Alma-Ata 1978: Primary Health Care: Report of the International Conference on Primary Health Care* (Geneva: World Health Organization, 1978).

**2.** *The State of the World's Children, 1984* (New York: Oxford University Press, 1983), Statistical Appendix.

**3.** *Health Sector Policy Paper* (Washington, D.C.: World Bank, 1980).

**4.** *World Development Report 1983* (New York: Oxford University Press, published for the World Bank, 1983).

**5.** Income projections assume that growth will not exceed 5.5 percent and that incomes now average less than $400 per capita in half the world. Current income levels were taken from *World Development Report 1983*.

**6.** The relationship between infant mortality and income was estimated using a simple linear regression analysis with data from *State of the World's Children* (The correlation coefficient (r) equals −.5 for countries with incomes less than $1,500 per capita, and −.4, between income and infant mortality above 100 per 1,000).

**7.** David Banta, deputy director of the Pan American Health Organization, Washington, D.C., private communication, February 1, 1984; Hector R. Acuna, *Toward 2000: The Quest for Universal Health in the Americas* (Washington, D.C.: Pan American Health Organization, 1983); *Health Conditions in the Americas 1977-1980* (Washington, D.C.: Pan American Health Organization, 1982); *State of the World's Children*.

**8.** *State of the World's Children*; and Pedro N. Acha and Boris Szyfres, *Zoonoses and Communicable Diseases Common to Man and Animals* (Washington, D.C.: Pan American Health Organization, 1981).

**9.** World Bank, Poverty and Basic Needs Series, "Water Supply and Waste Disposal," Washington, D.C., 1983; *Health Sector Policy Paper*; "Mortality and Health Policy: Highlights of the Issues in the Context of the World Population Plan of Action," draft report of the Population Division of the Department of International Economic and Social Affairs, United Nations, New York, May 4, 1983.

**10.** National Center for Health Statistics, *Health: United States, 1982* (Washington, D.C.: U.S. Government Printing Office, 1982); *State of the World's Children*.

**11.** *Health Conditions in the Americas 1977-1980*.

**12.** *State of the World's Children*; B. N. Ames, "Dietary Carcinogens and Anticarcinogens," *Science*, September 23, 1983; James B. Wyngaarden and Lloyd H. Smith, Jr., eds., *Cecil Textbook of Medicine* (Philadelphia: W.B. Saunders Company, 1982); Derek Bok, "Needed: A New Way to Train Doctors," *Harvard Magazine*, May-June, 1984; Richard Doll and Richard Peto, "Quantitative Estimates of Avoidable Risks of Cancer in the United States Today, *Journal of the National Cancer Institute*, November 1981.

**13.** Patricia W. Blair, ed., *Health Needs of the World's Poor Women* (Washington, D.C.: Equity Policy Center, 1981).

**14.** David Banta, private communication, February 1, 1984.

**15.** *State of the World's Children.*

**16.** *Cecil Textbook of Medicine.*

**17.** Daniel Yohalem, "American Children in Poverty," Children's Defense Fund, Washington, D.C., 1984; *State of the World's Children.*

**18.** John R. Evans, Karen Lashman Hall, and Jeremy Warford, "Health Care in the Developing World: Problems of Scarcity and Choice," *The New England Journal of Medicine*, November 5, 1981; Margaret Burns Parlato, *Primary Health Care: An Analysis of 52 AID-Assisted Projects* (Washington, D.C.: American Public Health Association, 1982); Rashid Faruqee, "Analysing the Impact of Health Services, Project Experience from India, Ghana, and Thailand," World Bank Staff Working Papers, Number 546, Washington, D.C., 1982; *State of the World's Children*; *Health Sector Policy Paper.*

**19.** W. Henry Mosley, "Will Primary Health Care Reduce Infant and Child Mortality? A Critique of Some Current Strategies, with Special Reference to Africa and Asia," draft report of the Ford Foundation to the United Nations Conference on Population.

**20.** *State of the World's Children.*

**21.** R. G. Whitehead, ed., *Maternal Diet, Breast-feeding Capacity, and Lactational Infertility* (Tokyo: The United Nations University, 1983).

**22.** "Findings of the World Fertility Survey on Trends, Differentials and Determinants of Mortality in Developing Countries," prepared by the Secretariat of the World Fertility Survey for the 1984 International Conference on Population, May 3, 1983.

**23.** Kathleen Newland, *Infant Mortality and the Health of Societies*, (Washington, D.C.: Worldwatch Institute, 1981); Galba Araujo et al. "Improving Obstetric Care In Northeast Brazil," *Bulletin of the Pan American Health Organization*, Volume 17, No. 3, 1983.

**24.** *Health Conditions in the Americas; State of the World's Children.*

**25.** *Cecil Textbook of Medicine.*

**26.** *Health Sector Policy Paper;* David Banta, private communication; James Heiby, Deputy Director, Health Services Division, U.S. Agency for International Development, private communication, June 8, 1984; *State of the World's Children.*

**27.** Lester R. Brown, *Building A Sustainable Society* (New York: W.W. Norton, & Co., 1981); *State of the World's Children; Health Sector Policy Paper*; "Findings of the World Fertility Survey on Trends, Differentials, and Determinants of Mortality in Developing Countries."

**28.** *Maternal Diet, Breast-feeding Capacity, and Lactational Infertility*; Lloyd Harbert and Pasquale L. Scandizzo, "Food Distribution and Nutrition Intervention: The Case of Chile," World Bank Working Papers, Number 512, Washington, D.C., May 1982; Pasquale L. Scandizzo, and Gurushri Swamy, "Benefits and Costs of Food Distribution Policies: The India Case," World Bank Working Papers, Number 509, Washington, D.C., 1982.

**29.** S. N. Tibrewala and K. P. Shah, "The Use of Arm Circumference as an Indicator of Body Weight in Adult Women," *Baroda Journal of Nutrition*, Vol. 5, No. 43, 1978, as cited in World Federation of Public Health Associations, "Maternal Nutrition: Information for Action Resource Guide," Washington, D.C., July 1983.

**30.** "Infant and Child Mortality in Rural Areas: Implications for Rural Development Programmes," prepared by the Food and Agriculture Organization of the United Nations for the 1984 International Conference on Population, May 3, 1983; R. V. Short, "Breast Feeding," *Scientific American*, April 1984; *State of the World's Children; Maternal Diet, Breast-feeding Capacity, and Lactational Infertility*.

**31.** F. C. Ramsey, "An Analysis of Breast-Feeding Findings in the Barbados National Health and Nutrition Surveys of 1969 and 1981, With Special Reference to the International Code of Marketing of Breast-milk Substitutes," *CAJANUS* 16(1):14-18, 1983, as cited in *Bulletin of the Pan American Health Organization*, Volume 17, No. 3, 1983.

**32.** *Cecil Textbook of Medicine*.

**33.** *Cecil Textbook of Medicine* and *State of the World's Children*.

**34.** *State of the World's Children*.

**35.** Rashid Faruqee and Ethna Johnson, "Health, Nutrition, and Family Planning in India: A Survey of Experiments and Special Projects," World Bank Working Papers, Number 507, Washington, D.C., February 1982. For descriptions of the Chinese primary care system, see Ruth Sidel, *Women and Child Care in China* (Baltimore: Penguin Books, 1974) and Wu Naitao, "The Healthy Growth of China's Children," *Beijing Review*, June 25, 1984.

**36.** "Modern Medicine is Quickly Gaining Acceptance," *New York Times*, December 28, 1983.

**37.** "Health, Nutrition, and Family Planning in India."

**38.** *Ibid*.

**39.** Warren L. Berggren, Douglas C. Ewbank, and Gretchen G. Berggren, "Reduction of Mortality in Rural Haiti Through A Primary-Health-Care Program," the *New England Journal of Medicine*, May 28, 1981.

**40.** "Health, Nutrition, and Family Planning In India."

**41.** "Primary Health Care: Progress and Problems, An Analysis of 52 AID-assisted Projects"; "Reduction of Mortality in Rural Haiti Through A Primary-Health-Care Program"; "Health, Nutrition, and Family Planning In India"; George Alleyne, director, Regional Programs Development, Pan American Health Organization, private communication, April 10, 1984.

**42.** Yusif Ali Fraj, "Point of View: No One Is Realistic About Family Planning," *World Health Forum*, Volume 4, No. 2 1983.

**43.** *Alma-Ata 1978: Primary Health Care*.

**44.** *World Military and Social Expenditures, 1983; Health Sector Policy Paper; State of the World's Children*.

**45.** *IDA In Retrospect: The First Two Decades of the International Development Association* (New York: Oxford University Press, 1982); *Security and Development Assistance*, Hearings before the Committee On Foreign Relations, United States Senate, February through March, 1983 (Washington, D.C.: U.S. Government Printing Office, 1983); *World Development Report 1983*.

**46.** *Toward 2000: The Quest for Universal Health in the Americas;* David Banta, private communication; George Alleyne, private communication.

**47.** *Amnesty International Report, 1983* (London: Amnesty International Publications, 1983).

**48.** Peter Bourne, president, Global Water, Inc., private communication, April 3, 1984.

**49.** *Safe Water and Waste Disposal for Rural Health; Health Conditions in the Americas; Health Sector Policy Paper;* "Water Supply and Waste Disposal."

**50.** John M. Hunter, Luis Rey, and David Scott, "Man-made Lakes—Man-made Diseases," *World Health Forum*, Vol. 3, No. 2, 1983; Paul L. Aspelin and Silvio Coelho das Santos, *Indian Areas Threatened By Hydro-electric Projects in Brazil* (Copenhagen: International Working Group for Indigenous Affairs, 1981); *Safe Water and Waste Disposal For Rural Health*.

CHAPTER 21

# Schistosomiasis and Water Projects: Breaking the Link

JONATHAN B. TUCKER

At the edge of the Shabelle River in southern Somalia, a brightly garbed woman collects some of the chocolate-brown water in a plastic bucket for drinking. Nearby, another woman squats barefoot in the shallows and does her washing. Both women are probably already infected with the parasitic disease schistosomiasis (also called bilharziasis, or snail fever), which afflicts an estimated 250 million people in the developing countries of Africa, Asia, and Latin America. After malaria, schistosomiasis is now the world's most widespread, serious infectious disease.

Ironically, many water-resource development projects in tropical areas have had the unintended consequence of markedly increasing the prevalence of schistosomiasis in the local population. Examples include the Aswan Dam in Egypt (see *Environment*, May 1981), the Gezira Irrigation Project in the Sudan (*Environment*, June 1983), Lake Volta in Ghana, and Lake Kariba in Zambia, among many others. In such cases, the economic benefit of the projects has been outweighed by the lowered human productivity caused by schistosomiasis and the increased costs of treatment and control of the disease. This adverse result is not inevitable, however, and there are a number of strategies through which it can be avoided.

## Natural History

In Africa, schistosomiasis occurs in two forms, urinary and intestinal, each caused by a different species of parasitic trematode worm or "blood fluke." Urinary schistosomiasis is produced by *Schistosoma haematobium*, which discharges its eggs from the human body in the urine; intestinal schistosomiasis is caused by *S. mansoni*, which distributes its eggs in the feces. Before the schistosomiasis parasite is capable of infecting man, it must first infect a particular species of fresh-water snail. As a result, the life-cycle of the parasite is intimately associated with slow-moving fresh water.

If the parasite eggs discharged in the feces or urine reach a body of fresh water within a month after excretion, they hatch almost immediately, releasing free-swimming, first-stage larvae that seek out and penetrate the appropriate snail host. (Each species of schistosome can only infect a corresponding species of snail.) After successful infection of the snail, the first-stage larvae multiply asexually into thousands of second-stage larvae, which emerge from the snails one or two months later as free-swimming, infectious agents. They now have a life span of approximately two days, during which time they must penetrate the skin of human beings who enter the water.

Once inside the human body, the second-stage larvae develop into young worms and migrate through the circulatory system to the liver, where they mature and mate. The adult male worm, measuring between one and two centimeters in length, has a central canal that enfolds the longer and thinner female during most of its life.

---



Jonathan B. Tucker has worked in Africa with CARE, and served on the board of editors of *Scientific American*. He specializes in science, technology and Third World development issues.

After mating, the worms migrate to the veins of their respective target organs: *S. haematobium* finds its way to the capillaries of the bladder and *S. mansoni* travels to the veins of the descending colon and rectum. When they can swim no further, the female worms deposit their eggs one by one—between 300 and 800 a day throughout their lifetime of three to five years. Some of these eggs burst through the lining of the bladder or the intestine and are carried out of the body with the urine or feces. Those eggs that reach slow-moving fresh water within a month's time hatch and begin a new reproductive cycle.

Although schistosomiasis is rarely fatal, it is often highly debilitating, causing chills, fever, and weakness, and reducing productivity. In severely infected individuals, the disease may result in permanent organ damage and, ultimately, death.

Disease due to schistosomiasis results from the incomplete excretion of parasite eggs, some of which are carried in the blood to other parts of the body. There they die off, causing inflammation and pathological lesions. *S. mansoni* infections may lead to fibrosis of the liver, and *S. haematobium* infections have been correlated with a high incidence of bladder cancer.[1] Since a single larva penetrating the skin of the human host develops into only one adult worm, the gravity of the disease depends on the number of larvae that infect the victim and the frequency of reinfection.

## Epidemiological Factors

Water contact is the most critical variable in the transmission of schistosomiasis. In most rural communities in Africa, it is difficult for villagers to avoid contact with infested water because they collect it for drinking, wash their clothes and cooking utensils in it, and bathe or swim in it. Men are often infected while washing cattle, fishing, planting rice, or engaging in *wadu*, the ritual washing that all observant Muslims must perform five times a day before praying.[2] Children tend to be infected by drinking contaminated water or by accompanying a family member to an infested river or pond; for young males, nearly all water contact is through bathing and swimming, at which time urination is common.

Peak water-contact activity tends to occur in the afternoon, which unfortunately coincides with the peak time of schistosome egg output in the urine and the shedding of the schistosome larvae by infected snails. As a result, the provision of adequate sanitation or safe drinking water is usually not sufficient to control schistosomiasis. In one study, children from different villages had much the same incidence of the disease regardless of whether or not their village had a piped water supply, simply because all of the children played in the heavily infested river nearby.[3]

Workers in a Sri Lanka rice paddy. Modernization of traditional tank irrigation and drainage systems, as well as overall improved water management, are essential to the control of schistosomiasis (World Bank photo by Tomas Sennett).

The prevalence of schistosomiasis increases throughout childhood and peaks in the second decade of life, after which—for reasons not well understood—it gradually declines. Since younger age groups make up 50 percent of the rural population in Africa, they are responsible for a very high proportion of the contamination of the environment with schistosome eggs, and hence the continued transmission of the disease.

## Effect of Water Projects

As mentioned before, the development of water resources for hydroelectric power or for irrigation has tended to intensify the transmission of schistosomiasis, as well as to extend it into new areas. In many parts of Africa the disease remained at low levels until the introduction of water projects, after which the snail vectors flourished and the prevalence of the disease soared.

There are four reasons for this phenomenon. First, irrigation of formerly arid areas creates additional habitats for the snail vectors beyond those already present in ponds and rivers. Second, defects in the design or engineering of water projects may also provide new habitats for the snail hosts. These sources of new vector habitats include poorly constructed canals that become rapidly fouled with silt or aquatic vegetation, inefficient drainage canals, and night-storage ponds filled with stagnant water. Third, the water projects have provided new opportunities for villagers to come into contact with the infested water, particularly when the projects are located close to villages in which no alternative water sources are available.[4]

Finally, the advent of perennial irrigation, in which the irrigation canals are in use year-round, has facilitated the multiplication of the snail vectors of schistosomiasis. Under natural conditions, the snail populations are unstable because of seasonal variations in water level. Lack of rainfall or low temperatures in winter tend to inhibit snail breeding or even stop it altogether. Most snails in Egypt, for example, are rid of parasites in winter due to chilling, so that there are practically no snails shedding infectious larvae between late December and the middle of May.[5] Under conditions of low rainfall, the canals dry out, forcing the snails to burrow into the mud; only when the rains return do the snails again become active.

Perennial irrigation has removed this natural check on snail populations, enabling the molluscs to multiply out of control, thereby increasing the rate at which schistosomiasis is transmitted to the human population. It seems likely that schistosomiasis transmission will continue to increase in developing countries as their exploding populations, attempting to extend irrigation into arid zones to grow urgently needed food, will create new habitats favorable to the snail vectors of the disease.

The schistosomiasis cycle begins with the parasite's infection of a particular species of freshwater snail. Second-stage larvae then infect humans (photo by Harold Royaltey, Project HOPE).

## Control of Schistosomiasis

Under the conditions prevailing in most developing countries, endemic schistosomiasis can rarely be eradicated. Egg production appears to be so great (relative to what is required for continued transmission of the disease) that even a small residual percentage of parasites would be sufficient to maintain it at a considerable level.[6]

Indeed, the actual number of snails infected in nature is surprisingly low. A survey in Egypt revealed that the highest rate of infection among *S. haematobium* snails was 0.3 percent, yet the incidence of schistosomiasis in the local human population was over 50 percent. Evidently, a few scattered snails can produce enough schistosome larvae to infect large numbers of people, because the amount of contact villagers have with the infested water is so great.[7]

As a result, the "breakpoint" in the transmission of schistosomiasis appears to lie so low as to be practically synonymous with total eradication. Given that this goal is unlikely with limited resources, a reasonable objective is to reduce the intensity of infections to the point where they do limited damage to the health of the local population.

Schistosomiasis can be controlled through a variety of measures directed against different parts of the parasite's life cycle. Each method has its advantages and drawbacks. The effectiveness of each depends largely on the conditions under which it is applied. A combination of methods is usually best able to achieve a threshold level of control below which intensive transmission of the disease will cease.

Although drugs are often used to treat schistosomiasis, the doses necessary to cure the disease may produce serious and occasionally lethal side effects. Most of the these agents are also expensive relative to the health budgets of developing countries. Moreover, although available drugs are fairly effective in killing the worms, they do not prevent reinfection. Continual administration of chemotherapy is therefore required to maintain a given level of incidence.[8] After a few years, programs that rely exclusively on chemotherapy tend to fail, either because of the development of drug resistance in the parasite or because efforts at control tend to relax when the incidence of the disease has fallen considerably.

A second way to limit the transmission of the disease is to kill off the snail vectors with poisons (molluscicides), which are applied at points of frequent water contact. Unfortunately, it is impossible to eradicate an entire snail population because some snails will burrow into the mud and survive, and because there is a continual influx of new snails from outside the area.[9] Although the use of molluscicides can reduce transmission, application must be continued for many years—and these chemicals possess toxicity for the environment, particularly for fish. The cost of such compounds is also quite high: estimates in the Philippines indicate that expenditures for molluscicides alone would have exceeded the entire annual budget of the country's Ministry of Health.[10]

In Africa, the emphasis has been on schistosomiasis control projects involving a combination of molluscicides and chemotherapy. Because of the drawbacks of both methods, however, there has recently been a growing interest in integrated programs of control comprising sanitation, provision of safe water supplies, environmental modification, and public-health education. This approach may turn out to be less costly and more advantageous because it provides multiple benefits to the local community.

## Engineering Strategies

Once snails have been introduced into an irrigation scheme, they are impossible to eradicate completely. It is therefore essential to plan control measures from the outset. Projects in Israel, the Philippines, Japan, and China have demonstrated that if irrigation projects are supplied with efficient drainage, good water management, and regular maintenance, it is possible to avoid an increase in the incidence of schistosomiasis.

In China, for example, schistosomiasis control has been an integral part of the economic plan. Peasants in areas where the disease is endemic have drained marshlands that were breeding grounds for the snails and turned them into productive farms. In addition, the farmers have rebuilt many irrigation canals, not only eliminating snails in the process but also creating a more efficient water network.[11]

Several design elements of irrigation systems are relevant to snail control. First, increased water velocity in the canals appears to prevent the transmission of schistosomiasis, since the snail vectors are rarely seen in streams or canals having an average flow rate of more than 30 centimeters per second.[12] Lining earthen canals with asphalt, concrete, or plastic and gravel will effectively reduce snail habitat by increasing water velocity, and also by reducing the growth of aquatic vegetation, preventing seepage to low-lying areas, and speeding the rate of drainage and drying. Although the cost of lining irrigation canals may seem excessive for a disease control program, it will yield major economic benefits by reducing seepage, which in unlined canals results in water losses of 25 to 50 percent.

Second, irrigation projects should be provided with effective drainage systems. Lack of drainage results in stagnant pools and seepages that create snail habitats. As long as the water is drained away within two to three weeks, the chances of creating new habitats are negligible.

Third, maintenance and water management in irrigation projects are essential for schistosomiasis control. Because aquatic weeds in canals and drains provide suitable habitats for the snails, it is important to control the weeds by removing them at regular intervals. Water in the canals should frequently be varied in depth, and the canals should be drained and allowed to dry out during the non-irrigation season. This technique will interfere with the life and reproduction cycles of the snails, and prolonged dessication may even kill them entirely.[13]

A second set of design modifications aims at limiting contact by the local population with the infested water. It is important to keep a distance of at least 500 meters between the canals and houses, and to prevent access to the irrigation system by means of walls and fences. Water contact can be further reduced by constructing bridges over canals, providing a public water supply, and building water facilities for bathing, washing, laundering, and public recreation.

Among irrigation engineers there is an increasing interest in closed systems such as concrete pipe, which require a higher capital investment but pay off in the long run through reduced transport losses and lower maintenance costs. Other approaches that decrease water contact include sprinklers and trickle-drip irrigation from plastic pipes, the latter of which has the additional advantage of conserving water.

## Role of Donor Agencies

The belated recognition by the major international lending agencies that an increase in schistosomiasis transmission can seriously jeopardize the real economic returns from an irrigation or hydropower project has led them to include schistosomiasis control as an integral part of plans for water-resources development.

The World Bank has been involved in schistosomiasis control since 1970, when all Bank-assisted projects were examined for their environmental and health effects. Although the Bank has not sought to eradicate schistosomiasis, it has tried to eliminate the risk of increased infection from the water projects it helps finance.[14]

The U.S. Agency for International Development (USAID) has also raised its awareness of the importance of schistosomiasis control in planning water projects. Since the mid-1970s, the agency has required an Initial Environmental Examination (similar to a domestic Environmental Impact Statement) to determine the possible deleterious effects of a development project on the environment and public health. This assessment attempts to strike a balance between economic and environmental concerns. Planners consider design alternatives but may still go ahead and build a system that is not optimal from the point of view of schistosomiasis control, provided that it has compensating benefits.

Although these efforts are a promising beginning, much more could be done to prevent the continuing rapid spread of schistosomiasis throughout much of tropical Africa and elsewhere in the Third World.[15] In areas where water-development projects are contemplated, public-health surveys should be done in advance, and measures necessary for protecting the health of the local population incorporated into the project design at the planning stage. The lending agencies, by requiring adequate measures for prevention of schistosomiasis as a condition of the loan, can ensure the necessary action.

## NOTES


1. N. Sherif, "A Water-Snail Shares the Blame," **World Health**, September-October 1981, p. 33.

2. Frederick E. McJunkin, **Water, Engineers, Development, and Disease in the Tropics** (Washington: USAID, July 1975), p. 80.

3. A. R. Walker and B. F. Walker, "Pure Water and Infections in Africa," **Lancet**, September 16, 1978, p. 639.

4. Patricia L. Rosenfield and Blair T. Bower, "Management Strategies for Mitigating Adverse Health Impacts of Water Resources Development Projects," **Progress in Water Technology**, 1979, Vol. II, No. 1/2, p. 286.

5. Henry van der Schalie, "Schistosomiasis Control in Egypt and the Sudan," **Natural History**, 1969, Vol. 78, No. 2, p. 65.

6. WHO Expert Committee, "Epidemiology and Control of Schistosomiasis," **WHO Technical Report Series**, No. 643, 1980, p. 32.

7. Van der Schalie, note 5 above, p. 65.

8. Rosenfield and Bower, note 4 above, p. 292.

9. Ibid., p. 292.

10. McJunkin, note 2 above, p. 54.

11. "Report of the American Schistosomiasis Delegation to the People's Republic of China," **American Journal of Tropical Medicine and Hygiene**, 1977, Vol. 26, p. 431.

12. McJunkin, note 2 above, p. 54.

13. Ibid., p. 72.

14. "The World Bank's Role in Schistosomiasis Control," World Bank Office of Environmental and Health Affairs (pamphlet), January 1978, p. 11.

15. WHO Scientific Working Group on Schistosomiasis, "Epidemiology and Control of Schistosomiasis: Present situation and priorities for future research," **Bulletin of the World Health Organization**, 1978, Vol. 56, No. 3, p. 362.

CHAPTER 22

# Ecological Effects of the War in Vietnam

GORDON H. ORIANS AND E. W. PFEIFFER

Wars are always destructive of environments, but never before have the ecological effects of a war been a major issue. For the past several years there has been widespread concern among scientists about the massive use of chemicals for defoliation and crop destruction in Vietnam. Because these chemicals have never before been used in military operations, there are no data upon which to predict the effects of such use. However, J. S. Foster, Director of Defense Research and Engineering, Department of Defense, has stated that the Department of Defense would not use these chemicals if it judged that seriously adverse ecological consequences would occur. The basis upon which this judgment was made is not clear in view of the fact that the report of the Midwest Research Institute (sponsored by the Department of Defense) on the ecological effects of herbicides (*1*) points out that predictions based on civilian uses are not valid. This is because the military application of herbicides in Vietnam is carried out under conditions that are not comparable to the civilian situation.

Recognizing that there were no data on the ecological effects of the military use of herbicides in Vietnam, the Department of State sent F. H. Tschirley, a U.S. Department of Agriculture plant ecologist, to Vietnam in March 1968, to make a 30-day assessment of the ecological effects of defoliation. His report (*2*) indicates that the defoliation program is having a profound effect on plant life in Vietnam. He was, however, unable to get first-hand data on many aspects of the problem, including effects on animal life. Accordingly, the Society for Social Responsibility in Science decided to sponsor a trip in March 1969, with the objective of supplementing Tschirley's observations with those of zoologists. Unfortunately both of these visits have been made in the dry season.

## Sources of Information

We gathered information and impressions from interviews with military personnel involved with both field operations and policy decisions. We traveled by helicopter over areas damaged by B-52 bombing raids, and we flew on spray missions with the C-123 aircraft which have been modified for spray application. We were also able to take a 2-hour, 40-minute (104 kilometers) trip by Navy patrol boat through the Rung Sat Special Zone, an extensive region of mangroves on the Nha Be River, which has been heavily defoliated. The main shipping channel to Saigon passes through the area and widespread defoliation has been used to reduce the incidence of rocket and mortar attacks on vessels coming up the river. We are grateful to the U.S. Embassy, Army, Navy, and Air Force, the Rubber Research Institute of Vietnam, Plantations Michelin, and the many Vietnamese biologists, both in governmental and nongovernmental positions in their country, for their cooperation and hospitality. All information which we requested from the Department of Defense that did not carry a security classification was made available to us.

Because rubber plantations are one of the most important sources of foreign capital in Vietnam and since the rubber tree *Hevea brasiliensis* is particularly susceptible to damage by defoliants, especially 2,4,5-trichlorophenoxyacetic acid (2,4,5-T) (*3*), we in-

Reprinted from *Science 168:* 544–554 (1970). 

Gordon H. Orians is director of the Institute for Environmental Studies at the University of Washington, Seattle, WA.

E. W. Pfeiffer is with the Department of Zoology, University of Montana, Missoula, MT.

terviewed plantation owners concerning defoliation damage. The planters themselves have not carried out systematic studies of the physiological effects of defoliants on rubber trees, but they have been very much interested in estimating their losses. The Rubber Research Institute of Vietnam, a private research corporation, has made careful observations of the nature and extent of damage to rubber trees and has carried out some experiments to find ways of minimizing the loss to defoliants. The data in the files and publications of the Institute, kindly made available to us by the director, Jean-Paul Poliniere, were invaluable to us in learning more about effects of defoliation on rubber trees. Also, during a visit to the research station of the Institute, we were able to observe recent damage to trees by defoliants and to view pictures of trees damaged and killed by previous defoliations. Officials of the Michelin Plantations also provided us with data from their files on the nature and extent of herbicide damage to rubber trees on one of their plantations.

The Faculty of Science, University of Saigon, and government agencies concerned with plants and animals, such as Ministries of Fisheries, Forestry, and Agriculture, are staffed with biologists trained primarily in France and the United States. These people are knowledgeable and concerned about the ecological effects of the war in their country. By means of interviews with them we were able to assess their concerns, find out what kinds of studies have been initiated, explore ways of helping them launch future studies, and to gather information they had collected which was relevant to our mission.

Wartime conditions prevented us from making ground observations in heavily defoliated forests, but we were able to discuss damage with B. R. Flamm, Chief, Forestry Branch, U.S. Agency for International Development, Saigon, and to examine photographs he took inside forests receiving one and two applications of defoliants. In addition, one of us (G.H.O.) visited some of the sites in Puerto Rico, which have been used to test defoliants under tropical conditions, in April 1969, for a closer look at vegetation recovery and animal populations.

Because previous work on the effects of defoliation in the field have dealt almost entirely with direct effects upon plants, we made a special effort to observe animals in all the areas we visited and to ask as many questions as we could about changes in the status of animals. Because our own knowledge was most extensive about birds we learned the most about them, but we did gather some information on other taxa through interviews. Because of the short duration of our visit we were unable to obtain definitive answers to some of the most important questions which have been raised by the American scientific community about the ecological effects of the war, but we feel that the material we gathered forms a significant contribution to continuing efforts to assess the impact of modern warfare upon the environment in which man must live.

## Operational Aspects of the Defoliation Program

Inasmuch as it is the widespread use of herbicides in Vietnam that has been of greatest concern to American scientists, we gave top priority to learning about the effects of the defoliation program in Vietnam. Defoliants have been used in Vietnam by the United States since 1962. The program started modestly but increased sharply after 1965 (Table 1). A peak was reached in 1967 followed by a slight reduction of total area sprayed with defoliants in 1968 as a result of the reassignment of equipment for other missions following the Tet Offensive (*4*). The bulk of the spraying is directed against forests and brush, but a significant proportion is directed against cropland in the mountainous parts of the country (*4*).

Table 1. Estimated area (1 acre = 0.4 hectare) treated with herbicides in Vietnam. Actual area sprayed is not known accurately because some areas are resprayed. Areas are estimated from the number of spray missions flown, the calibrated spray rates and the width of spray swath covered. [From Department of Defense data.]

| Year | Defoliation (acres) | Crop destruction (acres) |
|---|---|---|
| 1962 | 17,119 | 717 |
| 1963 | 34,517 | 297 |
| 1964 | 53,873 | 10,136 |
| 1965 | 94,726 | 49,637 |
| 1966 | 775,894 | 112,678 |
| 1967 | 1,486,446 | 221,312 |
| 1968 | 1,297,244 | 87,064 |

The U.S. military authorities believe the food grown in the mountainous areas is used to feed the forces of the National Liberation Front. They deny using defoliants on rice crops in the delta region. Much of the defoliation is along roads and rivers and around military establishments, and border areas (near Laos and Cambodia) are extensively defoliated. Forested regions north and northwest of Saigon in Tay Ninh, Binh Long, Binh Duong, Phuoc Long, and Long Khanh provinces have been very hard hit. This area contains some of the most valuable timber lands in the country. In most cases, broad forest areas have not been repeatedly defoliated, though possibly 20 to 25 percent of the forests of the country have been sprayed more than once. Roadsides and riverbanks are subjected to multiple defoliation at regular intervals.

Officially the defoliation program is a Vietnamese program with the assistance of the United States. The initial request for defoliation may be made by either a district or a province chief with the support of his American advisor. Included in the request must be the claim that the targeted area is under control of the National Liberation Front or of the North Vietnamese. The chief must also pledge to reimburse his people if there is any accidental damage to their crops by wind-blown spray or other causes. The request also must contain a promise to inform people in the target area that it will be sprayed, giving them the reasons for the spraying, and offering them the opportunity to change their allegiances if they so desire. Plans are supposed to be made in advance to handle any refugees which might result from the operation.

This request then goes to the division tactical zone commander and his American advisor, then to the Corps commander and his advisor, and then to the Vietnamese Joint General Staff and its American advisors in Saigon. In Saigon the request is circulated among a broad spectrum of groups dealing with pacification operations, intelligence, psychological warfare, and chemical warfare. Finally, permission must be given by the commanding general and the United States Ambassador to Vietnam.

Despite this formal arrangement, in Vietnam the program is generally considered to be an American one, and military justification of it is always given in terms of the American lives it saves. Moreover, there is evidence that the many precautions specified by the procedures are neglected regularly. For example, aerial reconnaissance of the target area prior to the decision to spray it, is omitted if the schedule is busy, and in enemy-held areas there is often no warning given.

To reduce transfer of herbicides by the wind and to improve the kill on the desired target, the military authorities have established regulations governing conditions under which defoliation may take place. Missions are to be flown only when the temperature is less than 85°F (29.4°C) and the wind is less than 10 knots. This restricts aerial spraying to morning hours, though usually an attempt is made to fly two successive missions each morning.

The defoliants used in Vietnam, the concentrations used, and those used in U.S. civilian operations, and the purposes for which they are best suited are given in Table 2. In the region of Saigon, where wind-blown and gaseous herbicides pose threats to cropland, agent White is now preferred because of its lower volatility and persistence, but in regions where there is little agriculture, Orange is the preferred agent because it is more economical. Presently in Vietnam, Orange constitutes about 50 percent of the total herbicide used, White 35 percent, and Blue 15 percent, the latter being used primarily against mountain rice crops (*4*). Approximate areas where extensive defoliation has been carried out are shown in Fig. 1.

## Effects of Defoliants on Trees

It was impossible for us to visit defoliated forests on foot or by means of ground transportation. We, therefore, are unable to add much to what has already been reported on the direct effects of defoliants on forest trees. We can confirm Tschirley's report (*2*) that the trees which are collectively known as mangroves are extremely susceptible to the action of defoliants and that one application at the normal rate employed in Vietnam is sufficient to kill most of the trees. Most of the areas we visited by boat on the Rung Sat Peninsula (Fig. 1) were still completely barren even though some of the areas had been sprayed several years earlier. Only in occasional places was there any regeneration of mangrove trees. We observed no growth of the saltwater fern *Achrosticum aureaum* which often invades mangrove areas.

Mangrove vegetation is floristically simple, the forests in Vietnam being dominated by *Avicennia marina, A. intermedia, Rhizophora conjugata, Bruguiera parviflora, B. gymnorhiza, Ceriops candoleana*, and *Nipa fruticans*, the latter species also forming dense stands along most rivers in the delta region where they are subject to tidal influence. The normal pattern of vegetation succession in mangrove areas has been reviewed by Tschirley (*2*) who suggests that about 20 years would be required for the reestablishment of the dominant *Rhizophora-Bruguiera* forest.

Table 2. Chemical composition, rates of application, and uses of military herbicides from data supplied by the U.S. Departments of Defense and Agriculture. One pound per gallon, acid equivalent (AE) equals 114 grams per liter. One pound per acre equals 1.12 kilograms per hectare.

| Agent | Composition (%) | Concentration (lb/gal AE) | Rate of application (lb/acre) | | Use |
|---|---|---|---|---|---|
| | | | Vietnam | U.S. | |
| Orange | *n*-Butyl ester 2,4-D 50<br>*n*-Butyl ester 2,4,5-T 50 | 4.2<br>3.7 | 27 | 2 | General defoliation of forest, brush, and broad-leaved crops |
| Purple | *n*-Butyl ester 2,4-D 50<br>*n*-Butyl ester 2,4,5-T 30<br>Isobutyl ester 2,4,5-T 20 | 4.2<br>2.2<br>1.5 | | | General defoliation agent used interchangeably with agent Orange |
| White | Triisopropanolamine salt, 2,4-D<br>Triisopropanolamine salt, picloram | 2.0<br>0.54 | 6<br>1.5 | <br>0.5–2 | Forest defoliation where longer term control is desired |
| Blue | Sodium cacodylate 27.7<br>Free cacodylic acid 4.8<br>Water, sodium chloride balance | 3.1 | 9.3 | 5–7.5 | Rapid short-term defoliation. Good for grass control and use on rice |

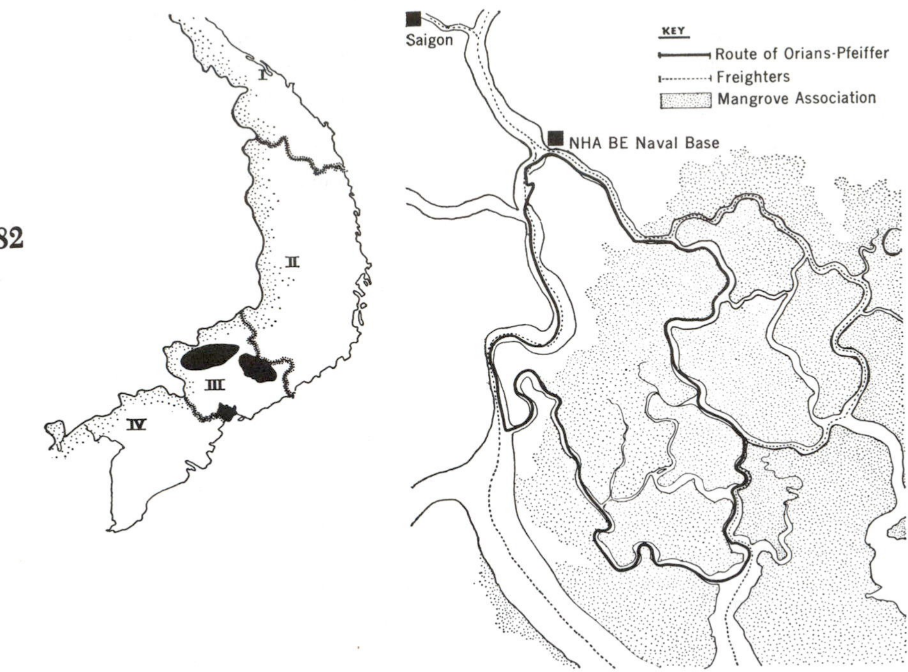


Fig. 1. (Above, left) Approximate areas of massive defoliation in III Corps, as indicated by Chemical Operations, MACV. (Above, right) Map of route through Rung Sat Special Zone. (Below) Defoliated mangrove association in Rung Sat Special Zone.

This estimate is based upon the assumption of immediate redistribution of seeds to the defoliated areas and the presence of suitable germination conditions when they arrive. Although our observations were limited to what we could see from the boats with binoculars, there is reason to believe that the timetable may be somewhat longer than this. Possibly conditions for seed germination are not now very good in the defoliated forests. The unusual soil conditions of mangrove forests may result in a failure of the herbicides to be decomposed. If the molecules remain bound to the soil particles they might influence seed germination for a long time. Alternatively, seed dispersal into the areas is difficult because of the large areas in which mature trees have been killed. Many of the areas, as a result of continued soil deposition under the trees, are flooded only at the highest of high tides, and seeds must be transported for long distances from the river channels under very unfavorable conditions. It cannot be excluded that reestablishment of the original forest may be impossible except along the edges of the river channels and backwaters.

Military operations in Vietnam provide an opportunity to study the effects of unusually high rates of application of herbicides. For example, before jet pods were installed in the C-123 aircraft, the planes were unable to remain aloft in case of engine trouble. In such a contingency, the crew could jettison the entire contents of the tank (1000 gallons; 3.79 kiloliters), in slightly less than 30 seconds, whereas normal spray time is about 4 minutes. Although such contingencies are said to occur less frequently now, they do continue to happen. On the spray mission which one of us (E.W.P.) accompanied as observer, the spray nozzles of one plane failed to work properly, and the entire tank was unloaded at the end of the target. Because the locations of targets are pinpointed very precisely, and because reports are made of all unusual activities during a spraying mission, it should be possible to keep a record of such occurrences. It is most important that all such incidents be recorded in order to enable biologists, in the future, to investigate the sites of concentrated defoliant applications.

## Effects of Defoliation on Upland Forests

Our observations on upland forests that were sprayed directly were limited to aerial reconnaissance. Regrettably we have nothing to add to the published studies about the short-term effect of defoliants on tropical forest trees after single applications of herbicides (*1*, *2*, *5*). The area in northern Long Khanh province that one of us (G.H.O.) observed from the air had been sprayed previously, and many of the trees on the actual target of the mission already appeared to be dead. Except for the wetter spots which were covered with bamboo, the ground was clearly visible in most areas from the low-flying aircraft. Many areas in War Zones C and D (Fig. 1) have been sprayed more than once, and this multiple spraying is also associated with coverage of wide areas. Vegetative recovery as judged from the air was limited to the growth of bamboo and understory trees rather than to refoliation of the canopy dominants.

Observations of defoliated upland forests were made from the ground by Tschirley (*2*) and Flamm (*6*). They visited defoliated forests near Special Forces camps in Tay Ninh and Binh Long provinces northwest of Saigon, a region of gray podzolic soils. According to these studies, after defoliation, on sites sprayed once, there appears to be a modest kill of canopy trees, but understory seedlings and saplings survive and forest regeneration begins fairly rapidly (Fig. 2). However, on sites that received two sprayings roughly 1 year apart, a heavy kill of all woody plants, including seedlings, is reported. Two or three spray applications may kill approximately 50 percent of commercially valuable timber in such forests. These areas are being invaded by grasses which are resistant to forest defoliants and which may arrest succession by preventing the reestablishment of tree seedlings for a long time. Even if this does not occur, it will take many decades before a mature forest grows. Subtle effects, such as changes in the species composition and forest physiognomy, may persist for much longer than that.

A year after spraying, timber is still in good condition, and could be harvested for commercial use, if equipment and markets are available. However shrapnel will be a serious problem for the Vietnamese lumber industry for many years. Most sawmills report that they lose from 1 to 3 hours each day because shrapnel in the logs severely damages the saw blades. The forestry program is looking for suitable metal detection equipment that might help to reduce this damage.

A variety of herbicides, including picloram, bromacil, isopropylamine, prometone, dicamba, divron, and fenac have been tested for their effects under tropical conditions in Puerto Rico since 1962 (*7*). The plots visited in April 1969 were located at an elevation of 540 meters in the Luquillo Experimental Forest in northeastern Puerto Rico. They had been sprayed in 1965 with a Hiller 12-5 helicopter which delivers the spray over a standard swath 35 feet (10.7 meters) wide. The plots were 60 by 80 feet (18 by 24 meters) separated by buffer zones 20 feet (6 meters) wide and there were three replications, ordered in a randomized block design, with 50-foot (15-meter) buffer zones between the strips. The extent of defoliation had been measured 1 year after treatment, the percentage of defoliation on each tree being estimated visually. Apparently there had been no recent ground visits to the sites because all the trails were overgrown and the boundaries of the plots were almost impossible to find. There has been little interest in the continuing effects of the herbicide treatments. This is unfortunate for some areas received very high rates of herbicide application [27 pounds acid equivalent per acre (30.2 kilograms per hectare)]. Assays of growth rate and germination of cucumbers, made in soils up to 1 year after application of the herbicides, revealed relatively high concentrations of picloram, although this technique does not provide precise quantitative measures.



There is a possible serious source of error in the visual estimates of the speed of refoliation in these Puerto Rican rain forest plots. At the higher rates of herbicide application, it was clear that most of the trees had been either killed or severely damaged. However, these plots had been invaded by vines which climbed the trunks of the dead trees and spread out over the for-

Fig. 2. Defoliated crowns of *Lagerstroemia* and *Pterocarpus* near Tong Le Chon. [Courtesy of Barry D. Flamm]

mer canopy. On some of the plots nearly all of the greenery above 3 meters was contributed by vines and not by refoliation of the original trees. Nevertheless, a quick visual estimate, particularly if it were made from a helicopter, might be taken to indicate that extensive refoliation of trees had occurred. The vine-choked plots will not return to their former state as rapidly as they might otherwise, because the dead trunks will probably collapse under the weight of the vines in a few years, creating a low, vine-covered mat through which regeneration could be very difficult. We urge that continued studies of vegetation succession on these and other Puerto Rican test plots be undertaken so that the time required to reestablish the original forest and the factors influencing the pattern of succession can be determined.

Some vine invasion was also characteristic of plots receiving lesser amounts of herbicides, but a severe setback in these forests did not appear to have taken place. Seedlings of mahogany, *Swietenia macrophylla*, and Caribbean pine, *Pinus caribaea*, which had been planted in some of the plots a month after defoliation were surviving quite well.

### Effects of Defoliants on Animals

Tschirley obtained no direct information on the effects that killing the mangroves had on animal populations, but he cited statistics that the fish catch in the Republic of Vietnam has been increasing. Because many factors influence total fish catch and because most of the fish are caught in regions not directly exposed to defoliation, the significance of these data is unclear. Therefore, we attempted to learn as much as we could about animal populations in the defoliated mangrove forests.

As might be expected, the almost complete killing of all of the vegetation of the mangrove areas by herbicides has had a severe effect upon the animals living there. During our tour of the defoliated areas we did not see a single species of insectivorous or frugivorous bird with the exception of barn swallows, *Hirundo rustica,* which are migrants from the north. Although no data regarding the bird populations in the Rung Sat prior to defoliation exist, our experiences in mangrove areas in tropical America indicate that there should have been large numbers of land birds. For example, in Panama as many species of birds were found in a pure red mangrove (*Rhizophora mangle)* forest as would be expected on the basis of the leaf height profile (density of leaves per unit volume as a function of height of forest) of the stand (*8*), and in a brief census of a similar mangrove forest (primarily *Rhizophora*) in Costa Rica, 44 species of land birds which appeared to be resident and breeding were recorded (*9*). Mangrove areas throughout the tropics are rich in bird species (*10*), many of them restricted to that type of vegetation, and the Southeast Asian mangroves are no exception.

Fish-eating birds seem to have suffered less severely, but even their numbers were much fewer than we expected. The species of birds and the number of individuals per species that we observed during a 2-hour period in the defoliated areas, are: oriental darter (*Anhinga melanogaster*), 2; grey heron (*Ardea cinerea*), 13; large egret (*Egretta alba*), 3; little egret (*E. garzetta*), 12; intermediate egret (*E. intermedia*), 1; javan pond heron (*Ardeola speciosa*, 6; stork· (*Leptoptilos* sp.), 2; black-winged kite (*Elanus caeruleus*), 1; osprey (*Pandion haliaetus*), 9; whimbrel (*Numenius phaeopus*), 3; little tern (*Sterna albifrons*), 10; and white-breasted kingfisher (*Halcyon smyrnensis*), 2. All except the kite, which feeds on small mammals, are fish-eating birds. This suggests, as would be expected, that aquatic food chains in the mangroves may have been less severely affected by defoliation than the terrestrial ones. The only other vertebrate we saw in the defoliated areas was a large crocodile *Crocodylus* on the bank of a small channel.

Of all the areas in Vietnam, the mangroves in the delta of the Saigon River have probably been most severely affected by defoliation. The area treated has been very extensive, covering many square kilometers, the vegetation is extremely sensitive to herbicides, and many of the species of animals inhabiting mangroves are restricted to that type of vegetation. These animals are therefore inhabitants of "islands" surrounded by unsuitable habitat and as such are expected to have higher rates of extinction even under normal conditions than species of more continuous habitats (*11*). These same properties make them more susceptible to local and complete extermination by disturbance and destruction of habitat than are species of upland habitats. Long-term studies of the ecology of the Rung Sat should be given a high priority, including investigation of the status of such invertebrates as crustaceans.

Birds were scarce in the heavily defoliated plots in Puerto Rico, but in the more lightly treated areas both species composition and general population density were comparable to that found in untreated areas in the general vicinity. There was not time to conduct a complete census, but it is doubtful whether such studies would be worth while since the plots are so small that they are less than the average size of most bird territories. Therefore, the effects of the tests on bird populations should in any event be mimimum. It is important to remember, however, that results from spraying of very small areas cannot be assumed to apply to extensively treated areas.

### Toxicity of Herbicides

The problem of the toxicity of herbicides to animals is not yet resolved.

Nearly all studies are short term, and results are contradictory. Some reports (*1*) suggest that at the prevailing concentrations herbicides are not directly toxic to animals, and Tschirley (*2*) states: "There is no evidence to suggest that the herbicides used in Vietnam will cause toxicity problems for man or animals." However, according to Holden (*12*) 2,4-dichlorophenoxyacetic acid (2,4-D) may constitute a potential danger to fish even in normal use. The $LD_{50}$ value for salmonids during a 24-hour exposure to 2,4-D is 0.5 part per million. Thus, a concentration of 4 pounds of active constituent per gallon (458 grams per liter) in a small [10 cubic feet (0.28 cubic meter) per second] stream would expose fish to about 100 times the $LD_{50}$. It should be recalled that the rate of application of 2,4-D in Vietnam is slightly greater than this. According to Holden, the toxicity of 2,4,5-T is about one-half that of 2,4-D.

Another possible source of toxicity to animals from defoliation is an indirect effect of the activity of 2,4-D in plants. Stahler and Whitehead (*13*) reported that there are several cases of cattle becoming ill or dying after eating certain species of weeds that had been treated with 2,4-D. These authors present data that clearly indicate that sublethal dosages of 2,4-D may markedly affect the metabolism of certain plant species so that toxic quantities of nitrates accumulate in the treated plants. In the animals the nitrates are changed to nitrites which are absorbed into the blood producing methemoglobin which results in oxygen deficiency to the tissues. This condition may cause death or illness resulting in abortion. Leaves of sugar beets that had been treated with 2,4-D were shown to have amounts of nitrate well above the minimum lethal concentration. A recent statement (*14*) by an American agricultural specialist emphasizes that "Dairy cows should not be grazed on irrigated pasture for seven days after application of 2,4-D at the one-half pound and over rate of application."

To our knowledge there are no studies of the effects of agent Orange on Vietnamese forage plants to determine whether these plants become toxic to animals due to nitrate accumulation following defoliation with Orange. Determination of nitrate concentration in leaves should be made in defoliated and control areas, and the hemoglobins of animals which feed on exposed plants should be studied.

A recent study of the teratogenicity of 2,4-D and 2,4,5-T (*15*) shows that the latter compound is highly teratogenic in rats and mice at dosages that are possible of ingestion by humans in Vietnam.

We uncovered little evidence of direct toxic effects on animals. The Tan Son Nhut air base in Saigon is sprayed by hand with agent Blue several times each year and nonetheless has a serious rat problem. A trapping crew every night puts out 100 snap traps and 30 live traps, baited with bacon. From 3 January 1969 to 19 March 1969, they had trapped 613 rats and 8 viverrids of at least two species. We netted and observed birds on a previously sprayed brushy area near Bien Hoa on two different mornings and found birds very common. We saw much territorial defense and singing as would be expected at the end of the dry season in the tropics.

We did receive one report of many sick and dying birds and mammals in forests following defoliation and two reports of death of large numbers of small pigs near Saigon, but were unable to follow up either report. The Ministry of Agriculture has received no bona fide claims of animal damage from defoliants. Nevertheless, we must not forget that habitat destruction, which defoliation regularly accomplishes, is in most cases the equivalent of death for animals. The widespread view that animals can move to other nearby areas is untenable because recent ecological evidence suggests that tropical forests hold the maximum number of individuals of most species that the resources will support. Reduction of forest habitats will decrease the populations of forest animals by an equivalent amount. Nor is it true that forest species can live successfully in the greatly modified conditions which prevail in even partially defoliated forests. Species characteristic of successional stages will, of course, be expected to move into the disturbed areas, but even they may have to wait until the basic food resources, such as insects and fruit, have built up again, and we do not know how long this will take.

A phenomenon that should be investigated immediately is a widespread sickness which appears at the beginning of the rainy season in commercially important freshwater fishes. The symptoms are many small, round, dark spots in the muscles. The taste of the fish is also adversely affected. Poor people continue to eat the fish even though they are diseased. This disease has always been characteristic of that time of the year in Vietnam, but the director of the Institute of Fisheries has received reports which suggest that the incidence is now higher than before. Conditions in the shallow water of the fields are ideal for concentration of herbicides. The Vietnamese fisheries people, who are qualified and presently have greater mobility in the country than Americans, are in a position to initiate such studies now. The Minister has already circulated a letter among his representatives in the provinces asking for any information they may have, and we agreed to help formulate a more detailed questionnaire for future circulation.

Some insight into the possible harmful effects of the herbicides now in use in Vietnam may be gained by consulting the labels which give directions for their uses. Dow Chemical Co., makers of agents Orange and White, warn that these chemicals should be kept out of reach of children and animals. The label on agent White states: "Do not allow material to contaminate water used for irrigation, drinking, or other domestic purposes." Dow Chemical Co. also recommends that no grazing be allowed on treated areas for 2 years after treatment and that some broad-leafed crops may show damage 3 years after application.

Ansul Chemical Co., makers of agent Blue, state that when an individual is exposed (to cacodylic acid) daily for extended periods, the inspection of skin sensitivity should be supplemented by monthly urinalysis for arsenic. Symptoms of acute poisoning from cacodylic acid are headache, vomiting, diarrhea, dizziness, stupor convulsions, general paralysis, and death. The dosage required to cause these symptoms may be as little as one ounce (28 grams) of cacodylic acid per human adult.

**Effects of Defoliants on Rubber Culture**

Most studies of the effects of defoliants on forest trees have been confined to observation of the percentage of defoliation after relatively short intervals following single applications of herbicides. Studies of the effects of defoliation on rubber trees have been initiated by the Rubber Research Institute of Malaya and by the Rubber Research Institute of Vietnam because of the economic importance of rubber trees to Vietnam and because of the widespread damage to plantations from military spraying. Although these studies contain the best available data, they have been limited by the shortage of funds and difficulties of field work in a country during wartime.

Damage to rubber trees in Vietnam has been extensive. During 1967–68, the Institute staff visited over 200 different plantations in the provinces of Bien Hoa, Binh Duong, Gia Dinh, Hau Nghia, Long Khanh, Phuoc Tuy, Tay Ninh, and Binh Long. (This covers most of the area between the rice-growing areas of the Saigon and Mekong River deltas and the mountainous central part of the country.) On this extensive area of approximately 130 by 40 kilometers, all plantations reported damage by defoliants. More than 40,000 hectares planted with rubber trees were defoliated at least to the extent of 10 percent. It is difficult to estimate the total amount of damage resulting from defoliation. Plantation owners might possibly submit exaggerated claims, but there is no doubt that the damage has been considerable. For example, Plantation de Dautieng of the Michelin Company has been affected by defoliants three times since 1965. In all cases, the defoliant has not been applied directly to the rubber trees, but has been carried by the wind from applications in the general area. No trees were killed, but, by measuring the drop in latex production due to stoppage of tapping, decreased yield of lightly damaged trees, and costs of cutting and trimming back partially killed trees, the company estimates that the damage amounted to $27,835 in 1965, $37,479 in 1966, and $27,844 in 1967. The areas of spraying, direction of the wind, and areas of the plantation affected are shown in Figs. 3 to 5.

The yield of rubber per hectare is decreasing. In 1960, rubber plantations in Vietnam yielded 1066 kilograms of dry rubber per hectare (on plantations of more than 25 hectares). In 1967, the yield had dropped to 793 kilograms per hectare. In contrast, in Malaysia the yield in 1960 was 758 kilograms of dry rubber per hectare, but had risen to 1007 kilograms per hectare in 1966. The decrease in yield in Vietnam is due to a combination of circumstances such as the cessation of tapping forced by military action, less experienced labor and less thorough control in the field, herbicide damage, lack of general upkeep of plantations, and the cutting of rubber trees along roads where about 3000 hectares have already been cut. The relative importance of each factor seems impossible to assess. It is a fact that they are all the consequence of the war.

The total yield of rubber in Vietnam has also declined. In 1960, 77,560 tons of dry rubber were produced. Rubber exports amounted to $48,000,000, which was 56 percent of South Vietnam's total exports for that year. In 1967, the yield had dropped to 42,510 tons of dry rubber, which, considering the devaluation of the piaster, amounted only to $12,800,000. Inasmuch as other exports suffered even more heavily, this diminished amount (26 percent of the 1960 exports) made up 72 percent of South Vietnam's exports, which had decreased to $17,800,000, or 20.8 percent of the 1960 exports (*16*).

If a rubber tree is completely defoliated by herbicides, the Institute recommends that planters stop tapping until its new leaves are fully grown. Because it takes a month for a new leaf to grow to full size from the time of breaking of bud dormancy and because dormancy is not usually broken immediately after defoliation, the minimum period of stopping is about 2 months. The maximum period of stoppage is, of course, permanent if the tree is killed. If tapping is not stopped while the tree is defoliated, there is competition between growth of new leaves and yield within the tree, and the future health of the tree is jeopardized. In a number of cases where trees were not killed, tapping has been stopped for as long as 1 year. If only some of the leaves are lost, tapping can be continued, but there is a drop in latex production after a lag of about 1 month. The loss, over a period of a year, has been estimated to be sometimes as much as 30 percent of the normal yield of latex. At current prices that amount of loss reduces profit from about $90 per hectare per year to nothing. As a consequence, most of the small plantations have been unable to stay in business.

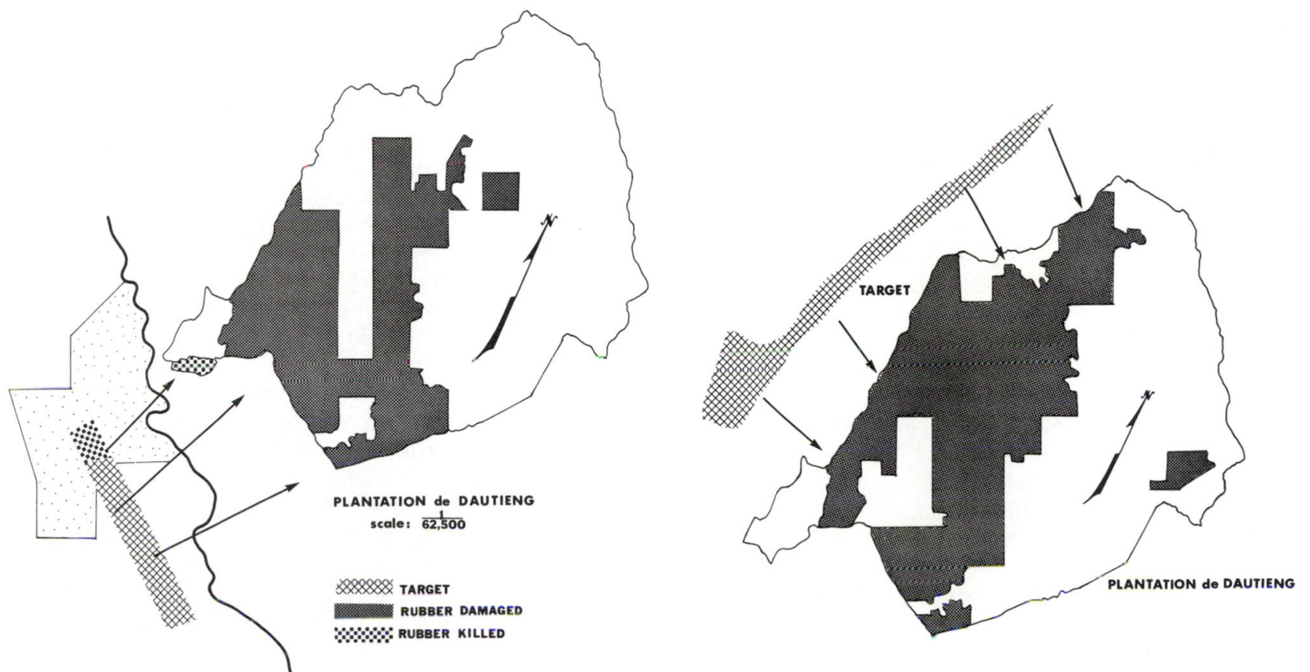


Fig. 3 (left). Target area and areas of rubber trees on Plantation de Dautieng affected by defoliation in February 1965, as indicated by Plantations Michelin. Wind direction is indicated by arrows. Fig. 4 (right). Target area and areas of rubber trees on Plantation de Dautieng affected by defoliation in December 1966, as indicated by Plantations Michelin. Wind direction is indicated by arrows.

Only the large planters, with solid financial backing, can afford to remain in operation despite the war.

According to studies by Dow Chemical Company (as reported to us by the Rubber Research Institute), the defoliant is absorbed through the leaves of the trees and is carried down through the phloem within 24-hours, and symptoms of defoliation appear within a few weeks after spraying. The distance the defoliant travels down the tree is a function of the dosage received, and the Institute people have assessed this by the simple device of cutting into the trunk of the trees at different heights to investigate the flow of latex. Necroses are also clearly visible in the sectioned trunks, many of which we examined in the laboratories of the Institute. As might be expected, the smaller the rubber tree, the more readily it is killed by defoliants. Research in Malaysia has

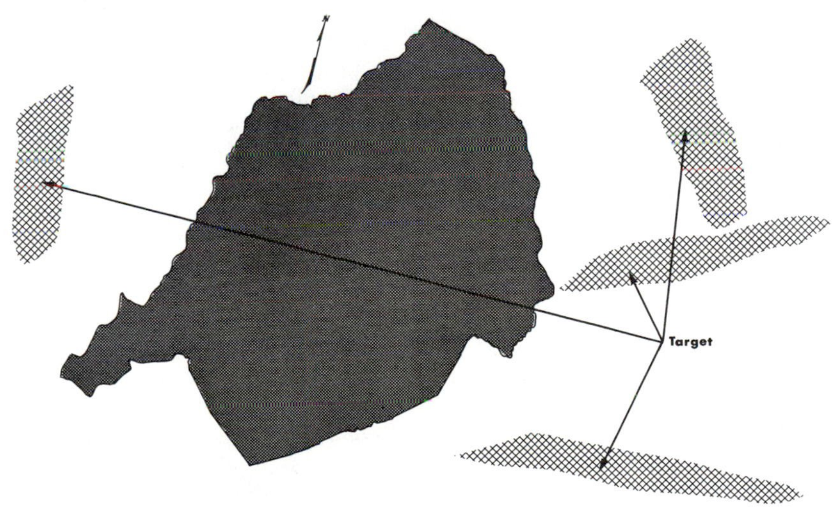


Fig. 5. Target area and areas of rubber trees on Plantation de Dautieng affected by defoliation in May 1967, as indicated by Plantations Michelin.

shown that a wide range of concentrations of the *n*-butyl ester of 2,4,5-T killed rubber seedlings in 6 weeks (*3*). Accidental defoliations in Vietnam indicate that trees less than 7 years old can be killed by the dosages used in military operations, but that older trees normally recover. Nevertheless, all trees on 100 hectares on Plantation Ben Cui were killed by herbicides in 1965, despite the fact that the trees were 33 years old. From such occurrences, the Rubber Research Institute concluded that repeated defoliations threaten the very existence of rubber culture in Vietnam (*17*).

In spite of such evidence, Chemical Operations Division, United States Army, claims that rubber trees cannot be killed by defoliants. According to our observations, although we do not claim expertise in this field, damage to rubber production is severe. The Rubber Research Institute, which does not itself maintain any plantations and cannot be accused of bias on that account, seems to be in an excellent position to conduct further research into the physiological effects of defoliants on trees. Funds are urgently needed for this purpose.

### Effects of Accidental Defoliation

The extent of damage resulting from wind-blown and gaseous herbicides has been much debated. Agent Orange is classified as a volatile herbicide by plant physiologists, but physical chemists regard it as nonvolatile. Under proper weather conditions nearly all of the spray is deposited on the vegetation or ground within a minute after release from the aircraft. Those vapors formed during fall of droplets subsequently diffuse according to the laws of gaseous diffusion. Therefore, it has been concluded that "The rate of downwind movement of vapors, and therefore the duration of exposure of plants to the vapors, is dependent upon wind speed in the first few minutes subsequent to spray release. *While no quantitative data are available*, it is our considered judgment, based on the above reasoning, that vapors arising during the actual spray operation, as usually carried out, can be dismissed as a source of herbicides for crop damage outside target areas" (*18*) (emphasis added by us). This assumes the existence of inversion conditions and that transport of the liquid spray droplets by the wind is negligible. Our direct observations and interviews suggest that the seriousness of this problem has been greatly underestimated.

We were able to observe defoliation damage to several species of trees far removed from target areas. On 25 March, in the village of Ho-Nai, we observed many fruit trees that had recently been damaged by defoliants. The characteristic sign was the presence of curled, dead leaves on the trees. Damage seemed excessive on the south side of the trees, which suggests that the spray was carried into the village by a southerly or southeasterly wind. Villagers informed us that spray had hit them about 1 week previously. Chemical Operations Division, United States Army, reported to us that a defoliation aircraft had had to jettison its chemicals at the time of takeoff from nearby Bien Hoa Air Base, approximately at the time when the Ho-Nai residents had observed the spray. The most severe damage was to jack fruit (*Artocarpus heterophyllus*, Moraceae) which is also a producer of a milky sap. The residents of Ho-Nai claimed to have been affected by defoliation missions seven times within the past year.

On 23 March, in a residential area between Saigon and the U.S. Air Base at Bien Hoa, we examined and photographed many diseased mango trees. The owner, a biologist trained in the United States, claimed that the trees suffered defoliation 3 years ago, after which they became infected and had not since flowered or produced fruit. In other areas we subsequently observed the same symptoms in mango and other trees. According to the Rubber Research Institute, latex-producing trees seem to be more susceptible to herbicide damage than other species.

Every Vietnamese biologist we talked to explained that actual herbicide damage has been frequent and regular over much of the delta region. In the Ministry of Agriculture we were shown photographs of damaged jack fruit, manioc, and rubber and were told that many guava trees had been killed. The Ministry has attempted in a preliminary way to assess the total damage reported and found it to be so extensive that adequate financial compensation to the owners of damaged trees would probably be impossible. The experimental station of the College of Agriculture of the University of Saigon at Tu Duc has been affected by wind-blown defoliants several times, usually with almost complete kill of vegetables.

It is difficult to determine the amount of claims actually submitted to or paid by the Vietnamese government. Funds for the payment of defoliation claims are provided by the United States, but the claims are handled by the Political Warfare Department of the Air Force of the Republic of Vietnam under the Military Civil Assistance Program. Damage claims are considered and paid by province officials under guidelines established by the central government. Everyone we talked with agreed that payments are minimum. We were told by Vietnamese that people who file claims with the government are often threatened with imprisonment if they continue to press their claims. Many others do not attempt to file claims because they feel it will be of no use. United States officials argue that most claims are fraudulent.

It is our opinion that significant quantities of defoliants are regularly carried by the wind over broad areas of cropland in the Republic of Vietnam. Even given the difficulties of making first-hand observations in a war zone, it would be possible for inde-

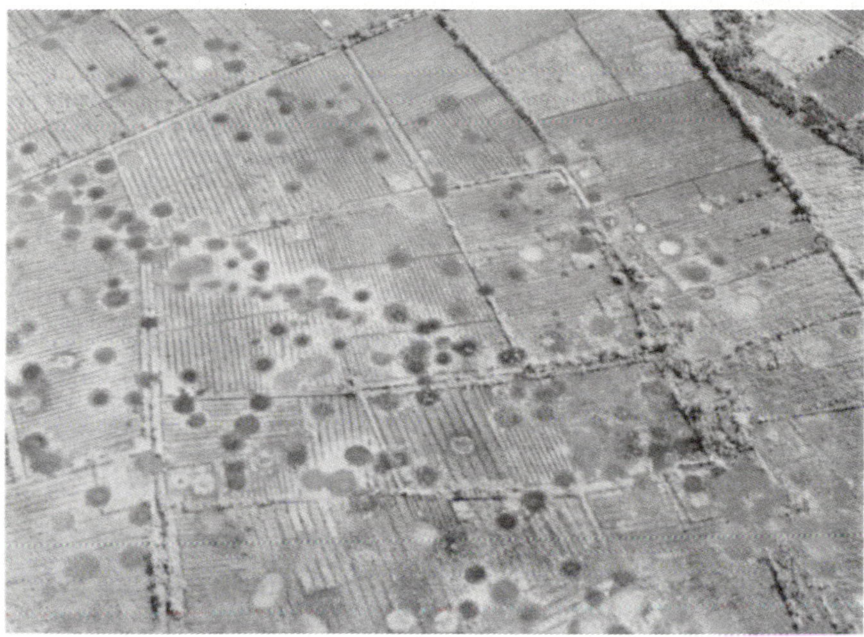

Fig. 6. Craters caused by bombs from B-52 aircraft in manioc fields about 20 miles northwest of Saigon.

pendent observers to verify or disprove many of the conflicting claims. Such a study is urgently needed. Rising damage claims in 1967 caused a serious review of the defoliation program at that time (*19*). Serious controversies over damage caused by wind-blown spray had arisen, and the psychological impact on the Vietnamese was great. It is noteworthy (and substantiates the claims of widespread crop damage) that the United States now has changed its policy and uses primarily agent White in the delta region because its volatility is lower than that of the other available agents. Nevertheless, we encountered many reports of very recent damage in that area.

## Crop Destruction

Direct and deliberate application of agent Blue to cropland has been restricted to the highland regions of the country which are held by the National Liberation Front. Consequently, scientists of the Republic of Vietnam and those of the United States are unable to make first-hand studies at present. American officials consider the program very successful because many captured soldiers from these areas are seriously undernourished; some to the extent of being stretcher cases at the time of capture. These reports might suggest that the "resource denial" program has been successful, but there are strong reasons for believing that food shortages affect women, children, and elderly people much more than they affect soldiers (*20*).

## Effect of B-52 Bombing

Although it has not attracted the concern of American scientists, the damage caused by raids with B-52 bombers is of considerable ecological significance. The 500- and 750-pound bombs dropped by these aircraft leave craters as much as 30 feet deep and 45 feet across. Most of these are filled with water even late in the dry season. The army does not disclose the total number of bombs dropped, and the total area affected cannot be calculated accurately. However, the magnitude of the effect can be estimated from the following facts. A standard load for a B-52 is 108 500-pound bombs or nearly 30 tons of explosives. Normally, a "mission" consists of 3 to 12 aircraft. In 1967, 982 missions were flown over the Republic of Vietnam. In 1968, 3022 missions were flown (Table 3). If one assumes an average of eight planes per mission, then one can estimate that about 848,000 craters were formed in 1967 and 2,600,000 craters in 1968. As one Vietnamese put it, we are making the country look like the surface of the moon (Fig. 6). Unless heavy earth-moving equipment can be brought to the sites to fill the craters they will remain a permanent feature of the Vietnamese landscape. Areas such as War Zones C and D, which have been heavily hit by B-52 attacks, are riddled by craters.

Since most of the attacks have occurred in militarily contested areas it has not been possible for scientists to investigate heavily cratered areas to determine the effects on local ecology. Obviously, they are potential breeding grounds for mosquitoes; they may possibly be fish-breeding ponds; they may also render many agricultural areas difficult to utilize.

## Miscellaneous Effects

The prolonged military activity in Vietnam is causing other ecological upheavals. Not the least are the major sociological changes that are taking place in the country, such as the

Table 3. Missions flown by B-52 bombers over Vietnam.

| Month | I Corps | II Corps | III Corps | IV Corps | DMZ | | North Vietnam |
|---|---|---|---|---|---|---|---|
| | | | | | N | S | |
| | | | *1967* | | | | |
| January | 18 | 14 | 27 | 1 | 2 | 7 | |
| February | 23 | 30 | 30 | 1 | 1 | | |
| March | 45 | 23 | 32 | | | | |
| April | 55 | 13 | 22 | 2 | 2 | 4 | |
| May | 55 | 27 | 23 | 1 | 1 | 3 | |
| June | 45 | 28 | 25 | 1 | | | |
| July | 44 | 31 | 22 | 3 | | 3 | 3 |
| August | 26 | 28 | 24 | | | 24 | 1 |
| September | 13 | 9 | 6 | | | 57 | 15 |
| October | 17 | 15 | 20 | | | 30 | 27 |
| November | 13 | 47 | 16 | 1 | | 12 | 9 |
| December | 6 | 19 | 22 | | | 27 | 8 |
| | | | *1968* | | | | |
| January | 59 | 22 | 23 | | | 4 | 1 |
| February | 204 | 53 | 34 | | | | |
| March | 222 | 58 | 27 | 4 | | 6 | |
| April | 173 | 66 | 19 | 7 | | 2 | |
| May | 71 | 123 | 27 | 10 | | 13 | 1 |
| June | 24 | 87 | 171 | 11 | 4 | 6 | |
| July | 28 | 34 | 152 | 26 | 7 | | 45 |
| August | 71 | 55 | 161 | 13 | 8 | 8 | 8 |
| September | 51 | 55 | 167 | 18 | 1 | 10 | 8 |
| October | 86 | 33 | 128 | 25 | 2 | 1 | 15 |
| November | 45 | 36 | 109 | 17 | | | |
| December | 53 | 22 | 125 | 17 | | | |

amazingly rapid rate of urbanization of the population. This results as people flee from war-torn countryside or are forcibly transported to the city. Within the last decade Saigon has changed from a quiet city of 250,000 to an overcrowded city of 3,000,000 inhabitants. The tremendous infusion of American capital has also resulted in rapid increase in the number of motorized vehicles in the streets. Japanese motor bikes and small cars of Japanese or Italian manufacture seem to be prevalent. Traffic accidents are common. Saigon's air pollution problem due to fumes from the mixture of gasoline and oil which serves as fuel is so severe that many trees along the major arterials in the city are dead or dying. (It is possible that the winddrift of defoliants has contributed to weakening the trees, but it is likely that the major cause is fumes from motor vehicles.) There are no immediate prospects for any improvement as the population of the city continues to grow and creation of an adequate municipal transportation system seems improbable.

A major cause of forest destruction in Vietnam today is fire. Some fires are started deliberately by the Vietnamese army and some are caused by artillery shells. Over 40 percent of the pine plantations in the country have been burned recently; the extent of destruction of the mixed forests is unknown. We were unable to estimate the total area involved.

Because of the war, all hunting in the Republic of Vietnam has been officially discontinued. Nevertheless, there are large numbers of armed men in the forest, many of whom are poorly nourished. Presumably, they regularly shoot all suitable food animals. Tigers, on the other hand, seem to have benefited from the war. In the past 24 years, they have learned to associate the sounds of gunfire with the presence of dead and wounded human beings in the vicinity. As a result, tigers rapidly move toward gunfire and apparently consume large numbers of battle casualties. Although there are no accurate statistics on the tiger populations past or present, it is likely that the tiger population has increased much as the wolf population in Poland increased during World War II.

## Summary and Conclusions

In Vietnam the chemical weapons of a technologically advanced society are being used massively for the first time in a guerrilla war. In this conflict there are no battle lines, no secure territory, and no fixed, permanent military installations which can serve as targets for attack. Rather, the military efforts are aimed at increasing the toll of fatalities, denying food to the enemy, and depriving him of the cover and concealment afforded by natural growth. This type of warfare is, therefore, extremely destructive, both of human lives and environment. Our own observations showed the profound effects of denuding the country of growth. The military is emphatic about the effectiveness of defoliation in reducing American casualties significantly. The demand for the services of 12th Air Commando Squadron greatly exceeds their ability to supply them. Although the total number of requests for defoliation missions was not disclosed, we were told that even if no further requests were made, the defoliation crews would be kept busy for years by the present backlog. The current extent of the defoliation program is not determined by military demand nor by any considerations of saving the ecology and viability of the land and natural resources of Vietnam, but solely by competition for equipment and personnel.

With general agreement among military experts that defoliation is a potent weapon in guerrilla warfare, it is to be expected that in any future wars of this nature more extensive use will be made of it. At the end of their war against the Vietnamese, the French discovered the usefulness of helicopters as field combat aircraft, but they had only about a dozen at their disposal. There are now several thousand helicopters in Vietnam as a major component of our offensive air power. Making a realistic appraisal of defoliation and its ecological consequences, we must, therefore, consider not only the present extent of use but also anticipate greatly expanded defoliation actions in the future.

We consider that the ecological consequences of defoliation are severe. Enough is now known to reveal that a significant fraction of mature trees in most forests are killed by single applications of herbicides and that almost complete kill, including destruction of seedlings and saplings, is to be expected if repeated sprayings are made. Because of military demands for respraying, we must expect virtual elimination of woody vegetation of defoliated sites as a common result of the military use of herbicides.

It is evident that the most stringent regulations for the application of defoliants cannot prevent the widespread dispersal of herbicides to areas far beyond those that were intended to be defoliated. We found abundant evidence of repeated moderate to severe defoliation of trees and herbs in areas many miles removed from sites of direct application. Every responsible Vietnamese person we met confirmed this. Moreover, a pilot in a war zone will jettison his load of defoliant, rather than jeopardize the safety of his crew and plane, and a spray plane will not return to its base with a full tank because its crew found the temperature or the wind velocity higher in the target area than anticipated. Military use of defoliants will inevitably result in herbicide damage to areas that are far more extensive than those specified as targets.

It is evident that the defoliation program has had tremendous psychological impact upon the Vietnamese people and has profoundly affected their attitude toward Americans. A farmer whose entire crop has been destroyed by herbicides, whose fruit trees do not bear fruit for 3 years, will inevitably be resentful. We were told repeatedly, though politely, that a significant deterioration of attitudes toward Americans has resulted from the massive use of defoliants. The claim that defoliation is more humane than other weapons of war because it does not directly cause human casualties, may appeal to those whose land has not been defoliated, but hardly to those whose food supply or property has been destroyed. A realistic assessment of the effects of defoliation must take into account the psychological effects upon the people.

The politically sensitive nature of effects of defoliation is fully recognized by the military authorities. Although they claim that defoliants produce no long-term effects on the environment, they have instituted the most stringent regulations to govern their use. The Army claims that it is more difficult to get permission for the defoliation of trees in Vietnam than for killing persons, and permission to spray rubber trees has never been granted, according to military sources, even when enemy forces were "known" to use plantations for concealment. It seems that preferential treatment of the politically powerful rubber interests in Vietnam has added to the hostility of the poorer Vietnamese.

The secrecy surrounding the use of defoliants in Vietnam has also contributed to the feelings we have reported above. The government of the Republic of Vietnam and American officials have not disclosed information to the Vietnamese about the agents used, areas sprayed, and the nature of the chemical action of defoliants and herbicides. The most concerned Vietnamese scientists did not know the chemical composition of the herbicides even though they have tried to ascertain it from their government.

## Recommendations

American scientists will want to know what investigations might be immediately possible to sift facts from among so many conflicting claims regarding the ecological effects of defoliants and to stem the tide of increasing mistrust between the Vietnamese and the Americans. Support for research projects should be initiated by the American scientific community without delay. In Vietnam there are scientists, well-trained at American and European universities, who are deeply concerned about the effects of the war on their country. They are eager to conduct research that is necessary for the rehabilitation of their ravaged land. The flora and fauna of the country are well known. The Rubber Research Institute of Vietnam continues to function, although it has once been displaced by military action. It is capable of expanded research into the physiological effects of defoliants on rubber trees and other species. Its staff is interested in investigating the possibilities of diversifying so that it can advise rubber planters on avoiding complete dependence upon rubber. A modest investment of funds for Vietnamese scientists is likely to produce important research results. It would also improve Vietnamese relations with American scientists.

Although long-term studies, such as following vegetational succession on heavily defoliated areas, would be impossible for Vietnamese (Saigon) or American investigators, there are no insuperable barriers to the investigation of fish diseases, of methods of minimizing herbicide damage to commercially important trees which have been deliberately or inadvertently sprayed, and of further studies of toxicity to animals. It should also be possible to gather soil samples from areas that have been subjected to different treatments to learn more about the fate of arsenical compounds, their effects on soil microorganisms, and possible accumulation in the soil of the more persistent herbicides such as picloram. We urge that such studies be initiated now rather than be delayed until hostilities cease, although obviously the difficulties are great. We recommend most strongly that the American Association for the Advancement of Science, in accordance with its resolutions of 1966 and 1968 (*21*), take the initiative in setting up an international research program on the long-range effects of the military use of herbicides in Vietnam. We believe that such action is necessary if United States scientists wish to maintain (or regain) the respect of scientists in Southeast Asia.

**References and Notes**


1. W. B. House, L. H. Goodson, H. M. Gadberry, K. W. Docter, *Assessment of Ecological Effects of Extensive or Repeated Use of Herbicides*, Midwest Research Institute Report No. 3102-B (1967).
2. F. H. Tschirley, *Science* **163**, 779 (1969).
3. F. Hutchinson, *J. Rubber Res. Inst. Malaya* **15-5**, 241 (1958).
4. Information supplied by officers in the Chemical Operations Division, Military Assistance Command, Vietnam.
5. R. A. Darrow, G. B. Truchelut, G. M. Bartlett, *OCONUS Defoliation Test Program, Tech. Rep. No. 79* (U.S. Army Biology Center, Fort Detrick, Maryland, 1966), p. 149.
6. B. R. Flamm, *A Partial Evaluation of Herbicidal Effects to Natural Forest Stands Principally in Tay Ninh Province* (U.S. Agency for International Development, 1968).
7. J. E. Munoz and L. W. Hill, *The use of herbicides for site preparation and their effects on tree survival, U.S. Forest Service Research Note No. IFT 12* (Institute of Tropical Forestry, Rio Piedras, Puerto Rico, 1967).
8. R. H. MacArthur, H. Recher, M. Cody, *Amer. Natur.* **100**, 319 (1966).
9. G. H. Orians, *Ecology*, in press.
10. G. D. Field, *Ibis* **110**, 354 (1968); R. P. Ffrench, *ibid.* **108**, 423 (1966); T. Haverschmidt, *ibid.* **107**, 540 (1965); E. M. Cawkell, *ibid.* **106**, 251 (1964); I. C. T. Nisbet, *ibid.* **110**, 348 (1968); B. C. Smythies, in *The Birds of Burma* (Oliver & Boyd, Edinburgh, 1953).
11. R. H. MacArthur and E. O. Wilson, *The Theory of Island Biogeography* (Princeton Univ. Press, Princeton, N.J., 1967).
12. A. V. Holden, *J. Inst. Sew. Purif.* **1964**, 361 (1964).
13. L. M. Stahler and E. I. Whitehead, *Science* **112**, 749 (1950).
14. T. Woiciechowski (Missoula County Agent), *Missoulian* (8 May 1969).
15. B. Nelson, *Science* **166**, 977 (1969).
16. From data supplied to us by Plantations Michelin, Saigon.
17. J. P. Poliniere, *Situation Phytosanitaire de Différent Sites Hévéicoles en Relation avec Apport Probable de Defoliants (à suivre)*, arch. 5/67 (Institut des Recherches sur le Caoutchouc au Viet-Nam, 1967).
18. C. E. Minarik and R. A. Darrow, *Potential Hazards of Herbicide Vapors* (Headquarters U.S. Military Assistance Command Vietnam, Chemical Operations Division, 1968).
19. E. Pond, *Christian Science Monitor* (25 and 29 Nov. and 8 Dec. 1967).
20. J. Mayer, *Scientist and Citizen* **9**, 115 (1967).
21. D. Wolfle, *Science* **167**, 1151 (1970).

# SECTION V
# SCIENCE, CONFLICT, AND HUMAN PROSPECTS

CHAPTER 23

# The Climatic and Biological Consequences of Nuclear War

HERBERT D. GROVER

Recent studies projecting the consequences of nuclear war estimate that from 750 million to 1.1 billion humans in the Northern Hemisphere could die from the blast, thermal, and radiation effects of a large-scale nuclear war. In addition, the number of individuals suffering serious injury and trauma, many of whom would not recover, could reach the hundreds of millions.[1]

Despite this devastation, perhaps 50 to 70 percent of the human population in both the Northern and Southern Hemispheres might survive the direct effects of a large-scale nuclear war. What would be the prospects for their survival? Would there be adverse indirect effects of nuclear war propagated throughout the global ecosystem that would place in jeopardy the survival of humans and other biota in regions far removed from the conflict? More important, what would be the long-term implications of changes that might occur in the character and degree of human reliance on natural ecosystems?

Projecting comprehensively the climatic and biological consequences of nuclear war brings into question the perceived security ascribed to policies of deterrence, as well as the efficacy of civil preparedness measures, particularly relocation plans. We all presume the probability that nuclear war will occur is low, but the potential consequences are so great that we must give the issue our full attention.

Nuclear war is the greatest environmental threat we face. Yet the problem should be solvable. It is especially important that environmentally concerned citizens and scientists apply their knowledge and understanding to the problem because many of the changes in environmental quality that concern us today, such as air and water pollution by a host of organic and inorganic agents, represent the principle mechanisms whereby life could be threatened in the aftermath of nuclear war.

---

Reprinted by permission from *Environment 26(4):* 6–13, 34–38 (1984). 

Herbert D. Grover is in the Department of Biology, University of New Mexico, Albuquerque, NM.

These are issues that over 100 physical and biological scientists recently considered in preparation for a privately funded conference, held in the fall of 1983,[2] in which I participated. The results of these various studies, elucidating what has become known as the nuclear winter effect,[3] as well as related work that has been performed more recently,[4] will be the focus of this article. Studies currently underway, such as the work being done under the auspices of SCOPE (Scientific Committee on Problems of the Environment) of the International Council of Scientific Unions, and the National Academy of Science, indicate that the next year to two years may see significant new evidence—possibly confirming, possibly radically different—emerging in this highly fluid field.

Basically, the nuclear winter effect refers to sunlight obscuration and corresponding reductions in air temperatures at the Earth's surface brought about by the combined influence of dust injected into the atmosphere by nuclear weapons and soot generated by urban and wildland fires resulting from nuclear war. Other studies have addressed the potential climatic effects of nuclear war,[5] concluding these to be minimal, but the significance of dust and soot was not fully appreciated until recently.[6] Scientists who had previously studied volcanic plumes and atmospheric circulation on other planets have in the last few years turned their attention to these questions.

Another reason for the delay in understanding the nuclear winter phenomenon is that the computer models necessary to simulate the dynamic effects of "nuclear" dust and soot in the atmosphere are only now attaining the level of sophistication required to perform such complex analyses.

Although in some ways imprecise, the calculations identifying the nuclear winter reveal first-order effects on climate that cannot be ignored. Indeed, consequence analyses of this sort have now gained a new importance in the debate regarding nuclear armament issues.[7]

## Nuclear Weapon Effects

With the exception of the Hiroshima and Nagasaki bombs, all of the explosives used in human warfare have depended upon chemical reactions for their expenditure of energy. TNT remains the most notable of chemical explosives and is used as the standard in expressing weapon yield.

Whereas TNT relies on the rearrangement of atoms of oxygen, hydrogen, and nitrogen for its energy release, nuclear weapons derive their energy from the forces controlling the structure and composition of atomic nuclei. These nuclear forces are several orders of magnitude greater in strength than the forces controlling the chemical behavior of atoms or molecules, and may be released through processes of fission or fusion.[8] Fission is the splitting of an atomic nucleus into two daughter nuclei, usually through bombardment by a free neutron. If the appropriate elemental isotopes are present under exacting conditions of purity and density, a single free neutron can initiate a chain reaction, thereby releasing in a few thousandths of a second more energy than was thought possible prior to the 1930s.

Uranium-235 and plutonium-239 are the two isotopes commonly used in fission weapons. The atomic bomb dropped on Hiroshima contained approximately 60 kg of uranium-235, of which only 700 g underwent fission. The yield of this weapon, expressed in tons of TNT equivalent, was approximately 12.5 kilotons (kT). The Nagasaki bomb contained about 8 kg of plutonium-239, with an equivalent yield of about 22 kT of TNT.

Nuclear weapons deriving their explosive yield from fusion use devices similar to the Nagasaki bomb as triggers. At the extremely high temperatures and pressures generated by a fission bomb (millions of degrees centigrade and millions of atmospheres pressure), the atomic nuclei of hydrogen (1 proton), deuterium (1 proton, 1 neutron), and/or tritium (1 proton, 2 neutrons) are caused to fuse with one another, forming atoms of helium (2 protons, 2 neutrons). Thus, fusion weapons are also known as hydrogen bombs, or thermonuclear bombs. The amount of energy released per unit mass of nuclear reactant is several times greater during fusion than in fission. Because of this, and because of advancements in weapons technology, the energy yield of fusion weapons can exceed several million tons of TNT equivalent, referred to as megatons (one megaton [MT] of TNT would fill a freight train 300 miles long).

The energy released by a nuclear weapon can take the form of blast or shock waves, thermal radiation (infrared light), and ionizing radiation. The proportion of yield expended to these respective categories is a function of weapon design (fission vs. fusion) and of the conditions of the medium (air, soil, or water) in which the weapon is detonated. Generally, nuclear devices detonated in the atmosphere (air bursts) will expend about 50 percent of their energy as blast and shock waves, about 35 percent as thermal radiation, and the remaining 15 percent as ionizing radiation. About one-third of the ionizing radiation is released directly from the fission or fusion reactions; the remaining two-thirds are released through the decay of radioactive debris known commonly as fallout. In a nuclear weapon detonated at or near the Earth's surface (surface burst), a greater proportion of its energy is transferred to the ground as blast and shock waves—about 60 percent, with less energy expended as thermal energy—and a moderate, but important, increase in the amount of radioactive fallout is produced.

The distance to which blast, thermal, and initial ionizing radiation effects are significant, either in a biological sense or because of their destructive potential, is not linearly related to weapon yield. For example, for air bursts the 5-p.s.i.[9] overpressure isopleth, which defines the "lethal area" for a nuclear weapon,[10] extends to about 2.5 km for a 100-kT weapon, 5 km for a 500-kT weapon, and about 5.8 km for a 1-MT weapon. By comparison, thermal irradiance sufficient to cause third-degree burns in exposed individuals extends to nearly 5 km, 9 km, and 10.7 km for 100-kT, 500-kT, and 1-MT weapons, respectively. Except at very low yields (1 kT or less), and then primarily because of special weapon design,[11] initial ionizing radiation does not extend beyond the range of blast and thermal effects. Therefore, in almost all cases, thermal effects are important over the greatest area, followed by blast effects, then initial ionizing radiation.

As mentioned above, the amount of radioactive fallout generated by surface bursts and near-surface bursts is substantially greater than for air bursts on account of the soil entrained by the nuclear fireball, some of which is made radioactive by neutron activation. In such cases, an appreciable fraction of the fallout produced is deposited locally within a few hours of the blast. For air bursts, fallout is produced from weapon debris only, and may remain suspended in the atmosphere for extended periods.

In either case, radioactive decay of fallout products begins immediately so that the potential dose rate for a given quantity of fallout decreases exponentially with time. One week after a nuclear explosion, almost 90 percent of the total radiation dose will have been expended. (Little relief can be found in this, however, if one is in a "hotspot" where fallout radiation doses may initially be so high that many weeks or months are required to reach safe radiation levels.)

The major climatic effects of nuclear weapons are caused by the particulates lofted into the atmosphere by the explosive force of the weapon and by the soot produced from fires ignited by the thermal radiation released. For surface and near-surface bursts, about $10^5$ tons of particulates may be lofted into the atmosphere for every MT of yield. Thermal energy exposures of 5 to 10 cal/cm$^2$ (calories per square centimeter) are sufficient to ignite many kindling materials, such as newspaper or dried leaves. This exposure level is met or exceeded at the 2 p.s.i. isopleth for weapon yields between 100 kT and 1 MT, with ranges from ground zero of about 6 to 12 km.

It is important to recognize the potential interaction between blast wave arrival time and the rate at which thermal energy is delivered, because the blast wave is accompanied by strong winds that can either extinguish or fan ignited materials. At the 2 p.s.i. isopleth, winds of 60 to 80 mph can be expected, arriving about 8 seconds after a 100 kT surface burst, or 13 seconds after a 1 MT surface burst (arrival times for air bursts are about double these values). Up to 70 percent of the thermal energy delivered at these distances, however, arrives within 2 to 5 seconds, giving many ignitions considerable time to become established. Although these interactions are important in dealing with wildland fires, the probability of "secondary" fires in urban areas is so great that concern for ignitions caused by weapon thermal energy exposure becomes moot.



## Scenario Development

The principles discussed in the preceeding section are indicative of the many variables that must be considered in an analysis of the consequences of nuclear war. Perhaps the most important variable is the scale of exchange presumed to occur. Generally, limited wars receive little attention because it is assumed that any use of nuclear weapons in warfare would result in rapid escalation to a full-scale exchange.[12] Regardless of the confidence one holds in this assumption, it is important to recognize that the course of international events in crisis situations is impossible to predict. Therefore, evaluations based on plausible worst case, or near worst case, scenarios follow the most conservative tact.

The accompanying sidebar describes the immense power contained in the world's nuclear arsenals. Nuclear war consequence-analysis requires, however, that the manner in

# *Dimensions of the World's Nuclear Arsenals*

The immense power contained in the nuclear arsenal of the world is overwhelming; yet it is important that we try to comprehend just how much potential energy is stored in the form of nuclear weapons. Figure 1 depicts the relationship between tons of TNT equivalent, the standard measure of weapon yield, and energy yield expressed in joules, where one ton of TNT equivalent is equal to 4.2 x $10^9$ joules. Dividing the total yield contained in the world's arsenals by the global human population reveals that about 3 tons of TNT equivalent are now present for every man, woman, and child. The largest World War II bombs, called blockbusters because they could level a city block, contained about 10 tons of TNT. Neutron bombs, also known as Enhanced Radiation Weapons, have yields of generally 1 kiloton (kT), or less. The Hiroshima bomb had a yield of about 12.5 kT, while the Nagasaki bomb and the Trinity test detonation yielded 20 to 22 kT.

A Poseidon submarine-launched ballistic missile (SLBM) can carry 10 to 14 independently targetable warheads of 40 kT each, for a total yield of 560 kT per missile. With 16 missiles per submarine, each can carry nearly 9 megatons (MT) of explosive power, or the equivalent of 720 Hiroshima bombs.

Underground tests are limited by agreement to 150 kT or less. These tests are conducted in the course of developing new technologies, and to determine whether stockpiled warheads are still functional. Worldwide during the past few years, an average of 40 to 50 test detonations have been conducted per year.

Each Minuteman III missile, (MM III), the current backbone of the U.S. intercontinental-ballistic missile (ICBM) forces, carries three independently targetable 170 kT warheads. There are presently 1,000 Minuteman missiles placed in silos throughout the Great Plains states, but not all of these carry multiple warheads. The MX (missile experimental) that the United States is now developing will be capable of delivering 10 independ-

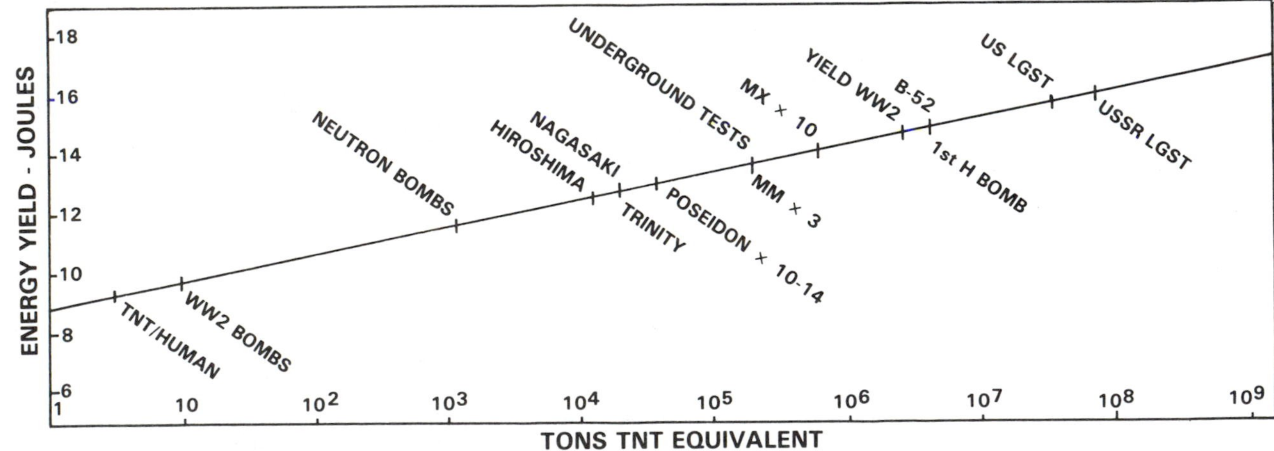


**FIGURE 1. The Relationship of TNT Equivalent and Energy Yield**

Several points of reference are illustrated here in terms of energy yield expressed in joules versus tons of TNT equivalent. Note that the vertical axis is scaled by factors of 10, with one ton of TNT equal to 4.2 × $10^9$ joules. Abbreviations are discussed in the accompanying text.

ently targetable warheads of 350 kT yield each. In addition to carrying a heavier payload, the MX or "Peacekeeper" missile has an accuracy projected to be such that 50 percent of its warheads can fall within 100 m of an intended target.

One of the more striking points revealed in Figure 1 is that a single B-52 is capable of carrying a payload with a yield almost double the cumulative explosive yield of World War II. Indeed, the first hydrogen bomb detonated by the United States had a yield of 5 MT, equal to the B-52 payload.

The largest nuclear test conducted by the United States (28 MT) was surpassed in yield by a factor of two in the largest Soviet test (58 MT). It is interesting to note that in the early 1960s, just after these massive atmospheric tests, many studies projecting the effects of nuclear weapons included 100 MT "superbombs" in their analyses. Clearly such weapons far exceed any conceivable military purpose. It is thought that very few weapons greater than 10 MT are now deployed, although the Soviets may have some warheads as large as 20 MT on some older, less accurate ICBMs.

These are staggering numbers. But even more staggering are the total yields estimated to be contained in the strategic and tactical nuclear arsenals of the superpowers and their allies. From Table 1, we learn that the United States and the Soviet Union control by far the greatest proportion of nuclear weapons in existence today. In both strategic and tactical categories, the United States leads in the number of warheads, while the Soviet Union leads in total yield. This information confirms that a nuclear exchange between the superpowers could conceivably involve several thousand warheads having a cumulative yield between 1,000 to 10,000 MT, or even greater. Extrapolating from Figure 1, an exchange within this yield range would release between $10^{19}$ and $10^{20}$ joules. This is an astounding quantity of energy, exemplified by the fact that, during a 24-hour period, the entire Earth's surface intercepts about $10^{19}$ joules of sunlight energy. Put in another frame of reference, if a 5,000 MT war were to occur using Hiroshima-size weapons, with one bomb dropped every second, 60 bombs per minute, the war would last 4.5 days.

**Table 1**
**REPRESENTATIVE ESTIMATES FOR THE NUMBERS OF WARHEADS AND TOTAL YIELDS OF THE WORLD'S NUCLEAR ARSENALS**

| | STRATEGIC WEAPONS[a] | | TACTICAL + OTHER[b] | | TOTALS | |
|---|---|---|---|---|---|---|
| | *Warheads* | *Megatons* | *Warheads* | *Megatons* | *Warheads* | *Megatons* |
| United States | 9,000 - 11,000 | 3,000 - 4,000 | 16,000 - 22,000 | 1,000 - 4,000 | 25,000 - 33,000 | 4,000 - 8,000 |
| Soviet Union | 6,000 - 7,500 | 5,000 - 8,000 | 5,000 - 8,000 | 2,000 - 3,000 | 11,000 - 15,000 | 7,000 - 11,000 |
| United Kingdom | | | | | 200 - 1,000 | 200 - 1,000 |
| China | | | | | 300 | 200 - 400 |
| France | | | | | 200 | 100 |
| | | | *Rounded totals* | | 37,000 - 50,000 | 11,000 - 20,000 |

**SOURCE:** A. I. Thunborg, *Nuclear Weapons: Report of the Secretary-General.* (Brookline, Mass.: Autumn Press, 1980), 223 pages.

[a]Strategic weapons here refer to those deliverable from the territory of one nation to another nation covering distances measured in thousands of kilometers.

[b]Tactical weapons, which here include other types of weapons not strictly considered to be strategic, such as some intermediate-range systems, are those generally designed for battlefield or theater use.

which this energy is dissipated be considered explicitly. Thus, it is essential that several targeting variables be delineated. For example, of great importance are: the numbers and kinds of military, industrial, and population centers struck; the yield of weapon(s) applied to each target; and weapon-burst heights.

Attacks on military targets are generally referred to as "counterforce" strikes, and may be preemptive or retaliatory, although timing in this sense is not usually important. "Countervalue" targets include primarily industrial centers, but it is impossible to separate most industrial targets from population centers. The U.S. Federal Emergency Management Agency has determined that 65 percent of the U.S. population lives within 20 miles of at least one prime military-industrial target, and 95 percent are within 100 miles of such targets.[13]

The explosive yields of nuclear weapons presumed to be used in a hypothetical war are important not just for estimating the area endangered by blast and thermal effects of the weapons. Yield also determines the quantities of dust and debris lofted into the atmosphere, and the height to which particulates and other products of the explosion (e.g., fallout, oxides of nitrogen) are delivered in the atmosphere.

The layer of the atmosphere in which we live, and in which weather processes occur, is called the troposphere. At approximately 10 km altitude, changes in atmospheric density and temperature form a boundary layer, the tropopause, which separates the troposphere from the overlying stratosphere. The stratosphere is important to living organisms in that it contains, under normal conditions, relatively stable concentrations of ozone that are effective in absorbing most of the harmful ultraviolet rays contained in incident sunlight. Nuclear weapons with yields approaching 100 kT produce mushroom clouds that can rise to altitudes exceeding 12 km at their tops, while weapons of 200 kT or greater yield produce mushroom clouds with at least half of their volume above the tropopause.

Injection of materials from nuclear explosions into the stratosphere is important for two reasons. First, radioactive debris and dust will remain suspended in the atmosphere for much longer periods of time if they are injected into the stratosphere. This is largely due to the tropopause, which restricts transfer of materials between the stratosphere and underlying troposphere. Precipitation that accompanies storms in the troposphere is very effective in scavenging particulates from the troposphere. Such scavenging occurs infrequently in the stratosphere.

The second reason that mushroom-cloud intrusion into the stratosphere is important involves the oxides of nitrogen produced in the nuclear fireball. About $10^{32}$ molecules of nitrous oxides are produced per MT of yield. In the troposphere, photochemical reactions act to convert nitrous oxides into what we commonly know as "smog." However, in the stratosphere, nitrous oxides catalyze reactions that destroy ozone, thereby allowing more ultraviolet radiation to reach the Earth's surface. In 1975 the National Academy of Sciences released a report estimating that a nuclear war involving 10,000-MT total yield would produce sufficient quantities of nitrous oxides to reduce stratospheric ozone concentrations by 30 to 70 percent.[14] As a result, ultraviolet light penetration to the Earth's surface could increase by 6 or 7 times and remain elevated for several years until ozone concentrations recovered. Such increases in ultraviolet exposure would prove detrimental to many plants and to most animal species.

Several other variables can be important in calculating the immediate effects of nuclear war. These include time of day when the attack occurs, season of the year, and weather conditions in the target zone. The *Ambio* study[15] is most notable. The study was the work of a committee of international scientists invited by the Royal Swedish Academy of Scientists to develop and discuss a nuclear war scenario. The resulting collection of papers was published in *Ambio*. It deals explicitly with each of these variables in presuming that a war occurs on a June day in 1985 at 11:00 A.M. New York time, 6:00 P.M. Moscow time. Weather patterns for a typical June day were used to characterize fallout plumes. The *Ambio* group was also very explicit in identifying targeting variables, such as the size and number of weapons assigned to military installations of different types, industrial centers, and population centers. Such specificity is uncommon; most studies assume that a certain number of weapons are detonated within, say, 30° and 60° north latitudes. As the following discussion will reveal, such variables can be very important to the outcome of a study.

## Climatic Effects of Nuclear War

With each new study performed over the past few decades, the consequences of nuclear war seem to take on new dimensions. Residual radioactivity was viewed for many years as the principle long-term effect of nuclear war. This concern arose after a hydrogen bomb detonated in the South Pacific in 1954 contaminated inhabitants of the Marshall Islands and a Japanese fishing boat with radioactive fallout.

In 1975 a National Academy of Sciences study[16] reviewed the climatic effects of volcanic eruptions to determine if the debris lofted into the atmosphere by a 10,000-MT nuclear war could result in significant temperature changes. Their findings suggested a minimal potential effect. However, it has since been pointed out that volcanic aerosols are less efficient in injecting particulates into the atmosphere than are nuclear weapons and that the aerosols can have very different reflective and absorptive properties than have "nuclear" dust and soot. Nonetheless, as was discussed in the preceeding section, the NAS study did discover a potentially significant effect of nitrous oxides produced by nuclear explosions on stratospheric ozone concentrations.

In contrast to the fallout problem, discovery of the effect that dust and soot could have on climate in the aftermath of nuclear war came not from direct experience but by extrapolating from the individual effects of many hundreds of nuclear explosions. As a result, several studies have recently shown that the amount of soot produced by fires ignited from nuclear weapons exploded over urban-industrial targets and over wildlands would be sufficient to block sunlight from reaching the Earth's surface, thereby initiating a series of climatic changes that could persist for months or years after the war.[17]

Seminal work on this "new" problem was done by Drs. Paul Crutzen and John Birks for the *Ambio* study.[18] They concluded from some very basic calculations that a nuclear war using between 5,000 and 6,000 MT could involve in fire about 4 percent of the Northern Hemisphere forests (one million km$^2$). The soot produced by wildland fires and the burning of industrial centers, oil refineries, etc., would be sufficient to reduce noon-time sunlight levels to that of a full moon.

These findings inspired several planetary and atmospheric scientists to apply their expertise to the problem. The result is a report now known as TTAPS (an acronym for the authors' names: R. P. Truco, O. B. Toon, T. P. Ackerman, J. B. Pollack, and Carl Sagan), in which the "nuclear winter" effect is elaborated.[19] In TTAPS, several nuclear war scenarios were constructed, each isolating certain variables of interest. For example, the total quantities and relative amounts of dust and soot produced were controlled by altering such target variables as the numbers of industrial versus military targets, sizes of weapons involved, and the proportion of air versus surface bursts. Total yields in these scenarios ranged from a 100-MT cities-only attack to a 25,000-MT future war, but the majority of theorized wars consumed a more plausible 3,000 to 5,000 MT.

On the basis of previously selected parameters, many of which were derived from weapons tests, the quantities of dust, soot, and nitrous oxides produced for each scenario were calculated. This information was then entered into computer models designed to simulate the effects of atmospheric particulates on sunlight transmittance and subsequent temperature changes at the Earth's surface. Finally, these models also projected how the dust and soot would be dispersed over the weeks and months following the war.

A scenario involving 5,000 MT was used as the baseline case in the TTAPS study. In this war, 10,400 warheads were utilized, having yields in the range of 100 kT to 10 MT. About 60 percent of these weapons were designated as surface bursts, with approximately 20 percent of the total yield falling on urban-industrial targets. It was conservatively estimated that such a war would involve about 240,000 km$^2$ of urban areas in fire. This represents about one-fourth of the developed area of the Northern Hemisphere, or about one-half of the urban centers in NATO and Warsaw Pact countries having populations over 100,000. Forest, shrubland, and grassland fires would consume an additional 500,000 km$^2$. As a result, an estimated 225 million tons of soot would be released to the atmosphere within a few days of the war. This is about 13 percent more soot than is produced in an average year by all natural and man-caused fires.

The nuclear "pall" described by these calculations would, within a week or two, encompass the entire Northern Hemisphere, and it is likely that extensions of these clouds would be spun off along the equator and intrude into the Southern Hemisphere as well. It is postulated that the soot in the nuclear clouds would be very effective in absorbing incident sunlight, so that almost none of this energy would reach the Earth's surface.

In the baseline case, for example, sunlight reaching the surface would be reduced by 96 percent within a few days of the war, and would remain at less that 50 percent of normal for nearly two months. By comparison, the 100-MT cities-only scenario, involving 100 cities receiving ten 100-kT warheads each, would result in a 95 percent reduction in sunlight penetration, with greater than 50 percent obscuration for about six weeks. In both of these cases, land-surface temperatures would drop from the hemispheric continental mean of 13°C, to −23°C. In the baseline case, temperatures would remain below freezing (0°C) for 95 days, whereas in the cities-only attack, about 70 days would be required for temperatures to return to freezing. Near normal temperatures would return in about 100 days in the city scenario, but would require more than 300 days to recover in the baseline case.

Overall, the range of scenarios considered in TTAPS revealed that soot is most important in reducing sunlight and atmospheric temperatures at the Earth's surface in the first few months after a nuclear exchange. Dust was found to be important in prolonging these effects beyond this period. Also, nitrous oxides produced would reduce stratospheric ozone levels so that, when the nuclear clouds dissipated, levels of ultraviolet light elevated by 2 to 5 times above normal would endanger exposed organisms.

It is important to note that the findings in TTAPS bear upon internal continental conditions; the thermal inertia of the oceans would ameliorate temperature reductions along coastal regions and on islands. Although this suggests that refugia could be found there, other impacts of nuclear war that were not quantified reduce the likelihood of survival anywhere in the Northern Hemisphere, even if climatic effects were less severe than projected.

For example, in addition to soot, the urban and industrial fires would produce large quantities of pyrotoxins as byproducts from the combustion of synthetic materials. Hits on chemical plants, nuclear power plants, or hazardous waste storage sites would contaminate large areas. On a broader scale, absorbance of sunlight energy in the upper strata of the nuclear clouds would alter convective air movement in the lower troposphere, thereby disrupting precipitation patterns, and the precipitation that did fall would likely occur as snow.

Other simulation studies using atmospheric models differing in structure and mathematical design from TTAPS have yielded similar results. Temperature decreases of about the same magnitude as projected in TTAPS were obtained by Soviet scientists V. Aleksandrov and G. Stenchikov.[20] Their models, however, also project a warming trend following several months of nuclear winter, with temperature increases sufficient to cause rapid melting in high-elevation snowpacks, endangering some areas with extensive flooding.

More recently, scientists at the National Center for Atmospheric Research (NCAR) completed modeling studies that incorporated aspects of atmospheric circulation that could not be handled by the TTAPS or Soviet models.[21] In this work, only the effects of soot generated by a 6,500-MT war were considered. In general, because of changes in air circulation patterns, dispersal of soot effects within the troposphere, and advective air movement from warm oceans to coastal regions, the NCAR study projects a less severe nuclear winter than described in TTAPS. For example, within 10 days of the exchange, temperature reductions of 15° to 20°C were projected for continental regions, or about half the reduction found in TTAPS.

The NCAR group also found that subfreezing temperatures would prevail over most of the land surface of the Northern Hemisphere within two weeks of a nuclear war. Some concern was expressed, however, about the ability to project climatic effects of nuclear war over the long term (beyond a few weeks) because air circulation patterns could be so profoundly altered by the soot clouds that our current understanding of aerosol dispersal may no longer apply.

Should we be concerned that the results of these studies differ by factors of two or three? Not really. What is important from these scientific exercises is not the precision with which nuclear winter is projected, but the fact that all of these studies are in agreement as to the qualitative changes that could occur in climate. We can say with some degree of confidence that nuclear war would generate substantial quantities of soot, and that this soot would be effective in reducing sunlight and atmospheric temperatures at the Earth's surface. These effects would be devastating for the survival of humans and other biota even if only a few-degree drop in average temperatures occurred.

In 1815, Mount Tambora erupted in what is now Indonesia.[22] Approximately 25 cubic miles of debris were injected into the atmosphere, changing climate in the Northern Hemisphere to the extent that 1816 is known as the "year without a summer." Severe crop failures were felt in the New England states and across much of Europe, causing famine in many areas. A pandemic of cholera in subsequent years may have been attributable to the crop failures as food export patterns were adjusted, spreading famine to colonies of the more influential nations. The drop in average temperatures responsible for this series of events was only about 1°C.

### Prospects for the Biosphere

The implications of the TTAPS findings for humans and other biota were explored by a team of biologists and ecologists with whom I served.[23] Although human deaths were not quantified for the entire Northern Hemisphere in these studies, it was clear from our analyses that the estimate of 750 million deaths in the *Ambio* report and 1.1 billion deaths projected by the World Health Organization are valid approximations of the nuclear holocaust, and may even be very conservative estimates in that they focused on immediate and short-term effects. The injuries inflicted upon humans would also number in the hundreds of millions, suggesting that one-third to one-half of the global human population could be affected directly by a full-scale nuclear war.

Moreover, the survivors, both injured and uninjured, would inherit a hostile environment, and would be forced to seek food and shelter in natural ecosystems that would also be reeling from the effects of nuclear war. Trying to characterize how natural biota would be affected by nuclear war and its after-effects was a major focus of our work, and continues to occupy the time of other natural scientists involved in studies, both national and international, that have not yet been completed.[24]

In a very real sense, each of the environmental insults depicted in TTAPS (light reduction, temperature reduction, pyrotoxins and other pollutants, enhanced ultraviolet radiation, and elevated ionizing-radiation exposure from fallout) would serve as selective biocidal agents, each driving susceptible species to precipitously low densities, even extinction.

For terrestrial and freshwater-aquatic ecosystems, temperature reductions would be the most devastating effect of the war. Many land plants can withstand seasonal fluctuations in atmospheric temperatures if given time to acclimate to changing conditions. This process is called "hardening," and may require seasonal cues other than temperature (e.g., day-length) for initiation of the physiological processes involved, which in turn may take several days or weeks to accomplish.

As one would expect, coniferous evergreen trees common in the farthest northern latitudes are most resistant to temperature reductions; many species are capable of withstanding air temperatures as low as −90°C before suffering injury. It is interesting to note that these northern species are also accustomed to extended periods of darkness since they regularly experience the long arctic nights. Most tree and shrub species of more temperate climes can withstand air temperatures of −10° to −20°C before being injured. Finally, subtropical and tropical plant species may be injured at temperatures of only 0° to 5°C. Indeed, more than half of all plant species are never exposed to sub-freezing temperatures, and presumably lack adaptive mechanisms for such conditions.

The effect of lowered temperatures on freshwater aquatic ecosystems is mechanistically different than for land systems. Here the major impact is caused by severe alterations in the physico-chemical condition of the aquatic habitat, as standing bodies of water would be subject to freezing to considerable depths under the conditions forecast in TTAPS.[25] In the baseline case, conditions reviewed earlier would result in ice depths of 50 to 90 cm, with 60 to 80 percent of the volume of smaller water bodies (1 $km^2$) converted to ice, while larger water bodies (1,000 to 10,000 $km^2$) may have only 5 percent of their volume solidified.

Changes such as these are significant because they entail not only a phase change in aquatic habitat but are also accompanied by an alteration in the chemical quality of the remaining liquid volume. As detritivores continued to consume oxygen during metabolism of dead organic matter, the ice cover would act to inhibit gaseous exchange with the atmosphere. In the baseline case, ice caps on many bodies of water would persist for an entire annual cycle, even if air temperatures recovered, because of a lag in melting. Severely anoxic conditions would result, decimating algal populations that form the base of the normal aquatic food chains.

Converting surface waters to ice presents special problems for terrestrial animals as well, humankind included. In most cases, wild animals would be unable to break through the ice covering lakes and ponds. Humans would have similar difficulty, and would only be able to exploit groundwater supplies if power were available to operate pumps. Hence, dehydration would serve as an additional threat to survival for several months after the war. This assumes, of course, that the water that could be obtained was not contaminated with hazardous chemicals or high concentrations of fallout.[26]

Species susceptibility to ionizing radiation is largely a function of chromosome volume, but it may also be affected by the age of an organism and its overall condition.[27] Feeding habits and socialness may also be important in determining exposure levels, as adults could carry contaminated foodstuffs to their more susceptible young or larvae. Sensitivity to radiation is expressed as the lethal dose effective for 50 percent of the exposed population ($LD_{50}$). For humans and many large mammals, the $LD_{50}$ falls around 400 to 500 rads.[28] Herbaceous plants, such as soybeans and corn, have $LD_{50}$s in the range of 1,000 to 3,000 rads, while perennial woody plants, such as oak, can withstand doses exceeding 5,000 rads. Interestingly, some conifers such as pine are about as susceptible as humans to ionizing radiation. Adult insects are among the most resistant organisms, with $LD_{50}$s of 10,000 rads or more.

However, it is important to note that sub-lethal effects of ionizing radiation can be incurred at much lower doses than cited here. Cataracts formed several months to years later in some Hiroshima and Nagasaki survivors receiving doses as low as 100 rads, and temporary sterility at similarly low doses are observed in males.[29] Similar effects can be expected in natural biota, with consequent reductions in long-term survival of some species and delayed recovery of other species.

In contrast to terrestrial and freshwater ecosystems, marine ecosystems would suffer to a greater degree because of light attenuation rather than from temperature reductions. This is because of the thermal inertia of the ocean waters and the short generation times of the phytoplankton upon which marine food chains depend. Many phytoplankton have life cycles of only a few days, or even hours. Sustenance for zooplankton and carnivorous fishes further up the food chain depends upon the rapid growth and reproduction of these small organisms. Elimination of sunlight for two to three weeks would exterminate phytoplankton populations, not only by natural mortality and suppressed reproduction but also because of consumption by starving consumers. This effect would cascade up the food chain to larger fishes as well.

Some consolation can be found in the recoverability of marine ecosystems; horizontal mixing of ocean waters would probably restore phytoplankton to affected areas in a relatively short period of time. However, marine phytoplankton are sensitive to ultraviolet light, suggesting that recovery could be delayed until stratospheric ozone levels return to near normal.

Our current understanding of interhemispheric circulation patterns in the atmosphere suggests that the effects on climate would impact tropical latitudes and portions of the Southern Hemisphere. Although temperature and light reductions would probably be less severe (perhaps less than a 5°C drop in average temperature) and of shorter duration, tropical and sub-tropical ecosystems are least adapted to variation in environmental conditions. Intrusion of cold air masses into the Central Americas or into northern South America and Africa could have unprecedented consequences for the biota and on human populations otherwise not involved in the nuclear conflict.

Indeed, our global population currently depends on the Northern Hemisphere, and North America in particular, for much of its food and technology. Many developing nations in the Southern Hemisphere have in recent years experienced an influx of peoples to urban areas and now import more than half of their foodstuffs, as well as critical agricultural supplies (e.g., pesticides, fertilizers, seeds) from North America. Eliminating commerce between the Northern and Southern Hemispheres would force a chaotic return to sustenance agriculture in regions of the globe where current human populations can no longer be supported without food imports and advanced agricultural technologies.

Even if no climatic perturbations were incurred below the equator, exploitation of these already stressed natural ecosystems would prove devastating, and would result in massive starvation for these peoples. Given the calculable and incalculable impacts depicted in TTAPS and associated studies, the recovery of human civilization from nuclear wars of even moderate intensity cannot be assured.

## Policy Implications

In spite of comments occasionally made by world leaders implying the "winnability" of nuclear war, there has probably never been greater concensus over an issue as now exists across all levels of society regarding the apocalyptic nature of nuclear war. We know from experience with single nuclear weapon detonations that the use of these leviathans in warfare would have consequences for humankind far beyond any calamity ever witnessed.

The findings reported in TTAPS and associated studies indicate that the direct and indirect consequences of nuclear war would render the biosphere permanently and severely altered. Many thousands of species would face extinction or would have their ranges drastically changed. The global ecosystem would continue to function, might even "recover" in some sense, but would be very different in composition and organization from the world in which we now live. Indeed, the extinction of humankind cannot be excluded as a possibility. But even if our demise were not absolute, our cultures and societies would be laid to waste, and some believe that our current level of technological development could never be regained.

As this review has pointed out, the critical factor in initiating the "nuclear winter" effect is the unprecedented quantities of soot that could be produced by urban fires (recall the 100-MT cities-only scenario). Therefore, the number of warheads takes on great importance, as even relatively low-yield nuclear weapons are capable of igniting massive conflagrations over certain kinds of targets.

Is there a threshold, some number of weapon detonations that, if surpassed, would ensure climatic suicide for the aggressor and defender alike? Perhaps so. One of the TTAPS authors, Carl Sagan, concludes that the use of between 500 and 2,000 warheads could induce severe climatic changes. This represents less than 10 percent of the approximately 18,000 warheads now contained in the world's stockpiles of strategic nuclear weapons.

Among the measures that could be taken to reduce the threat of nuclear war, the following seem most critical. First, the numbers of nuclear warheads could be reduced to subthreshold levels if such a level could be determined. The mechanisms for such reductions have been proposed to our government, and numerous lay and professional organizations have adopted resolutions calling for this kind of action. Second, to the extent that nuclear arsenals are deemed necessary for defense, the configuration should be adjusted to be less provocative. This is the subject of much policy discussion at present, and concerns such things as the accuracy of delivery systems, the number of nuclear warheads, verification, the likelihood of accidental war initiated by computer paranoia or some other technological malfunction, confidence-building measures, and improvements in international communications.

The conclusion that the use of nuclear weapons could create an uninhabitable world leads one to question the rationale for equating national security with military strength, and military strength with greater numbers and yields of nuclear weapons. Regardless of how serious we perceive the Soviet "threat" to be, evidence now available forces us to consider the potentially devastating and permanent consequences of permitting the nuclear dilemma to go unresolved.

## NOTES

1. J. Peterson, ed., *The Aftermath: The Human and Ecological Consequences of Nuclear War* (New York: Pantheon Books, 1983), 196 pp. (note: originally appeared as a special issue of *Ambio* II(1982):76–176, and is referred to as such throughout the text); and S. Bergstrom et al., *Effects of Nuclear War on Health and Health Services*, WHO Publ. A36.12 (1983).

2. "The World After Nuclear War, Conference on the Long-Term Worldwide Biological Consequences of Nuclear War," October 31–November 1, 1983, Washington, D.C.

3. R. P. Turco et al., "Nuclear Winter: Global Consequences of Multiple Nuclear Explosions," *Science* 222 (1983):1283–1292.

4. C. Covey et al., "Global Atmospheric Effects of Massive Smoke Injections from a Nuclear War: Results from General Circulation Model Simulations," *Nature* 308(1984):21–25.

5. R. U. Ayres, *Environmental Effects of Nuclear Weapons* (Harmon-on-Hudson, N.Y.: Hudson Institute, Inc., HI-518-RR, 1965); and A.O.C. Nier et al., *Long-Term World-Wide Effects of Multiple Nuclear Weapons Detonations* (Washington, D.C.: National Academy of Sciences Press, 1975), 213 pp.

6. P. J. Crutzen and J. W. Birks, "The Atmosphere After a Nuclear War: Twilight at Noon," *Ambio* 11(1982):115–125.

7. Carl Sagan, "Nuclear War and Climatic Catastrophe: Some Policy Implications," *Foreign Affairs* No. 62202 (Winter 1983/84):257–292.

8. S. Glasstone and P. J. Dolan, *The Effects of Nuclear Weapons.*, 3rd ed. (Washington, D.C.: prepared for U.S. Department of Defense, and Energy Research and Development Administration, 1977), 653 pp.

9. The term "p.s.i. overpressure" refers to atmospheric pressures above ambient (14.7 lb/in$^2$) brought about by the blast wave.

10. "Lethal area" refers to the area within which the number of survivors equals the number of fatalities outside that area. For nuclear explosions, the 5 p.s.i. isopleth is accepted as defining the lethal area.

11. Fusion weapons can be shielded with reactive uranium-238 to reflect neutrons back into the weapon core, and to increase yield. Enhanced radiation weapons, also known as neutron bombs, lack this shield so that large quantities of high-energy free neutrons are released to the surrounding environment.

12. P. Sharfman et al., *Effects of Nuclear Weapons* (Washington, D.C.: U.S. Congress Office of Technology Assessment, Pub # OTA-NS-89, 1979), 151 pp.

13. *FEMA Attack Environment Manual* (Washington, D.C.: U.S. Federal Emergency Management Agency, CPG 2-1A1, May 1982), chap. 1.

14. Nier et al., note 5 above.

15. *Ambio*, note 1 above.

16. Nier et al., note 5 above.

17. Report of Committee on Atmospheric Effects of Nuclear Explosions (Washington, D.C.: National Academy of Sciences, in press); and V. V. Aleksandrov and G. L. Stenchikov, *On the Modeling of the Climatic Consequences of Nuclear War*, The Computing Centre of the USSR Academy of Sciences (1983); see also R. P. Turco et al., note 3 above; and C. Covey et al., note 4 above.

18. P. J. Crutzen and J. W. Birks, note 6 above.

19. R. P. Turco et al., note 3 above. There are several assumptions in the TTAPS report that are crucial to the conclusions. The most important is the targeting; TTAPS assumes that there is no overlap in strategy so that each warhead is falling on virgin territory. This may overstate the case for dust and soot, since a warhead falling on an area already burning would do nothing to increase the amount of soot pro-

duced—and may even reduce the amount by snuffing out the fires. However, since the selected parameters are conservative, such overlapping is not a major quantitative factor.

20. V. V. Aleksandrov and G. L. Stenchikov, note 17 above.

21. C. Covey et al., note 4 above.

22. H. Stommel and E. Stommel. "The Year Without a Summer," *Scientific American* 240(1979):176–186.

23. P. R. Ehrlich et al., "Long-Term Biological Consequences of Nuclear War," *Science* 222(1983):1293–1300; and M. A. Harwell, "The Human and Environmental Consequences of Nuclear War," (in prep.)

24. See T. F. Malone, "Preventing Nuclear War," letter to the editor in *Science* 223(1984):223.

25. M. A. Harwell, note 23 above.

26. See K. G. Wetzel, "Effects on Global Supplies of Freshwater," *Ambio* 11(1982):126–132.

27. See G. M. Woodwell, "The Biotic Effects of Ionizing Radiation," *Ambio* 11(1982):143–148.

28. Rad refers to radiation-absorbed-dose.

29. E. Ishikawa and D. L. Swain, trans., *Hiroshima and Nagasaki* (New York: Basic Books, Inc., 1982), 706 pp.

C H A P T E R 24

# Third World in the Global Future



PETER H. RAVEN

THE PROBLEMS associated with the Third World affect us all. If we are wise enough or lucky enough to avoid nuclear war, how we deal with those problems will largely determine our future, and that of our children and grandchildren.

What was already, in 1950, a record human population tripled during the course of a single human lifetime. This represents an extraordinary and unprecedented situation. The challenge presented to the productivity capacity of the earth by this increase is neither "normal" nor a circumstance that we can expect to deal with by applying the standard behaviors of the past.

The worldwide distribution pattern of this population growth ought also to be a major cause of concern. During the past 34 years, well over two billion people have been added in the less developed countries alone. This number equals the entire world population as recently as 1932. During the same period "only" 300 million people were added in the developed countries, including those of the Near East and Korea which are not tropical. About 90 percent of such countries do lie wholly or partly in the tropics, a relationship that is important to understand.

During our single hypothetical lifetime, from 1950 to 2020, the proportion of people living in developed countries will fall from about a third of the total world population to about a sixth. Over the same period, the population of the mainly tropical, less developed countries will grow from approximately 45 percent of the total to more than 64 percent. In sum, the plurality of people live in the tropics, and their percentage of the world population is rapidly increasing, while ours is rapidly falling. Small wonder that we hear more and more about El Salvador, Nicaragua, the Philippines, Africa and other tropical regions and that we are steadily becoming more and more concerned with the development of appropriate policies to pursue in these hitherto unfamiliar parts of the world.

The worldwide rate of economic growth has fallen substantially from the 4 percent characteristic of the third quarter of the twentieth century. For the next two decades, many estimate that it may be no higher than 2 percent, as it has been for the past several years. This sort of economic environment, in the context of the much-discussed debts of countries like Brazil, Mexico, Nigeria and Kenya, makes it difficult to imagine how these countries will be able to meet the ordinary needs of their people, much less be able to improve conditions in the future.

The developed countries, with less than a quarter of the people in the world and an average per capita income of more than $9,000, control some 80 percent of the global economy. In stark contrast, the less developed countries, with an average per capita income of less than $1,000, control only about 17.6 percent of that economy. Further, the developed countries consume about 80 percent of the total world supply of energy, the less developed countries about 12 percent. And, as a final index to the disproportionate distribution of wealth, the consumption of iron, copper and aluminum by developed countries ranges from 86 to 92 percent, by the less developed countries, even including China, from 8 to 14 percent.

---

Reprinted by permission from *The Bulletin of the Atomic Scientists*, a magazine of science and world affairs Vol. 40(9): 17–20. 

Peter H. Raven is the director of the Missouri Botanical Garden in St. Louis and Engelmann Professor of Botany at Washington University, St. Louis, MO.

A quarter of the world's population (a proportion that is rapidly dropping) controls more than four-fifths of the world's goods, while a majority of the population (rapidly increasing in size) have access to no more than a sixth of any commodity involved in the world's productivity. Can this relationship be sustained as the disproportionate distribution of people becomes ever more extreme? The consequences of population growth in Kenya today are absolutely different from those in Europe or the United States of a century ago, and a direct comparison between the two situations is invalid.

An associated global problem is that of the rapid destruction of the forests and other potentially renewable tropical resources. In 1981, the Tropical Forest Resources Assessment Project of the U.N. Food and Agriculture Organization (FAO) estimated that 44 percent of the tropical rainforests had already been disturbed. The study estimated that about 1.1 percent of the remainder was being logged each year at that time. The total area of the remaining forest amounted to approximately the size of the United States west of the Mississippi River, with an additional area about half the size of Iowa being logged each year. If the clearing were to continue at this rate, all of the tropical rainforests would be gone in 90 years—a minimum estimate of the time necessary for their disappearance.

This estimate only begins to suggest the gravity of the problem. First, clearcutting is merely the most extreme form of forest conversion. Norman Myers, in his outstanding book *The Primary Source* (Norton, 1984), estimates the overall rate to be two to three times as great as that suggested by the FAO figures, or more than 2 percent per year. At that rate, even with no acceleration, the forests will all have been converted in less than 50 years. And other kinds of forest conversion are also threatening the extinction of species.

In the next 36 years alone, the population of the tropical countries will approximately double from its present level to about five billion. The governments of these countries are already faced with staggering debts, a sluggish world economy and the rapid loss of the productive capacity of their lands. For these governments to be able to expand their economies rapidly enough to continue to care for the needs of their people at 1984 levels clearly would be an unprecedented economic miracle. But even if they were able to do so, the numbers of their people living in absolute poverty would continue to increase as rapidly as their populations as a whole. Poor people would obviously continue to destroy their forests more and more rapidly with each passing year.

The rate of destruction and deterioration of tropical forests is by no means uniform. Three large forest blocks—in the western Brazilian Amazon, in the interior of the Guyanas and in the Congo Basin—are larger, less densely populated, and therefore being exploited more slowly than the remainder. Some of the forests in these three regions might actually persist in a relatively undisturbed condition for another 40 years or so, until the surging populations of their respective countries finally exhaust them. But all of the remaining forests in other parts of the tropics will surely be gone, or at least profoundly altered in nature and composition, much earlier. For the most part, these forests will not remain undisturbed beyond the end of the present century. This process of destruction is apparently irreversible, and it is accelerating rapidly. The tropical forests certainly will never recover from this onslaught.

The uneven distribution of wealth is one major factor in the destruction of tropical forests. The World Bank estimates that, of the 2.5 billion people now living in the tropics, one billion exist in absolute poverty. This term describes a condition in which a person is unable to count on being able to provide food, shelter and clothing for himself and his family from one day to the next. According to the World Health Organization, between 500 and 700 million people, approximately one out of every four living in the tropics, are malnourished. UNICEF has estimated that more than 10 million children in tropical countries starve to death unnecessarily each year. Worse, many millions more exist in a state of lethargy, their mental capacities often permanently impaired by lack of access to adequate amounts of food.

Shifting cultivation and other forms of agriculture generally fail quickly in most tropical regions. The reason lies in some of the characteristics of the soils and plant communities that occur in these areas. Although tropical soils are extremely varied, many are highly infertile. They are able to support lush forests, in spite of their infertility, because most of the meagre amounts of nutrients present actually are held within the trees and other vegetation. The roots of these trees spread only through the top inch or two of soil. Quickly and efficiently, the roots recover nutrients from the leaves that fall to the ground, transferring them directly back into the plants from which they have fallen.

Once the trees have been cut, they decay or are burned, releasing relatively large amounts of nutrients into the soil. It is then possible to grow crops on this land successfully for a few years, until the available nutrients are used up. If the cut-over areas are then left to recover for many years, and if there is undisturbed forest nearby, the original plant communities may eventually be restored. This process normally takes decades, even centuries, depending upon the type of forest involved. But rarely will it be allowed to reach completion anywhere in the world in the future. There are simply too many people and consequently too little time. The relentless search for firewood, the most important source of energy in many parts of the tropics, is one reason that the forests usually cannot recover.

Shifting cultivation, particularly under circumstances where the time of rotation must be short, virtually guarantees continued poverty for the people who practice it. Agricultural development in the tropics without proper management of the soil is not successful. Cultivation can be sustained on the better tropical soils under ideal conditions, involving fertilization, but such conditions lack meaning for the roughly 40 percent of the people who make up the rural poor—those who actually are destroying most of the forests. Trees generally make more productive crops in the humid tropics than do other kinds of plants; and agroforestry, the combination of annual crops and pastures with trees, is probably the most suitable form of agriculture for many of these regions. Unfortunately, very little research is being done in this area, and the practical options are few.

The FAO estimates that a 60 percent increase in world food production will be needed by the year 2000 if the world's population is to be fed. It does not appear that current efforts will lead to our even beginning to approach this goal, although some optimistic estimates have suggested that we might achieve half of it. In *World Indices of Agricultural and Food Production*, issued last year, the U.S. Department of Agriculture calculated that there has been little progress since 1973 in raising food consumption per capita for the world as a whole. Only greatly expanded efforts might offer the hope of improving this record significantly in the near future.

IN SUB-SAHARAN Africa, the problem is worse than elsewhere. There, per capita food production has declined every year since at least 1961; the Department of Agriculture estimated that in 1982 it was 11 percent less than in 1970. Currently, food production in this region is growing at about 1.3 percent per year, the population at about 3 percent. Even worse, food production *per acre* has been declining in recent years, despite some $8 billion of international aid spent annually in Africa.

For tropical countries, only sustainable local agricultural productivity—not food exports—will lead to stability. There are indications that other regions, including northeastern Brazil, the Andean countries, Central America and the Indian subcontinent, may soon face the same difficulties in food production that Africa does now, if their rapid population growth, soil erosion and underinvestment in agriculture continue unchecked. Only about 8 percent of the food eaten in tropical countries is imported, and it is highly improbable that this total could be increased significantly, especially in the face of these nations' staggering debts and their rapid population increases.

UNLESS WE recognize and address these problems, we can expect the instability now characteristic of so many tropical countries to spread and to become increasingly serious. About two-thirds of the people in these countries are farmers, and this number includes most of the truly poor, many with very little land, or none at all. Only if we can find better ways to use tropical land productivity for human benefit, concentrating on the areas that will be most productive and on the rural poor, shall we be making a genuine contribution to peace and harmony for those who come after us.

The population of Central America is about 23 million. In concentrating on that area, we are concerning ourselves with only one percent of all the people in the tropics—and their total number is projected to double in the next 30 years or so! To attain our political and economic goals throughout the world, we must find some way to help to alleviate the plight of the billion people in the tropics who are at the edge of starvation. If we cannot collectively find the means to eliminate rural poverty in these regions, as many experienced observers have concluded, these poor people will soon topple any government, be it friendly or unfriendly to us.

Yet despite these realities, the authors of the Kissinger Report on Central America pay very limited attention to the ecological problems that underlie the complex difficulties confronting that region. Although the Report explicitly recognizes the contribution of widespread poverty and population pressures to the difficulties of that area, it fails to connect them with their underlying ecological causes. Only by coping with all these factors can true regional stability be attained, and we can begin to secure U.S. interests there.

It is no coincidence that El Salvador is ecologically the most devastated of all the countries of Central America. For well over a decade, the relationship between its degraded environment, the lot of its people and its persistent internal conflicts has been stressed by virtually everyone concerned with that country's future. Throughout Central America and the other tropical regions, the best land is held by relatively few people; half of the farmland in El Salvador is owned by 2 percent of the population, for example. In practice, such a pattern tends to force the peasants to shifting cultivation in, and consequent permanent destruction of, the productivity of marginal lands.

Today, about a sixth of all American manufactured exports go to tropical countries, exports which support over 600,000 jobs in the United States alone. We also send nearly half of our agricultural exports to these areas and obtain many of our most important commodities from them. Such commerce will not be possible in the kind of world that is rapidly developing. The instability spreading throughout the tropics, including the constant threat of war, arises in many cases because of the prevalent extreme poverty and resource depletion. This poverty has brought about massive emigration. The U.S. Immigration and Naturalization Service in 1982–1983 apprehended over a million illegal immigrants at the Mexican border alone and estimates that 30 to 40 million more Latin Americans may enter the country illegally by the end of the century, in addition to the number who enter legally. Such a pattern occurs precisely because after 40 years of sustained economic growth, fully half of the population of Mexico still lives in poverty.

BEYOND THE SOCIAL and political consequences of the exhaustion of tropical resources, however, is a still more fundamental problem. It is the extinction of a major fraction of the plants, animals and microorganisms during the lifetime of a majority of people on earth today.

Approximately 1.5 million kinds of organisms have been named and classified, but these include only about 500,000 from the tropics. The total number of species of organisms in temperate regions is estimated to be approximately 1.5 million, but in relatively well-known groups of organisms—birds, mammals and plants—there are about twice as many species in the tropics as in the temperate regions. It may therefore be estimated that at least three million species exist in the tropics. Of these, we have named, and therefore registered, no more than one in six.

Many tropical organisms are very narrow in their geographical ranges and are highly specific in their ecological and related requirements. Thus, tropical organisms are unusually vulnerable to extinction through disturbance of their habitats. More than half of the species of tropical organisms are confined to the lowland forests. In most areas, these forests will be substantially altered or gone within the next 20 years.

Nearly 20 percent of all the kinds of organisms in the world occur in the forests of Latin America outside the Amazon Basin; another 20 percent occur in the forests of Asia and Africa outside the Congo Basin. All of the forests in which these organisms occur will have been destroyed by early in the next century. What would be a reasonable estimate of the loss of species that will accompany such destruction?

The loss of half of the species in these forests would amount to at least 750,000 species, about most of which we know nothing. This amounts to more than 50 species a day—fewer in the immediate future, more in the early part of the next century. And, because of the subsequent destruction of the remaining large forest blocks, there will be a continuing acceleration in the rate of extinction. The ultimate possibility is that of reaching stability after the human population does so, but only after many additional organisms have become extinct.

WHAT WE HAVE in the tropics, therefore, is a record and explosively growing human population, already well over twice as large as it was in 1950 and projected to double again in size in the next 30 years or so. More than one out of every four of these people are malnourished, many of them actually living at the edge of starvation. These people are dealing with the natural resources of their countries largely without regard to their sustainability, since no other options are available to them.

A human population with these characteristics will certainly exterminate a major proportion of the living species of plants, animals and microorganisms on earth before it begins to approach stability. For those unfamiliar with ecology and tropical biology to ignore or attempt to minimize the importance of events of this magnitude is to court disaster for themselves and for all the rest of us.

The extinction event projected within our lifetimes and those of our children may be about as extensive as that which occurred at the end of the Cretaceous Period 65 million years ago. For that time, David Raup, professor of geology at the University of Chicago, has estimated very approximately that about 20 to 30 percent of the total number of species may have disappeared permanently. There has been no comparable event since.

With the loss of organisms, we give up not only the opportunity to study and enjoy them, but also the chance to utilize them to better the human condition, both in the tropics and elsewhere. The economic importance of wild species, a tiny proportion of which we actually use, has been well documented elsewhere. Suffice it to say that the entire basis of our civilization rests on a few hundred species out of the millions that might have been selected, and we have just begun to explore the properties of most of the remaining ones.

The process of extinction cannot be reversed or completely halted. Its effects can, however, be moderated by finding the most appropriate methods of utilizing the potentially sustainable resources of tropical countries for human benefit. The explicit relationship between conservation and development was well outlined in the World Conservation Strategy, issued jointly in 1980 by the International Union for the Conservation of Nature and Natural Resources, the World Wildlife Fund and the United Nations Environmental Program.

Beyond the extinction of species, we are participating passively in the promotion of unstable world conditions in which it will no longer be possible to enjoy the benefits of civilization as we know them. It may seem comforting, temporarily, to use unwarranted scepticism and inadequate understanding of ecology as a basis for offering false reassurance to our leaders. To do so, however, is to offer them exceedingly bad advice at an extraordinarily dangerous time.

CHAPTER 25

# Science and Technology in a World Transformed

DAVID A. HAMBURG

My father was born in 1900 and died in 1984. In those years, he participated in one of the most drastic transformations any species has ever experienced. It is difficult to comprehend the extraordinary changes that people have witnessed in this century as a result of advances in science and technology. We are so deeply embedded in the present that it takes a difficult mental effort to comprehend the time scale of human life on Earth and the recency of the kind of world that we live in now. Human ancestors have been separate from the apes for about 5 to 10 million years. For almost that entire time, there were fewer than 1 million people on earth, subsisting by hunting and gathering in small, nomadic groups. Agriculture and large, settled populations have existed for much less than 1 percent of that epoch, and our technical world has been present for a mere moment in the time scale of human evolution. The way we live today is, in many important respects, a novelty for our species.

*Summary*. In this era of rapid, far-reaching transformation, our way of life is in many respects a novelty for our species. Opportunities arising from profoundly enhanced capabilities in science and technology are felt in every sphere of life from health to communication, yet each advance has side effects that take time to appear. Grave institutional inadequacies are manifested in the prevalence of totalitarian governments, proliferation of devastating weapons, failure of educational institutions to prepare most people for the modern world, failure to use what we know to prevent damage to a large proportion of the world's children; and the weakness of international institutions to deal with global interdependence in the face of persistent ethnocentrism and prejudice. The American scientific community can usefully become more deeply engaged with these great problems.

Reprinted from *Science 224:* 943–946 (1984). 

David A. Hamburg is president of the Carnegie Corporation, New York, NY 10022, and is president of the American Association for the Advancement of Science.

Our ancestors—prehuman, almost human, and distinctly human—lived in small groups in which they learned the rules of adaptation for survival and reproduction. They used simple tools to cope with the problems of living and struggled to obtain more control over their own destiny. For the most part, they were vulnerable to the vicissitudes of food, water, weather, predators, other humans—whatever nature might bring. Their world began to change with the onset of agriculture about 10,000 years ago. But the most momentous changes occurred with the industrial revolution two centuries ago, and above all with its pervasive implementation in the 20th century.

Much of the technology that structures American lives today, in ways we largely take for granted, is of extremely recent origin. In 1900 there were few automobiles or household telephones; motion pictures were just getting under way; there were no household radios, no airplanes, no televisions, no computers. Today it is almost impossible to imagine a world without these technologies—and in this country, without their presence everywhere. What a difference a century makes—indeed, even a decade as events move now.

## Opportunities and Complications

The opportunities arising from our profoundly enhanced capability in science and technology are visible in every sphere of human life—in medicine and public health, in agriculture and food supply, in transportation and communication, and elsewhere. Every advance has brought side effects—like a new medicine whose benefits are clear but whose complications take considerable time to appear. But complications there are: extreme population growth in much of the world, drastic urbanization with its crowding of strangers beyond any prior experience, environmental damage, resource depletion, the immense risks of weapons technology, and new patterns of disease—all are largely products of changes that have occurred only in the most recent phase of human evolution. We have rapidly changed our technology, our social organization, our diet, our activity patterns, the substances of daily use and exposure, patterns of reproductive activity, tension-relief, and human relationships. These changes are laden with new benefits and new risks, and the long-term consequences are poorly understood.

Many of the technological changes are exceedingly attractive since they free our species from hardships and dangers; they provide gratifications that were beyond reach at least for most people in the past. In many respects, ordinary citizens live today as kings of an earlier time never could.

The automation of the household has drastically reduced the requirement of physical labor at home. Its social implications have been far-reaching. Similarly the revolution in telecommunications has come close to making this country a single large community in some respects and may one day have a similar effect on the world as a whole.

Technological innovation is now associated with far-reaching and extremely rapid changes in the nature and scope of work available in this and other countries. The pervasive mechanization of work appears to be tangible on the horizon, not as a distant prospect, but as a powerful current gathering momentum and affecting the entire economy in far-reaching ways. Benefits in productivity are clearly visible. Concomitant social dislocations are not as visible but are as likely to occur.

### Growing Pressures on Resources

There are in many parts of the world today strong tensions between population pressures and available resources. These tensions have explosive potential within countries and also for international conflict. It has become a matter of our enlightened self-interest as well as of decent human concern for us to try hard to understand our species in its worldwide interdependence, paying just as much attention to the Southern Hemisphere as to the Northern Hemisphere. For our own sake as well as theirs, we need to strengthen our ties and work cooperatively with people in developing countries toward the reduction of poverty, ignorance, and disease.

Since the sciences provide our most powerful problem-solving tool, it is essential that we bring their strengths to bear on these problems to the maximum extent possible. The task requires an intensive effort now to learn what can be extracted from the efforts at economic and social development during the past several decades, sorting out failures from successes, looking for strengths on which to build future efforts of practical value.

The problems that face modern industrial nations are related to sociotechnical conditions that have appeared very recently in the evolution of the human species: the magnitude and rate of change make it difficult for us to devise and implement solutions to these problems. In developing countries, the shift from old to new ways has occurred even more rapidly, and change has been partly imposed from outside. These nations confront exceedingly difficult problems. Across much of Africa, Asia, and Latin America, explosive population growth and abject poverty have contributed to extremely severe health problems that stand as an enormous obstacle to sustained development and social progress. Their burden of early death and long-term disability is exceedingly heavy. Infectious diseases take the lives of a great many infants before they reach 1 year of age and handicap for life many of those that survive. Susceptibility to a wide range of diseases is heightened by the marginal character of subsistence.

### Global Interdependence

This set of facts raises ethical questions for countries with strong scientific capability. In recent decades, the United States has given little attention—in research, education, and practice—to some of the most important disease problems in the world today. How can we and the other more developed nations help improve health in developing countries? One way in which we are especially able to help is through research—including capacity-building in developing countries so that they can tackle their own problems. Priority areas are: (i) epidemiological assessment of specific needs; (ii) applying molecular biology to parasitic diseases; (iii) devising a wider array of fertility-control methods with special reference to cultural acceptability and feasibility of use; and (iv) clarifying relations of health and behavior, with special reference to breast-feeding, nutrition, child care, sanitation, water use, and family planning. What is fundamentally needed is a heightened awareness within the scientific community of the opportunities that exist, since even a modest shift of attention to such problems could yield major benefits.

Difficult as these problems are, and crying out for a larger place in the work of the scientific community, another aspect of our global interdependence is even more urgently in need of attention. The overriding problem facing humanity today is the threat of nuclear holocaust. Humanity's capacity for destruction has radically outstripped its institutional capacity to control intergroup violence. That violence is rooted in the nature of the human species. Mass expression of violence—war, terrorism, and genocide—persists throughout the world, and no people should be considered incapable of it. But the invention and deployment of nuclear weapons represents a qualitative break in the history of violence. It is now possible to destroy human life on Earth. Both the United States and the Soviet Union probably have the capacity to do that, or at least to make the human condition unbearable. In 1983, new evidence of the incredible devastation—immediate, long-term, and permanent—of nuclear war was brought to light.

Former Secretary of Defense Harold Brown points out that in the first half-hour of a nuclear war there might well be 100 million deaths each in the United States, in the Soviet Union, and in Europe. It is plausible that a billion people would die in just a few weeks. In point of fact, there is nothing in our history as a species to prepare us to comprehend the real meaning of such devastation.

Human societies have a pervasive tendency to make distinctions between good and bad people, between heroes and villains, between ingroups and outgroups. It is easy for most of us to put ourselves at the center of the universe, attaching a strong positive value to ourselves and our group, while attaching a negative value to certain other people and their groups. It is prudent to assume that we are all, to some extent, susceptible to egocentric and ethnocentric tendencies. The human species is one in which individuals and groups easily learn to blame others for whatever difficulties exist. But in the present predicament around nuclear conflict blaming is at best useless and most likely counterproductive.

A new level of commitment of the scientific community is urgently needed to reduce the risk of nuclear war. This requires a mobilization of the best possible intellectual, technical, and moral resources in a wide range of knowledge and perspectives. A science-based effort is essential to maximize analytical capability, objectivity, and respect for evidence—the outlook that is characteristic of the scientific community worldwide.

These efforts should bring together scientists, scholars, and practitioners in order to clarify the many facets of avoiding nuclear war. To generate new options for decreasing the risk, we need analytical work by people who know the weaponry and its military uses, people who know the Soviet Union, people who know international relations broadly, people who know the processes of policy formation and implementation, people who understand human behavior under stress, and people who understand negotiation and conflict resolution. Such analytical studies are likely to be more useful if they take into account policy-makers' perspectives, and policy-makers can benefit greatly from ready access to new ideas, a wider range of options, and deeper insights.

**Commitment to Science Education**

The rapid acceleration of technological and social change in recent decades sharply heightens the importance of our educational institutions. Indeed, during the past year we have experienced a sort of national rediscovery of education, particularly with reference to science and technology. We must address fundamental challenges in education.

1) How can we give all our children, regardless of social background, a good opportunity to participate in the modern technical world? In this time of high unemployment we must especially consider employment opportunity.

2) What constitutes a decent minimum of literacy in science and technology that should be part of everyone's educational heritage?

3) Given the rapidity of sociotechnical change, how can we make lifelong learning a reality so that people can adjust their knowledge and skills to new circumstances?

4) Since educational institutions will more than ever be trying to hit a moving target as they prepare people for unpredictable circumstances, how can we prepare for change itself?

5) How can we enlarge that talent pool so that we can find promising people for science-based careers, regardless of their socioeconomic background?

6) How can we broaden the spectrum of the sciences so that modern education will become increasingly informative with respect to the human experience?

7) How can the educational system foster a scientific attitude that is useful in problem-solving throughout the society, in relating scientific principles to the major issues on which an informed citizenry must decide?

8) How can we achieve an informed worldwide perspective in an era of profound interdependence?

The American dilemma in science education involves a remarkable paradox. We have the largest and probably the most respected scientific community in the world, yet our precollegiate science education is at a low ebb. Is there a way to resolve this paradox creatively?

We must seek mechanisms to link the science-rich sectors of our society and the science-poor sectors—that is, connect the scientific talent of universities, colleges, corporate laboratories, and national laboratories with the elementary and secondary schools, thereby strengthening national capability for broad education in the sciences: physical, biological, and behavioral.

The linking of science-rich to science-poor sectors as partners is a means to both teacher education and curriculum development. We can learn from the successes and failures of the curriculum reform movement that followed Sputnik. Improvements in education can flow from the collaboration of classroom teachers with subject matter experts (for example, in physics, chemistry, or biology) and also with scholars in the field of human cognition and learning. Such collaborations could be an important step toward incorporating teachers of science into the scientific community. While major efforts must be made to improve the salaries of teachers, and especially those of science teachers, it is equally important to find ways to bolster respect for their profession and strengthen their morale. Participation in the scientific community would be helpful in this regard.

Closer links of elementary and secondary schools with colleges, universities, corporate laboratories, and government laboratories could include summer institutes for teachers, Saturday activities for teachers throughout the school year, summer jobs in science for teachers, and the preparation of curricular materials. Leadership from different sectors of society will be necessary for the major upgrading that is required. The schools are central to the effort, but they alone do not have the resources or the clout to do the job.

**The Transformed World**

The rapid, pervasive, and truly unprecedented transformation resulting from science and technology is a central fact of our lives. It calls for the strengthening of institutional capability for objective, scholarly analysis of critical issues based on a broad foundation of knowledge and experience.

Colleges and universities, academies, and free-standing institutes can mobilize a wide range of talent to address the great issues of our time in a sustained, fascinating, and effective way. They can give us a better chance to get the complex facts straight and to clarify the most promising options—all this in a way that is credible and even intelligible to nonspecialists. Such efforts can be helpful to open-minded policy-makers, but also—and perhaps more importantly in the long run—to the education of a broadly informed public on the great issues of our time and the policy choices available to us.

Surely the problems associated with the social and economic concomitants of technological change are not insuperable. The opportunities provided by the advancing technology, if judiciously utilized, suggest that it is worth a lot of trouble and hard work to find ways to make a decent social adjustment. We will require broad, multifaceted analytical work to understand more deeply what is going on and to anticipate insofar as possible the likely consequences of major technological trends.

To deal effectively with real-world policy problems requires novel conjunctions of knowledge and talent. Typically, many different facets of a complex issue must be taken into account. The great problems do not come in packages that fit the traditional disciplines or professions, however excellent they may be. Organizations such as universities, scientific academies, and research institutes can make a greater contribution than they have in the past if they can organize effectively to share information, ideas, and technical abilities widely across traditional barriers and systems. Of course, most advances in knowledge require specialization. Yet for crucial social purposes these pieces must somehow be related to each other.

A particularly valuable undertaking is the intelligible and credible synthesis of research related to important policy questions. What is the factual basis drawn from many sources that can provide the underpinning for constructive options in the future? This is especially significant in view of the fact that pertinent information is almost always widely scattered. Moreover, it is very difficult for the nonexpert and sometimes even for the expert to assess the credibility of assertions on emotionally charged issues. In the current process of world transformation, such studies are needed to tackle vital and complex issues in an analytical rather than a polemical way.

Whether it be toward avoiding nuclear war, strengthening education in the sciences, or fostering human resources in developing countries, there is a precious resource in the great scientific community of the United States—and its links to the worldwide scientific community—which can be brought to bear on these crucial problems. This involves activation of a wide range of the sciences and an unusual degree of cooperation among them. Furthermore, it involves linking analytical work with education in a variety of modes, for the general public must take an informed part in the decisions that affect all our futures. The scientific and scholarly community can deepen its contribution to pressing social concerns if it is informed and stimulated by those on the firing line, whether the latter be engaged in teaching poor children, struggling with policy dilemmas, or coping with international tensions. There can be a mutually beneficial interplay between social concerns and basic inquiry.

**Conclusion**

There is little in our history as a species to prepare us for this hypermodern world that we have so rapidly made. The transformation of the world will press us toward transformation of our institutions—to keep up with events, to understand below the surface, to look ahead and prepare, to enjoy the fascination of deep insights, to make wise use of technology, to relieve poverty and disease, above all to resolve the deadliest of conflicts. In the historic effort to avoid unmitigated disaster and fulfill the potential of our species, science can help profoundly and in novel ways. But to do so, the sciences must transcend their traditional boundaries and achieve an unprecedented level of mutual understanding, innovation, and cooperation. These efforts will have to go across disciplines, across sectors of societies, and across nations.

Let us hope that fundamental values of freedom, curiosity, opportunity, diversity, excellence, and human decency will guide our institutions as they evolve. These enduring values can make it possible for us to cope with the great problems of our time, to work steadily towards the humane uses of science and technology, and to take advantage of unprecedented opportunities that are emerging for the benefit of people everywhere.

# Index

INDEX